D1327096

Karl J. Niklas

Plant Allometry

The Scaling of Form and Process

The University of Chicago Press
Chicago and London

Karl J. Niklas is professor of botany at Cornell University. He received the 1993 Michael A. Cichan Award, presented by the Botanical Society of America, for *Plant Biomechanics: An Engineering Approach to Plant Form and Function,* also published by the University of Chicago Press.

The University of Chicago Press, Chicago 60637
The University of Chicago Press, Ltd., London

Printed in the United States of America
03 02 01 00 99 98 97 96 95 94 1 2 3 4 5

ISBN: 0-226-58080-6 (cloth)
0-226-58081-4 (paper)

Library of Congress Cataloging-in-Publication Data

Niklas, Karl J.
Plant allometry : the scaling of form and process / Karl J. Niklas.
p. cm.
Includes bibliographical references and index.
1. Plant allometry. I. Title.
QK731.N54 1994
581.3′1—dc20 94-2418
CIP

⊗ The paper used in this publication meets the minimum requirements of the American National Standard for Information Sciences—Permanence of Paper for Printed Library Materials, ANSI Z39.48-1984.

Contents

Prologue

The overarching goal of this book is to examine size-correlated variations in plant form, metabolism, reproductive effort, and evolution. The basic question asked throughout is, *What is the relation between organic size and plant form or process?* Engineers, physicists, and chemists routinely ask this type of question, because the absolute and relative magnitudes of different measurements of size, such as length, cross-sectional area, volume, and mass, invariably influence the behavior of structural, mechanical, or chemical systems. Consequently structures differing in size must be built to different relative proportions whenever they are designed to sustain their own weight, mechanical devices differing in size must be reproportioned to accommodate perfunctory dynamic loadings, and differences in the absolute and relative concentrations of molecular species must be adjusted to maintain the rates of chemical processes.

In the physical sciences, the size-dependent behavior of various systems is evaluated by means of scaling analysis. In this context, *scale* refers to the proportion that a representation of an object or system bears to the prototype of the object or system. Although living things are neither mechanical "objects" nor chemical "systems," they undeniably evince size-correlated variations in organic form and process. For this reason the behavior of living things is not only amenable to the approach taken by the physical sciences, but potentially accountable in terms of the consequences of absolute and relative size. In the biological sciences, the study of size-correlated variations in organic form and process traditionally is called *allometry* (Greek *allos,* "other," and *metron,* "measure"; Huxley 1932; Gould 1966; Reiss 1989). Although the title of this book reflects this tradition, I have elected to eschew the term wherever possible in favor of *scaling analysis*—hence the subtitle, *The Scaling of Form and Process.* My reason for this tactic is that the word allometry has been used to describe very different things (see Schmidt-Nielson 1984; LaBarbera 1989), leading to avoidable confusion in a literature already unnecessarily obtuse and complex. In its general usage, allometry has two meanings: (1) the growth of a part of an organism in relation to the growth of the whole organism or some part of it, and (2) the study of the consequences of size on organic form and process. Allometry has yet another, albeit much more restrictive, meaning. It may be used to connote any departure from a condition called geometric similitude, which results when geometry and shape are conserved among a series of objects differing in size. As we shall see, geometric similitude conventionally serves as the null hypothesis in empirical as well as analytical treatments of allometric phenomena. Geometric similitude is a prominent concept, therefore, that must be thought of in a precise and unambiguous way. A more egregious fault of the term allometry is that it is sometimes used to imply that size is the driving force underlying organic size-correlated changes in form and process. This is a highly dubious notion because the correlations among size, shape, structure, chemical composition, and other biological variables reflect the way different levels of organization are integrated.

My purpose in writing this book, however, is not to entertain terminological debate, nor is it to espouse the view that organisms can be fully reduced to structures, mechanical devices, or chemical processes. Rather, I want to share my excitement about a point of view that embraces virtually every aspect of biology, from the molecular traffic within cells to the seemingly erratic historical flow of plant life, in terms of the effects of absolute and relative size on organic form and process. The diversity of size-correlated botanical phenomena evinced by the relations treated in this book is vast (there is a brief summary of them later in this prologue). Indeed, the diversity is so great that it may be perceived in two polarized ways—either as prima facie evidence that scaling analysis leads to an unnecessarily refractive view of plant biology, or as a vivid indication that every level of biological organization is responsive to and in turn influences absolute and relative size. If there is "truth" to be found here, it occupies the middle ground rather than the ends of this dialectic view of scaling analysis. Truthfully, taken in isolation, scaling analysis often is an intellectually trivial exercise—just another way of describing how far biological variables correlate with some measure of size. But correlations and the phenomenological descriptions that emerge from them lie at the heart of every scientific Weltanschauung, and therefore scaling analysis has a legitimate and meaningful place in the technological approaches of the biological as well as the physical sciences by virtue of its capacity to quantitatively describe the behavior of complex systems.

My purpose, however, is to show that scaling analysis transcends this role when employed to account for myriad superficially unrelated features by means of a comparatively few basic principles. In his justly famous book *On Growth and Form,* D'Arcy Thompson made the important point that, as they grow in size and change in form, organisms make use of physical and chemical processes whose underlying phenomenology extends the raw genetic capacity of an organism to orchestrate changes in its external shape (morphology) and internal structure (anatomy). In this manner, every living thing draws upon and responds to immutable physical principles as well as mutable genetic programs that propel ontogeny and ultimately dictate features of evolutionary pattern. If D'Arcy Thompson's perspective is true, as I think it is, then growth in size and the attending expression of organic form depend on two types of information, one contained in physical and chemical "laws" and another obtained from genetic "algorithms." It is undeniable that physical and chemical processes are size dependent. It seems reasonable, therefore, to assume that the operation of physical laws and the procession of ontogenetic and phylogenetic patterns are in resonance and that the harmonics of their interplay can be exposed when viewed in terms of the absolute and relative scales at which they operate. Also, it is not presumptuous to think that the perspective gained from scaling analysis can transfigure a basically phenomenological approach into one that permits a greater depth of understanding of the mechanisms linking the animate to the inanimate world.

The chameleon-like influence of absolute and relative scale is evident whenever we draw upon an equation derived from the physical sciences to describe a biological feature. Much of this book is devoted to these equations. Nonetheless, the view taken here is that the pristine and sometimes unassuming equations of physical sciences become animated and informative only when applied to tangible numbers obtained from real organisms. It is for this reason rather than mere pedagogy that I have striven to balance a treatment of the principles of the physical sciences as they relate to biological scaling analysis with the application of these principles to real data. As we shall see, the assumptions and the equations inspired by the physical sciences isolate the variables that are influenced by (and in turn influence) changes in organic size. It is for this reason that I draw on a broad variety of equations to illustrate the diversity of size-correlated changes in form and process bearing on plant biology. Note, however, that the equations presented throughout this book are extremely deceptive—they evince a level of precision from which we may infer an accuracy rarely achieved in biology. Living things respond to relative as well as absolute quantities. And for this reason their behavior tends to be "sloppy," often forcing us to accept qualitative rather than quantitative answers to the complex questions we ask. In many cases the physical sciences offer little or no guidance on the appropriate numerical values (or even the magnitudes) of the panoply of biological variables that typically confront us. Indeed, even the variables themselves may be poorly understood or simply not known. Since many of the equations used in this book are extremely sensitive to the values assigned to variables as well as to the variables actually elected for consideration, much of what appears here serves to illustrate how data may be manipulated rather than to establish categorical statements of what is biologically "true" or "false."

This philosophy—one of illustration rather than affirmation—applies equally to statistical analyses, which are invariably required for scaling analysis. Statistical inference and the mathematical tools it engenders are a prominent part of this book. Curiously, although regression analysis and correlation analysis (which are not the same thing) are the first steps in any scaling study, at a very fundamental level, which one among competing regression models is the most appropriate for a particular study remains a matter of active debate among statisticians as well as biologists. For this reason the statistical results obtained by different methods of regression analysis (least squares regression and reduced major axis regression) are provided for each of the scaling relations treated. Fortunately, in most cases the results from these two regression methods do not differ significantly, but in some important cases they do. As the author, I am obliged to choose the results of one over those obtained from alternative methods of analysis, but each choice is subject to legitimate, well-reasoned differences of opinion.

By way of offering some guidance as well as to expose the nature of current debates over which statistical technique is most appropriate, an appendix is provided. The topics covered in this appendix are referenced throughout the main body of the text to draw attention to particular matters of statistical inference and methodology. Numerical examples of each type of analysis, employing sets of data discussed in different chapters, are provided as well. After many years of teaching, I have found that the equations used in statistics can be needlessly intimidating yet are easily understood once they are worked through with "real numbers." Be that as it may, those unfamiliar with the fundamental differences between least squares and reduced major axis regression (the differences between model I and model II regression analysis) or the principles of multiple regression analysis, the methods of analysis of variance, nested analysis of variance, and the like are strongly advised to read the appendix before engaging the discussion offered in the main text.

The appendix is preceded by five chapters, each serving as an admittedly idiosyncratic essay on a particular topic whose main points are summarized at its end. The overarching theme binding all five chapters together is evolution. I see no way to truly understand biology in general or size-correlated phenomena in particular other than in the historical context that holds together otherwise ecologically and taxonomically diverse organisms. In this regard it is only fair to tell you that I am an "adaptationist"—I subscribe to the notion that most form-function relations are the consequence of natural selection and adaptive evolutionary changes. For this reason, like many other authors, I strongly depart from the approach of D'Arcy Thompson, who eschewed mentioned of natural selection. In particular, I argue throughout this book that scaling analysis assumes real importance because the functional obligations of organisms are dependent upon absolute and relative scale. Since the levels of performance of these obligations define an organism's fitness, size-correlated phenomena are a modus operandi of natural selection.

Chapter 1 introduces the general topic of scaling analysis in the context of botany, discussing the hypotheses of geometric similitude and functional similitude. Here I introduce the notion that an organism is required to change shape or structure whenever the levels of performance of its various functions must remain comparatively constant regardless of absolute size. In terms of function, the principal focus is on the interplay between the size dependency of photosynthesis and the size-correlated variations attending plant growth and development. Although the emphasis compares and contrasts plants with animals, my purpose here is not simply to argue that plants differ from animals but rather to expose how and when these differences become important when we use scaling analysis to understand the meaning of size-dependent variations among plants. Chapter 2 treats the scaling of aquatic plants and discusses how the physical properties of water, particularly its ability to absorb light and apply dynamic mechanical forces, influence the biology of aquatic plants. In this chapter much emphasis is placed on the transport laws, the equations that predict the exchange of mass or energy between any time of organism and its external fluid medium. In apposition, chapter 3 deals with the scaling of terrestrial plants, and there I discuss the importance of water conservation and the influence of gravity on plants that have evolved onto the terrestrial landscape. Once again, this chapter is an extension of an evolutionary theme, the historical transition of life from the aquatic to the terrestrial habitat that marks one of the most important adaptive events in the history of life on earth. Chapter 4 extends the theme of scaling to cover size-correlated variations in plant reproduction, placing particular emphasis on the allocation of metabolic resources to reproductive and vegetative functions. Finally, chapter 5 brings together much of the previous material in an effort to deal with the consequences of absolute and relative size on the pageantry of plant evolution. This chapter, which is neither an exposition of paleobotany nor a detailed treatment of the plant fossil record, is the most speculative, the least mathematical, and the most narrative in design.

The principal faults of any book reflect the intellectual gifts of friends and colleagues ill used. It is for this reason that I must acknowledge and apologize to the following talented and generous individuals: Tom L. Phillips, who first introduced me to the joys of paleobotanical research and the need always to think of fossil as well as living plants as organisms that functioned in a complex physical and biological environment; Dominick J. Paolillo Jr. (Cornell University), who provided many insightful comments and stimulating discussions about plant growth and development; Bruce H. Tiffney (University of California), who generously covered early drafts with an apparently inexhaustible quantity of red ink to draw my attention to many inadequacies; Sharon B. Emerson (University of Utah), who read the penultimate version of this book and helped me focus on and deal with the recent insights provided by the zoological and statistical literature in terms of scaling analysis; and finally, Susan E. Abrams (University of Chicago Press) without whose encouragement and consummate professional skill this book would never have been completed. The gifts bestowed by these exceptionally kind individuals constitute a great debt, which I begin to pay back by offering my sincerest apologies for the sins of omission and commission that undoubtedly remain.

Symbols Frequently Used

a	Major semiaxis of an ellipse
A	Cross-sectional area
α_{LS}	Scaling exponent of ordinary least squares regression
α_{RMA}	Scaling exponent of reduced major axis regression
b	Minor semiaxis of an ellipse
β	Scaling coefficient
C_D	Drag coefficient
d	Characteristic dimension
D	Diameter; diffusion coefficient
δ	Thickness of boundary layer
e	Eccentricity of an ellipse
E	Young's modulus
EI	Flexural stiffness
g	Gravitational acceleration
Gr	Grashof number
H	Heat flux density; height
I	Second moment of area
J	Mass flux density
L	Length
ln	Natural (base *e*) logarithm
log	Common (base 10) logarithm
M	Body mass
Nu	Nusselt number
P	Mechanical load
Pe	Péclet number
Pr	Prandtl number
ρ	Density
r	Correlation coefficient
r^2	Coefficient of determination
R	Radius
Re	Reynolds number
$\Re$	Aspect ratio
S	Surface area
Sc	Schmidt number
Sh	Sherwood number
t	Thickness of fluid layer
$\tilde{t}$	Thickness of "rind"
U	Fluid speed
μ	Dynamic viscosity
V	Volume
υ	Kinematic viscosity
W	Width of plate
x	Distance
Z	Section modulus

Some Scaling Relations

Prokaryotes
Cell biomass $\propto$ (cell volume)$^{0.999}$
($r^2 = 0.99$, $N = 7$ genera; $\alpha_{RMA} = 1.0 \pm 0.001$)

Unicellular and multicellular plants
Body mass $\propto$ (body length)$^{2.86}$
($r^2 = 0.94$, $N = 66$ species; $\alpha_{RMA} = 2.95 \pm 0.09$)

Unicellular and multicellular animals
Body mass $\propto$ (body length)$^{2.79}$
($r^2 = 0.97$, $N = 67$ species; $\alpha_{RMA} = 2.84 \pm 0.09$)

Unicellular algae
Cell biomass $\propto$ (cell volume)$^{0.794}$
($r^2 = 0.96$, $N = 45$ species; $\alpha_{RMA} = 0.81 \pm 0.03$)

Cell surface area $\propto$ (cell volume)$^{0.682}$
($r^2 = 0.98$, $N = 45$ species; $\alpha_{RMA} = 0.69 \pm 0.01$)

Growth rate $\propto$ (cell biomass)$^{0.787}$
($r^2 = 0.92$, $N = 53$ species; $\alpha_{RMA} = 0.82 \pm 0.03$)

Weight-specific growth rate $\propto$ (cell biomass)$^{-0.213}$
($r^2 = 0.45$, $N = 53$ species; $\alpha_{RMA} = -0.32 \pm 0.03$)

Chlorophyll *a* concentration $\propto$ (cell biomass)$^{0.740}$
($r^2 = 0.86$, $N = 47$ species; $\alpha_{RMA} = 0.80 \pm 0.04$)

Chlorophyll *a* concentration $\propto$ (growth rate)$^{0.891}$
($r^2 = 0.95$, $N = 47$ species; $\alpha_{RMA} = 0.91 \pm 0.03$)

Volvocalean total surface area $\propto$ (volume)$^{0.745}$
($r^2 = 0.98$, $N = 13$ genera; $\alpha_{RMA} = 0.76 \pm 0.04$)

Leaves
Volume fraction of spongy mesophyll $\propto$ (leaf thickness)$^{-0.49}$
($r^2 = 0.87$, $N = 28$ species; $\alpha_{RMA} = -0.53 \pm 0.08$)

Lamina mass $\propto$ (length of petiole)$^{1.84}$
($r^2 = 0.87$, $N = 193$ leaves; $\alpha_{RMA} = 1.97 \pm 0.04$)

Petiole flexural stiffness, $EI \propto$ (petiole length)$^{3.09}$
($r^2 = 0.91$, $N = 193$ leaves; $\alpha_{RMA} = 3.24 \pm 0.05$)

Leaf area $\propto$ (stem diameter)$^{1.84}$
($r^2 = 0.86$, $N = 46$ species; $\alpha_{RMA} = 1.98 \pm 0.05$)

Wood
Density-specific strength $\propto$ (density-specific stiffness)$^{0.45}$
($r^2 = 0.53$, $N = 60$ species; $\alpha_{RMA} = 0.62 \pm 0.14$)

Breaking strength $\propto$ (Young's modulus)$^{0.99}$
($r^2 = 0.75$, $N = 60$ species; $\alpha_{RMA} = 1.14 \pm 0.09$)

Density $\propto$ (breaking strength)$^{0.82}$
($r^2 = 0.80$, $N = 60$ species; $\alpha_{RMA} = 0.92 \pm 0.05$)

Density $\propto$ (Young's modulus)$^{0.71}$
($r^2 = 0.45$, $N = 60$ species; $\alpha_{RMA} = 1.06 \pm 0.09$)

Plant height
Overall $\propto$ (stem diameter)$^{0.896}$
($r^2 = 0.95$, $N = 670$ species; $\alpha_{RMA} = 0.92 \pm 0.01$)

Nonwoody $\propto$ (stem diameter)$^{1.29}$
($r^2 = 0.97$, $N = 190$ species; $\alpha_{RMA} = 1.32 \pm 0.02$)

Gymnosperm-angiosperm tree height $\propto$ (trunk diameter)$^{0.535}$
($r^2 = 0.54$, $N = 480$ species; $\alpha_{RMA} = 0.73 \pm 0.02$)

Dicot tree height $\propto$ (trunk diameter)$^{0.474}$
($r^2 = 0.47$, $N = 375$ species; $\alpha_{RMA} = 0.69 \pm 0.03$)

Conifer tree height $\propto$ (trunk diameter)$^{0.430}$
($r^2 = 0.25$, $N = 105$ species; $\alpha_{RMA} = 0.87 \pm 0.08$)

Moss seta length $\propto$ (seta diameter)$^{1.10}$
($r^2 = 0.97$, $N = 40$ species; $\alpha_{RMA} = 1.12 \pm 0.04$)

Pteridophyte height $\propto$ (stem diameter)$^{1.69}$
($r^2 = 0.85$, $N = 16$ species; $\alpha_{RMA} = 1.83 \pm 0.06$)

Dicot herb height $\propto$ (stem diameter)$^{1.26}$
($r^2 = 0.74$, $N = 117$ species; $\alpha_{RMA} = 1.46 \pm 0.09$)

Palm height $\propto$ (stem diameter)$^{1.76}$
($r^2 = 0.94$, $N = 17$ species; $\alpha_{RMA} = 1.82 \pm 0.09$)

Reproduction
Number of moss spores $\propto$ (capsule length)$^{2.11}$
($r^2 = 0.76$, $N = 75$ species; $\alpha_{RMA} = 2.42 \pm 0.05$)

Spore volume $\propto$ (capsule length)$^{-0.26}$
($r^2 = 0.68$, $N = 75$ species; $\alpha_{RMA} = 0.32 \pm 0.02$)

Seed number per ovulate cone $\propto$ (ovulate cone mass)$^{1.06}$
($r^2 = 0.94$, $N = 5$ genera; $\alpha_{RMA} = 1.09 \pm 0.04$)

Total seed mass per cone $\propto$ (ovulate cone mass)$^{1.67}$
($r^2 = 0.91$, $N = 5$ genera; $\alpha_{RMA} = 1.75 \pm 0.03$)

Biomass of androecium $\propto$ (biomass of filament)$^{0.852}$
($r^2 = 0.97$, $N = 56$ species; $\alpha_{RMA} = 0.86 \pm 0.02$)

Biomass of gynoecium $\propto$ (biomass of perianth)$^{0.914}$
($r^2 = 0.87$, $N = 39$ species; $\alpha_{RMA} = 0.98 \pm 0.06$)

Biomass of androecium $\propto$ (biomass of perianth)$^{0.978}$
($r^2 = 0.92$, $N = 39$ species; $\alpha_{RMA} = 1.02 \pm 0.04$)

Individual fruit mass $M_{IF} \propto$ (individual seed mass, M_{IS})$^{0.78}$
($r^2 = 0.55$, $N = 139$ species; $\alpha_{RMA} = 1.05 \pm 0.06$)

$M_{IF}/M_{IS} \propto$ (number of seeds per fruit)$^{0.84}$
($r^2 = 0.73$, $N = 139$ species; $\alpha_{RMA} = 0.98 \pm 0.06$)

Total average seed mass $\propto$ (individual fruit mass)$^{0.93}$
($r^2 = 0.84$, $N = 139$ species; $\alpha_{RMA} = 1.10 \pm 0.04$)

Total reproductive biomass $\propto$ (stem diameter)$^{2.9}$
($r^2 = 0.94$, $N = 13$ species; $\alpha_{RMA} = 2.99 \pm 0.05$)

Diaspore settling velocity $\propto$ (diaspore biomass)$^{0.18}$
($r^2 = 0.50$, $N = 66$ species; $\alpha_{RMA} = 0.26 \pm 0.05$)

1 Growth and Metabolism

1.1 Introduction

As it grows in size and develops, every kind of organism, plants as well as animals, undergoes changes in external shape (morphology), internal structure (anatomy), and process (metabolism and reproductive effort). These changes appear to unfold in an organized fashion during ontogeny and, when undisturbed, tend to produce the adult organic proportionalities characteristic of the species to which each individual belongs. The tendency for growth and development to move adult form toward the norm of a species yet permit variation among adults in a manner apparently molded by the external conditions attending growth and development affords strong circumstantial evidence that all levels of plant organization are both highly integrated and attuned to the physical environment.

As I noted in the prologue, this book deals with this interrelatedness from the perspective of the consequences that size has for organic form and process. Throughout, I shall ask, *Why do plants persistently attain size-correlated variations in their form and process?* Clearly, there is no reason to assume a priori that size, shape, structure, chemical composition, and the like are correlated with one another for all manner of things. Grains of sand come in a variety of sizes, shapes, and mineral compositions. Clouds all have the same chemical composition yet are mercurial in size and shape. By the same token, many of the objects we fabricate for enjoyment, like toy cars and boats, have the same shapes as their prototypes but are structurally and often chemically very dissimilar to their larger counterparts. These and other examples show that size, shape, structure, and composition are not invariably correlated for a great many things in our everyday experience. Yet living things suffer or die when externally manipulated so that they fail to achieve their normal proportions. To my mind the reason this is so—why living objects persistently attain their characteristic proportions—is function. Every living thing must perform a number of functions to survive, grow, and ultimately reproduce. When functional obligations are imposed on an object, all manner of things evince size-dependent variations in form and process. This holds true for the machines we build as well as for plants and animals as they grow in size. The logical inference drawn from this—and on which this book is predicated—is that organic proportionalities often reflect the consequences of natural selection operating on the relation between form and function.

This "adaptationist" point of view is not a precondition for an interest in scaling analysis. Even those who espouse the view that scaling relations appear as a consequence of historical accident employ scaling analysis because simple phenomenological description of the relations among variables is vital to understanding biology in general and botany in particular. The first objective of any empirical scaling analysis is to establish the functional (mathematical) relation and significance (degree of correlation) between one or more biological variables and organic size. This is accomplished by regression analysis and correlation analysis, which lie at the heart of scientific analysis. At this level of inquiry, scaling analysis and many of the standard statistical analyses employed by the biologist are essentially indistinguishable.

The importance of regression and correlation analyses to virtually every aspect of biology is so well known that it hardly seems necessary to draw attention to it. Yet the "obvious" becomes so only when it is pressed into service. In the case of scaling analysis, for example, we can estimate the value of a biological variable for the specified value of the variable (typically, but not invariably, size) against which it is regressed. Consider the scaling of a hypothetical species whose overall height, denoted by H, and overall leaf surface area, represented by S, obtain the proportionality $S \propto H^2$ (i.e., the overall surface of leaves is proportional to the square of plant height). As a statistical consequence of the circumstances of growth, therefore, a taller plant is predicted to have a much larger leaf surface area than its shorter counterpart. It is evident that the statistically verifiable relation $S \propto H^2$ makes no claim regarding the adaptive value, if any, of this scaling relation and that the status of $S \propto H^2$ as a hypothesis relies solely on its predictive capacity in a statistical sense. It also should be clear that different relations among the same variables are possible, such as one that obtains for plants with vestigial leaves or, in extreme cases, the leafless condition ($S \propto H^{-2}$). Although it is easy to speculate on the environmental (adaptive) circumstances that would favor $S \propto H^{-2}$ (dry or "xeric" conditions) or $S \propto H^2$ (wet or "hydric" conditions), these suppositions involve hypotheses that transcend statistical inference.

One of the objectives of this book is to show that, in contrast to the empirical (statistical) approach to the relations between absolute and relative size and between organic form and process, scaling relations can be analytically derived. That is, the way biological variables correlate with size may be adduced from "first principles," leading to a hypothesis very different from that obtained by statistical inference. This is so because analytically derived scaling relations are based on some argument of similitude. The first recorded argument of similitude is found in Galileo Galilei's *Discorsi e dimostrazioni mathematiche* (1638; see English translation by Crew and De Salvio 1914). In this book Galileo demonstrated from first principles that, to maintain comparable breaking strengths, the shapes of prismatic support members (columns, beams, and shafts) must change as size increases. Based on the size-dependent changes in the ratio of surface area to volume, Galileo also argued that shape must change among related organisms differing in size whenever physiological similarity is required. Arguments of similitude reverberate through the works of Herbert Spencer (1868), Simon Schwendener (1874), D'Arcy Thompson (1917), Alfred Lotka (1956), Sir Julian Huxley (1958), Nicolas Rashevsky (1960), and other theorists of more recent vintage. Thus the intellectual influence of these arguments is undeniable.

Arguments of similitude will be presented not because they are historically important, although this is sufficient justification, but because they can provide important insights into biology whenever verified by data drawn from phyletically or ecologically diverse organisms. The appearance of similar scaling relations among very different and unrelated organisms is not likely to occur by chance alone and therefore provides strong circumstantial evidence for convergent or parallel evolution. Consider geometric similitude, which assumes that geometry and shape are size-independent properties. This assumption obtains for a number of mathematically a priori relations among properties like surface area, volume, and mass. For example, as shown by Galileo, the surface area S of objects differing in size but sharing the same geometry and shape unavoidably scales to the 2/3 power of volume V: that is, $S \propto V^{2/3}$. If the rate of a metabolic process associated with V is assumed to depend on S, then the rate of the process must also scale as $V^{2/3}$. The additional presumption that the bulk density of organisms differing in size is size independent leads to the additional prediction that the metabolic rate will scale to the 2/3 power of an organism's mass. Regardless of their apparent elegance, however, scaling hypotheses derived from first principles are notoriously fragile, often shattering in unexpected ways when confronted by hard data obtained from the real world of biology. Galileo's suppositions about the relations among surface area, volume, and mass, although mathematically true, generally do not hold for living things as they grow in size and develop, because organisms differing in size do not subscribe to the same shape, geometry, or bulk density and therefore do not fulfill the preconditions of geometric similitude. The delightfully brutal confrontation between hypothesis and empirical observation, which is a hallmark of every science, provides one of the major themes running through the discussions of scaling analysis that follow.

The agenda of this chapter is to develop the major themes whose variations emerge in subsequent discussions by exploring the unassuming mathematical tools essential for scaling analysis and applying them to a variety of size-correlated phenomena attending the growth in size and the development of external form and internal structure of plants.

1.2 Growth and Development

The concepts of growth and development are understood intuitively by every biologist, although it is undeniable that precise definitions for these terms are elusive, given the diversity of life. Operational definitions for growth and development are required, however, if only because they make communication less ambiguous. Figure 1.1 diagrams some of the relationships between plant growth and development. Some of these relationships apply equally to animals, others do not. For my purposes, in its strictest sense, growth is defined as any permanent increase in size, while development is defined as any orderly change in external shape (morphogenesis) or internal structure (histogenesis). Unlike most animals, in plants morphogenesis is influenced by the physical presence of cell walls. An increase in the volume of the protoplast contained within the cell wall requires a relaxation of the elastic properties of the wall by means of the physiological activity of the protoplast. Permanent cell wall yielding is required for growth. Likewise, differential plastic deformation is required

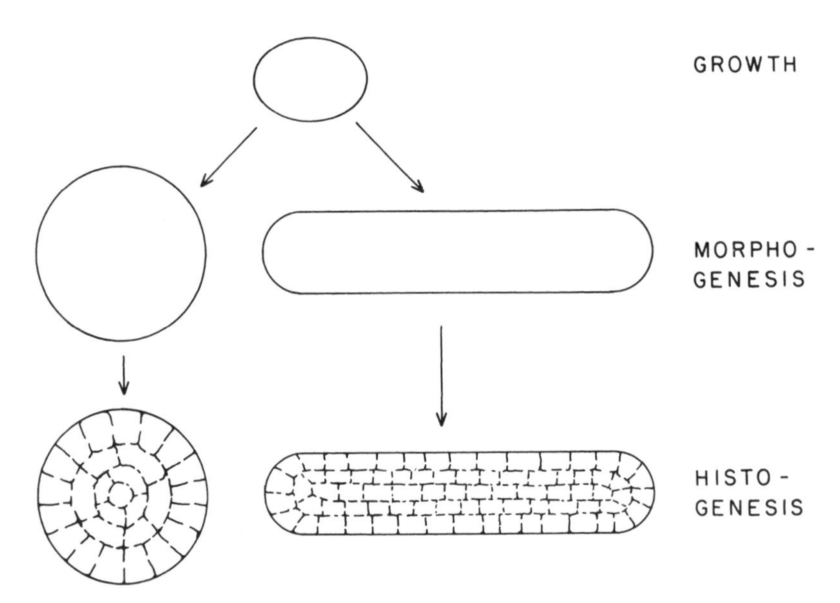

Figure 1.1. Schematic relations between growth and development (morphogenesis and histogenesis).

to change cell shape; that is, the cell wall must exhibit anisotropic mechanical behavior when subjected to an isotropic internal hydrostatic force. For a multicellular plant, organogenesis requires duplicitous alterations in the shapes and sizes of cells. Plant morphogenesis typically requires growth in size regardless of whether the plant is unicellular or multicellular in construction. This condition contrasts with that seen for animals, which, in the absence of cell walls, can modify overall shape by means of cell migration, translation, or the like. The production of the cell wall, which is a characteristic of most plant species, constitutes a significant distinction between plants and animals in the way morphogenesis is accomplished.

Neither morphogenesis nor growth in size need attend histogenesis. Cell walls can be deposited within a plant without growth in size. Growth, however, typically precedes anatomical changes, as shown by the fact that tissue differentiation occurs in closer proximity to actively growing (meristematic) regions as the rate of growth declines. Nonetheless, a protoplast can secrete internal cell walls without changing external shape or absolute size (fig. 1.1), as in the development of multicellular endosporic gametophytes.

The distinctions drawn between growth and development and between morphogenesis and histogenesis draw attention to the possibility that potentially independent processes attend the expression of plant internal and external form. Circumstantial evidence for this comes from observing unicellular and multicellular plants grown under natural or experimental conditions. Dependent on the forces exerted by the mechanical action of waves, naturally growing plants of *Caulerpa laetevirens* and *C. racemosa*, which are unicellular species, take on different external appearances and vary in how much their protoplasts are structurally reinforced with internal cell wall extensions called trabeculae (fig. 1.2). In high-energy wave environments, growth in height is reduced, and the trabecular extensions of the cell wall become thickened and more numerous. The horizontal portions of the unicellular plant develop numerous excrescences and protuberances and become felted together. These modifications in morphology and internal structure are functionally advantageous. They strengthen the cell wall, reduce the exposure of vertical portions of the cell to the shearing effects of waves, and enhance the plant's ability to remain attached to its substrate. The organs of multicellular plants likewise have the potential for size-dependent changes in shape and structure, as is shown by sun and shade leaves, all of which begin their development as small multicellular protrusions (leaf primordia) extending from the apical meristematic dome of a shoot yet differ in size, shape, and anatomy when mature.

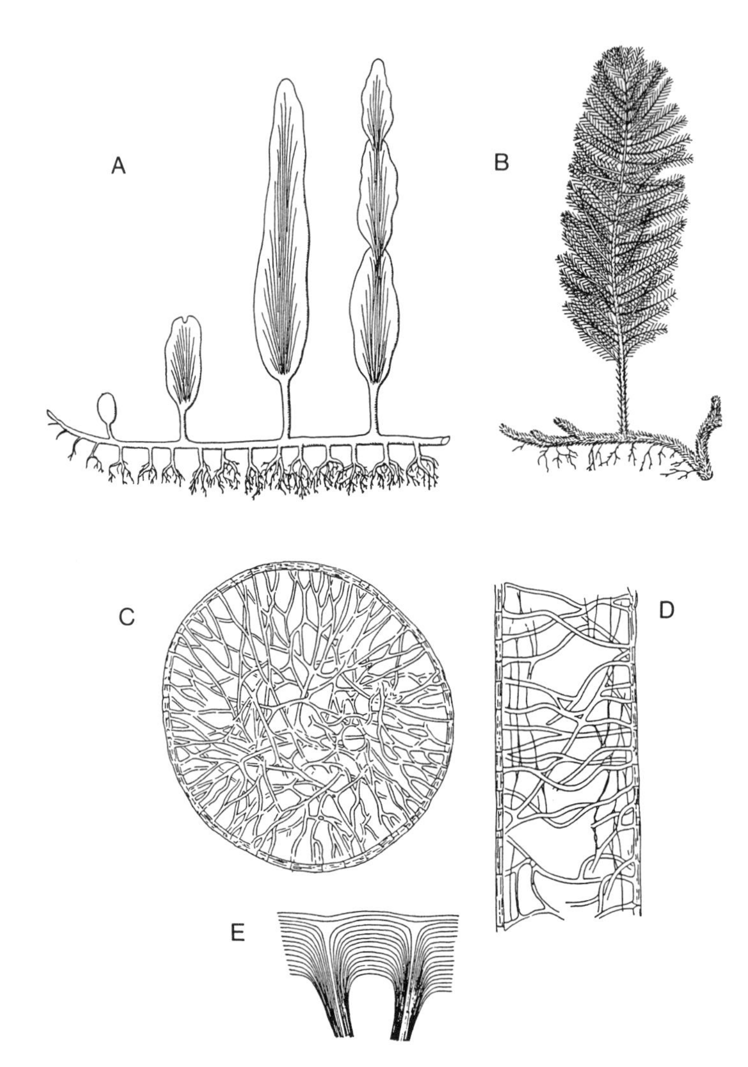

Figure 1.2. Morphology (A–B) and internal structure (C–E) of the siphonous (unicellular) green alga *Caulerpa*. (A–B) The external appearance of two species of *Caulerpa* evincing leaf-, stem-, and rootlike structures. (C–D) Schematics of a transverse and longitudinal cross section through a *Caulerpa* plant, showing internal extensions of cell walls (called trabeculae). (E) Diagram of the superimposed cell wall layers at the base of two trabeculae extending into a *Caulerpa* plant.

Experimental manipulation of plants also shows that patterns of histogenesis and morphogenesis can proceed independent of one another (e.g., Tenopyr 1918; Sinnott 1939; Golub and Wetmore 1948; Poethig 1987; Kaplan and Hagemann 1991). Also, progressive transformations in the shape and size of an organ can occur without benefit of cell division (Haber and Luippold 1960; Haber, Carrier, and Foard 1961, reviewed by Kaplan and Hagemann 1991). The possibility that external and internal form are under separate genetic control was formalized by Haber (1962) in the following scheme:

specifies → form (morphology)
genetic information →
specifies → cell shape and size (anatomy).

This scheme resonates with the distinction drawn by Paul Green (1976) between *surface extension* and *cell partitioning* in reference to the traditional concepts of *growth in form* and *growth in cell division*. Shape and structure therefore may be separate expressions of genetic information, as suggested by Green (1976) and others (Kaplan and Hagemann 1991). Nonetheless, changes in shape and internal structure must correlate with one another by virtue of the influence of the cell wall on plant morphogenesis and growth in size. Indeed, Haber's scheme identifies cell form, shape, and size as components of a single block of genetic information. When stressed or grown under significantly different conditions, a genotype may have the capacity to relax this correlation, but it cannot abandon it entirely except in traumatic and likely lethal circumstances. For example, the morphology and cell wall structure of *Caulerpa* plants differ as a function of environment, but conspecifics are identifiable as such by the correlations among cell shape and cell wall structure. Likewise, the size, shape, and anatomy of foliage leaves borne on the same plant often differ, but these properties evince strong correlations (see Wylie 1951 for an early and informative study of sun and shade leaves). In summary, the available evidence shows that each plant has two developmental routes (morphogenesis and histogenesis) with which to respond to size-dependent phenomena, but that each of these processes is dependent on and therefore correlated with the other.

1.3 Multicellularity and Modular Construction

We now turn to multicellularity, which provides numerous physiological and mechanical benefits by virtue of the compartmentalization of the living biomass of the plant body into cells. The formation of cells increases the overall surface area through which metabolites diffuse. At the tissue level, this is perhaps best seen in transfer cells, whose inner walls are highly folded and therefore increase the contact area among neighboring cells. Multicellularity, which increases the surface area of plasma membranes as a whole, also permits the formation of preferred routes for the translocation of metabolites and water by modification of the cell wall infrastructure. For example, as cell walls are secreted, the frequency distribution of plasmodesmata among neighboring cell walls can be adjusted to create preferred routes for the active transport of nutrients (e.g., sieve tube members of phloem tissue; see Biao et al. 1988). Likewise, the coordinated death of the protoplasts within mature elongated cells can produce intracellular conduits designed to offer little resistance to the passage of water (e.g., vessel members of xylem tissue). For a comparatively small plant whose relative surface area is large, these physiological benefits are not crucial. As the comparative size of an organism increases, however, surface-dependent functions (respiration, photosynthesis, and in some cases locomotion) and volume-dependent functions (translocation) suffer from the tendency of the ratio of surface area to volume to decrease. This phenomenology led Ryder (1893) to explicitly suggest that size-dependent variations in physiological processes are the formal cause of cell division. Clearly other factors are important, but it is fair to say that scaling influences optimal cell size.

Multicellularity confers mechanical benefits as well. The secretion of internal cell walls provides an infrastructure of *beams* and *columns* that offers resistance to the internal mechanical forces generated by the influence of gravity on the living biomass (self-loading) and by the movement of the ambient fluid (water or air) against obstructions like leaves and stems that result in dynamic loadings. From first principles, we know that the magnitude of self-loading depends on the size and orientation of a plant as well as the density of the fluid medium in which it grows. Small prostrate terrestrial plants experience little or no compression due to the influence of gravity on their biomass. Large vertically oriented plants experience proportionally larger compressive forces as they grow in height. By contrast, even very large aquatic plants experience little or no self-loading, because the bulk density of plant tissues is not much greater than the density of water. Water, however, is roughly a thousand times denser than air. Aquatic plants living in rapidly moving water therefore often experience higher magnitudes of dynamic loadings than their terrestrial counterparts of equivalent size. Recollecting how *Caulerpa* plants elaborate their trabeculae in response to wave activity shows that mechanical support is a design consideration for aquatic as well as terrestrial plants.

In summary, the adaptive value of multicellularity is complex, because the benefits provided by partitioning an organism into cells depend on the size of the organism and the environment in which it fulfills its functional obligations. This complexity is emphasized by exploring how the physical properties of a particular environment influence physiological and mechanical phenomena and how growth and development influence the way physical processes operate on the organism (see chaps. 2 and 3).

Among multicellular plants (metaphytes), changes in size, shape, and structure occur in a curious developmental milieu that allows the individual plant to increase in size indefinitely by periodically and cyclically adding additional organs (e.g., leaves) or functional groups of organs (e.g., shoots), many of which are determinate in their growth. Because of this pattern of growth and development, most plants have a modular construction. Each module experiences a single episode of morphological and anatomical modification until it reaches its functionally mature size. Modular construction confers a number of useful features. To some degree, each module is expendable and can be replaced with a new one when damaged. As each module grows, its morphology or anatomy can be modified in response to local environmental conditions attending organogenesis. Over the course of evolution, each module can be developmentally modified to assume different functional roles and can be combined and modified to produce novel compound structures. Well-known examples suffice: the abscission of senescent leaves; the formation of sun and shade leaf morphologies on the same plant; the assumption of a variety of functions by the developmental products of leaf primordia (photosynthetic foliage leaves, protective bud scales and spines, digestive organs among carnivorous plants, etc.); and the formation of compound reproductive structures (flowers, cycad strobili, pinecones, etc.). An additional advantage is that the scaling of shape and structure may be dealt with at the level of the organ (whose growth is typically determinate) rather than the individual plant (whose growth is typically indeterminate). This is highly effective because it is considerably easier to developmentally adjust shape to a final "anticipated" size than it is to continuously modify the overall shape and structure of an individual plant that increases indefinitely in size.

1.4 Size and Shape

The terms size and shape have been used rather glibly, even though, as with the terms growth and development, most biologists have an intuitive grasp of their meaning. As we shall see, however, intuition often is highly inadequate. Indeed, the definition of size and shape can be as elusive as the definition of growth or development.

Following the terminology of Ipson (1960), it is useful to begin this topic by noting that biological as well as nonbiological features are described in terms of *substantial* and *natural* variables. Substantial variables are expressed in terms of physical quantities that provide the units of measurement. In science, seven fundamental physical quantities are agreed upon, as listed in table 1.1. The kilogram and the meter are the fundamental physical quantities used to express mass and length, respectively. Substantial variables also can be expressed in terms of derived physical quantities based on combinations of the seven fundamental physical quantities. There are seventeen derived physical quantities; some of the more important are listed in table 1.1. Regardless of whether a substantial variable is measured in terms of a fundamental or a derived physical quantity, its magnitude must be associated with the units of measurement.

Size is a substantial variable. It can be expressed in units of a fundamental physical quantity, like mass, or in units of a derived physical quantity, like volume. The size of any object can be measured in a variety of useful ways. Nonetheless, the manner of measurement typically is operationally contingent on the object of a particular study. Confusion often results because of the failure to note that fundamental physical quantities, and thus different measurements of size, are not interchangeable. The size of a leaf can be measured in terms of its mass, surface area, or volume,

Table 1.1. Units and Dimensional Symbols of Fundamental and Some Derived Physical Quantities

Quantity	Unit	Dimensional Symbol
Fundamental		
Length	meter	L
Mass	kilogram	M
Time	second	T
Temperature	kelvin	K
Amount of substance	mole	S
Light intensity	candela	I
Electric current	ampere	A
Derived		
Volume	cubic meter	L^3
Force	newton	MLT^{-2}
Density	kilogram per cubic meter	ML^{-3}
Acceleration	meter per second squared	LT^{-2}
Production rate	watt	ML^2T^{-3}
Power	watt	ML^2T^{-3}
Pressure	pascal	$ML^{-1}T^{-2}$

or the ratio of surface area to volume. Each of these measurements has magnitude, and each is expressed in units of a fundamental or derived physical quantity. But we cannot convert measurements of mass into measurements of surface area, volume, or the ratio between these two variables without additional information (i.e., density). As a consequence, the results of different studies based on different methods of quantifying size are often difficult to interrelate.

Natural variables are not based on fundamental or derived physical quantities: their descriptions do not require reference to external, artificial standards. Rather, they derive their meaning from the physical system they describe. Although a natural variable has magnitude, it lacks units. Unlike measurements of size, shape is a dimensionless quantity "from which all information about position, scale, and orientation has been drained" (Bookstein 1978, 8). One of the most convenient methods of describing shape is the construction of an aspect ratio, henceforth denoted as $\Re$. The ratio of the length to the unit radius $L{:}R$ of a terete cylinder, which is called the *slenderness* or *fineness* ratio, is an example of $\Re$. Note that this $\Re$ describes the shape of a cylinder without reference to an external, artificial physical quantity. The ratio of the cube of surface area to the square of volume $S^3{:}V^2$ also describes the shape of a terete cylinder or any other three-dimensional object. Thus a variety of natural variables can be constructed to describe the shape of the same object.

Size provides an ordering principle for a population of related objects, while shape may be used to distinguish among objects, although it does not often lend itself to ordering them according to size. When dealing with scaling, our concern is to order measurements of size and shape into a progression that either reflects growth and development (if our concern is the scaling of the individual) or reflects size-correlated differences among individuals drawn from a population (if our concern is an intraspecific scaling relation). In either case, ordering objects differing in size and shape is never a simple task. Consider the triangle, whose size can be measured by area, length of longest side, or something else, and whose shape may be expressed in terms of its internal angles, although other expressions for shape are easily contrived. In geometry, a population of triangles possessing the same angles constitutes a single category of shape regardless of the range of size it contains. Conversely, a population of triangles that all have the same area but possess different angles is judged to belong to a single category of size but different categories of shape. A collection of triangles varying in size and shape, therefore, contains different categories of both.

It is worth emphasizing that all triangles belong to the same category of geometry (three-sided planar figures), but not all triangles have the

Table 1.2 Mensuration Equations for Surface Area S, Volume V, and $S{:}V$ and Aspect Ratios $\Re$ of Some Common Geometries

	S	V	$S{:}V$	$S^3{:}V^2$	
Sphere	$4\pi R^2$	$\frac{4}{3}\pi R^3$	$\frac{3}{\Re R}$	$\frac{36\pi}{\Re}$	where $\Re = 1$
Cylinder	$2\pi RL$	$\pi R^2 L$	$\frac{2}{\Re L}$	$\frac{8\pi}{\Re}$	where $\Re = R/L$
Cone	$\pi R_1(R_1^2 + L^2)^{1/2}$	$\frac{\pi}{3}R_1^2 L$	$\frac{3}{\Re L}(R^2 + 1)^{1/2}$	$\frac{729}{\Re}(\Re^2 + 1)^3$	where $\Re = R_1/L$
Spherical segment	$2\pi RL$	$\frac{\pi}{3}L^2(3R - L)$	$\frac{6\Re}{L(3\Re - 1)}$	$\frac{216\Re^3}{(3\Re - 1)^2}$	where $\Re = R/L$
Frustum of cone	$\pi(R_1 + R_2)[L^2 + (R_1 - R_2)^2]^{1/2}$	$\frac{\pi}{3}L(R_1^2 + R_2R_2 + R_2^2)$	$\frac{3(\Re + 1)[1 + \Re_0^2(\Re - 1)^2]^{1/2}}{L\Re_0(\Re^2 + \Re + 1)}$	$\frac{729(\Re + 1)^6[1 + \Re_0^2(\Re - 1)^2]^3}{\Re_0^2(\Re^2 + \Re + 1)}$	where $\Re = R_1/R_2$ $\Re_0 = R_2/L$
Prolate spheroid	$2\pi b^2 + \frac{2\pi ab}{e}(\sin^{-1} e)$	$\frac{4}{3}\pi ab^2$	$\frac{3}{4\Re a}\left[2\Re + \frac{\sin^{-1} e}{e}\right]$	$\frac{1.69}{\Re}\left[\Re + \frac{\sin^{-1} e}{e}\right]^3$	where $\Re = b/2a$ $e = (1 - 4\Re^2)^{1/2}$
Oblate spheroid	$2\pi a^2 + \frac{\pi b^2}{e}\ln\frac{(1 + e)}{(1 - e)}$	$\frac{4}{3}\pi a^2 b$	$\frac{3}{4\Re a}\left[1 + \frac{2\Re^2}{e}\ln\frac{(1 + e)}{(1 - e)}\right]$	$\frac{1.69}{\Re^2}\left[2 + \frac{\Re^2}{e}\ln\frac{(1 + e)}{(1 - e)}\right]^3$	where $\Re = b/2a$ $e = (1 - 4\Re^2)^{1/2}$

Note: a, b = major and minor semiaxes of elliptical cross section; e = eccentricity of ellipse = $(a^2 - b^2)^{1/2}/a$; L = length; R unit radius; R_1 = bottom radius; R_2 = top radius; ln = natural logarithm.

same shape. This is somewhat counterintuitive but nonetheless a simple fact of geometry. Shape and geometry are not synonymous terms, nor are shape and symmetry (for a beautiful discussion of the latter, see Weyl 1952). The failure to distinguish between shape and geometry (or symmetry) produced much confusion in the early literature. Table 1.2 illustrates the difference between shape and geometry for some simple solid geometries. It gives the equations for calculating size (measured as surface area S, volume V, or $S:V$) and shape (measured in terms of $S^3:V^2$) for spheroids, terete cylinders, and the like. Each equation contains a manifestation of $\mathfrak{R}$, thereby emphasizing that shape can vary within each category of geometry. Unfortunately, simple measurements of shape such as $\mathfrak{R}$ fail to address the question, "Whence and what art thou, execrable shape?" posed by John Milton (*Paradise Lost,* 2.681).

Unlike a collection of triangles consisting of objects separated in space, all the categories of size and shape resulting from the growth and development of an organism are expressions of the same object separated by intervals of time. Each category is logically, though not biologically, independent of those that temporally precede or follow, provided we are ignorant of the underlying mechanism responsible for their ontogenetic relationship. A continuous progression of size and shape may in fact be the result of numerous processes whose operations are discontinuous but overlap in time. In terms of growth and development, what we call biological structures may legitimately be viewed as slow processes of long duration, and what we call functions often are appropriately viewed as quick processes of short duration. This is not metaphor, nor is it a refuge for ignorance. The measurement of by-products is not the measurement of mechanism. And we must constantly be on the lookout to avoid the danger of confusing effect with cause.

Figure 1.3 illustrates some of the points raised so far. The data represent measurements taken from twenty-one internodal cells of the green alga *Chara corallina.* Each of these cells is geometrically approximated by the terete cylinder with unit radius R and length L. Shape therefore can be expressed in terms of two natural variables ($\mathfrak{R} = R:L$ or $S^3:V^2$), each of which has magnitude but lacks units. Size is given in terms of L measured in units of meters (a fundamental physical quantity). Inspection of figure 1.3 shows us that no two cells possess the same shape as defined by either natural variable. Clearly, both cell size and shape provide reasonable ordering principles, although we could not have ordered cells precisely according to their size or shape. More important, nothing in the structure of these data gives insight into why cell shape evinces size dependency. This shows us that we can order objects without any understanding of their biological relationships. The paleontologist may wish to consider what effect this has on the interpretation of phylogeny from morphological differences among species for which we have no genetic information.

1.5 Scaling Principles

(Readers unfamiliar with statistics should now read the appendix.)

Scaling refers to any size-dependent variation in organic form or metabolism (Huxley 1932; Gould 1966; Reiss 1989). Figure 1.3 illustrates the size-dependent variation in the shape of *Chara* internode cells. It distinguishes between two categories of scaling analysis that, for convenience, I call empirical and analytical (see sec. 1.1). The former results from any statistical regression analysis of two or more variables, and the resulting regression equation identifies an observed size-dependent relationship. This is a purely descriptive (statistical) approach to the consequence of variations in size on organic form or physiology. For example, ordinary least squares (henceforth denoted as LS) regression of the data plotted in figure 1.3 obtains the formula $S^3{:}V^2 = 25.3\ L^{-3.21}$. This for-

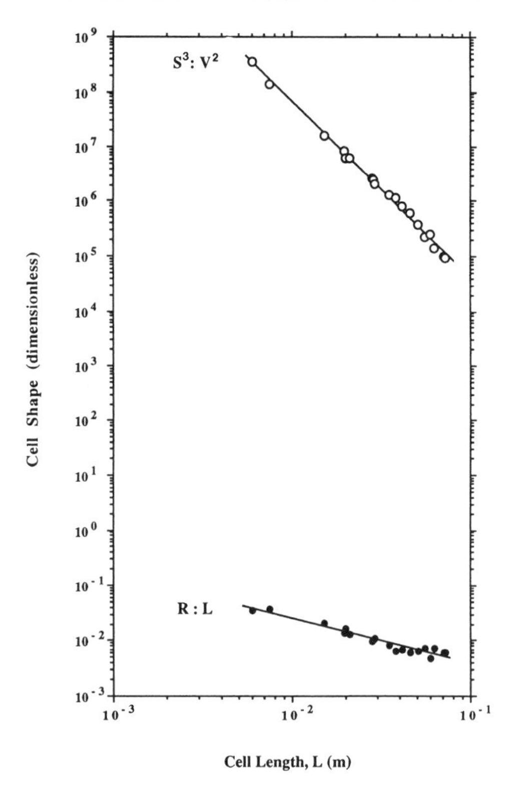

Figure 1.3. Log-log (base 10) plot of cell shape (defined in terms of two aspect ratios, $S^3{:}V^2$ and $R{:}L$) versus cell length L based on measurements taken on the internodal cells of the green alga *Chara corallina*.

mula, whose complete rendition requires the statement of the standard errors of the regression coefficient and regression exponent as well as the confidence intervals for the latter, merely describes the statistical tendency of twenty-one randomly selected internode cells. And, with considerable caution, we may employ it to crudely estimate $S^3{:}V^2$ for a specified value of L. By contrast, an analytical argument for a particular scaling relation a priori predicts size-dependent variations in form or metabolism. It adduces why these variations are obtained for all or most organisms. Such a scaling equation is derived from first principles and therefore must subsequently be tested against observations. Analytical scaling relationships are designed to probe statistically verifiable relations so as to understand why differences in size may require changes in form or metabolism.

An analytical approach to the data plotted in figure 1.3 might start by noting that the growth of *Chara* internode cells, which are geometrically similar, depends on passive diffusion for the exchange of water and other nutrients between cells and their external environment. The surface area of objects differing in size but sharing the same geometry and shape must scale as the 2/3 power of volume. Accordingly, the relationship between $S^3{:}V^2$ and L for *Chara* might be expected in light of the desirability of keeping physiological rates constant regardless of cell size. That some size-correlated differences in proportion are required is proposed by this approach. Whether these differences are the result of unavoidable relationships among size, shape, and metabolism must be examined empirically. Although the juxtaposition of the two types of analyses is necessary whenever an analytical approach is taken, the distinction between them must never be lost. Observed scaling relationships may be *required,* or they may be *permitted* within the biological context in which they are observed. The resolution of which of these two possibilities legitimately explains a biological phenomenon is not a simple task, yet it lies at the heart of a variety of hypotheses.

The most commonly used scaling equation is the power function equation:

$$Y_1 = \beta Y_2^{\alpha}, \qquad \text{(1.1a)} \qquad \text{(power function equation)}$$

where Y_1 is the variable of particular interest, Y_2 is the variable measuring size, and α and β are parameters describing the functional (mathematical) relation between Y_1 and Y_2. As noted in the appendix, scaling studies rarely deal with independent variables, because size is subject to natural variation and measurement error. Therefore, although the literature treating "allometry" typically gives eq. (1.1a) as $Y = \beta X^{\alpha}$, we should avoid the implication that size is an independent variable. The roles played by

α and β in defining a relationship are seen explicitly when eq. (1.1a) is transformed into linear form:

$$\log Y_1 = \log \beta + \alpha \log Y_2, \tag{1.1b}$$

which indicates that α is the slope of the log-log plot (i.e., the change in Y_1 with respect to Y_2) and β is the value of Y_1 when $Y_2 = 1$. In the scaling literature, α is referred to as the scaling (= regression) exponent, and β is called the scaling (= regression) coefficient. I shall return to the importance of these parameters shortly. For now, however, note that logarithmic transformations of data tend to diminish the differences among large numbers and accentuate the differences among small numbers. They are useful when the untransformed data do not have a normal frequency distribution, for example, when we are dealing with interspecific comparisons for which small species typically outnumber larger species. Log transformation also reduces the statistical problems resulting from a number of "outlying" data points. A mathematical consequence of dealing with log transformation, however, is that the mean log *Y* is the median *Y* of the distribution of untransformed values of *Y*. This introduces a systematic bias in the regression parameters α and β. Mathematical techniques permit correction of this bias (see appendix, eq. A.11). In the final analysis, logarithmic transformations are no more subject to abuse than any other system of computation.

The units of Y_2 depend on how size is defined, since this definition subsequently determines the standard of measurement (e.g., kilograms, meters). As noted, when the size of a plant is defined in terms of cell volume *V*, then size can be expressed in units of cubic meters per cell (m^3 $cell^{-1}$) or convenient multiples or submultiples of meters. If size is defined in terms of cell mass, then size can be expressed in terms of picograms of carbon content per cell (pg C $cell^{-1}$) or the like. Y_1, however, can be any measurement of size, any measurement of shape (see table 1.1), or any measurement of metabolic rate (e.g., growth in units of pg C $cell^{-1}$ hr^{-1}). Thus Y_1 and Y_2 can be any substantial or natural variables, although caution must be exercised when both are measurements of shape, since changes in size need not attend changes in shape.

How closely Y_1 and Y_2 are correlated obviously determines how well we can predict the value of one variate from the value of the other. Ordinary LS regression analysis and correlation analysis generally are adequate when the objective of a scaling analysis simply is to predict Y_1 based on Y_2. When the objective of a scaling analysis is to determine the numerical value of the scaling exponent α in order to test the predictions of an analytical scaling relation, however, LS regression generally is inadequate, particularly when the coefficient of correlation is small. Alternative regression techniques are discussed in the appendix. For convenience and uniformity—both admittedly problematic reasons—I have elected to use reduced major axis regression, denoted henceforth as RMA regression, to determine the scaling exponent for data against which analytical arguments are tested. Throughout this book, the scaling exponent obtained by RMA regression is reported as α_{RMA}. It is useful to know that the scaling exponent determined by RMA regression equals that calculated by LS regression divided by the coefficient of correlation ($\alpha_{RMA} = \alpha_{LS}/r$).

As noted in the appendix, we require the confidence intervals for α_{RMA} to test an analytically derived scaling relation as well as to determine whether two empirically determined scaling relations differ from one another. Thus we need to know the standard error, *SE*, for α_{RMA}, which incidentally is the same for α_{LS}. Typically, confidence intervals are determined for the 95% level (see appendix, eq. A.7). With very few exceptions, therefore, the numerical value, the standard error, and the lower and upper 95% confidence intervals for α_{RMA} are reported throughout this book.

Generally, the scaling exponents predicted by geometric similitude serve as the null hypothesis. Geometric similitude obtains whenever a series of objects differing in absolute size share the same geometry and the same shape (when the two linear dimensions of geometrically identical objects differ by the same multiplier). Geometric similitude is evinced by an isometric (= allometric) relation between Y_1 and Y_2. As we shall see, however, an isometric relation between two variables may not be taken as prima facie evidence for geometric similitude (see Prothero 1986, 275, fig. 5). With this caveat in mind, isometry is indicated when the scaling exponent of the log-log plot of Y_1 versus Y_2 equals unity ($\alpha = 1$), provided Y_1 and Y_2 are measured in the same units. That is, allometry indicates that Y_1 varies in direct proportion to Y_2. When Y_1 and Y_2 do not have the same units, the value of α for anisometric relationship is easily determined from a dimensional analysis (see sec. 1.6). For example, volume V and surface area S are expressed in the same units (length L) but differ in the exponents for L (volume has the dimensions of L^3, and surface area has the dimensions of L^2). Thus when Y_1 is volume and Y_2 is surface area, the value of the exponent α for geometric similitude is given by the ratio of the exponents of the ratio L^3:L^2 (i.e., $\alpha = 3/2$). Likewise, when $Y_1 = L$ and $Y_2 = V$, then L^1:L^3 and $\alpha = 1/3$ or 0.333, and so forth. When $\alpha \neq 1$, the relation between Y_1 and Y_2 sharing the same units of measurements is anisometric. That is, Y_1 and Y_2 do not vary in direct proportion and therefore geometric similitude is violated.

Not all scaling relations need take the form of the power function equation. In many cases, for example, in many intraspecific comparisons, a simple linear relation is obtained when Y_1 is regressed against Y_2. In the case of the simple linear function,

$$Y_1 = b + mY_2, \qquad \text{(simple linear equation)} \tag{1.2}$$

where b is the Y_1-intercept (the scaling coefficient) and m is the slope (the scaling "exponent") of the linear relationship. For a simple linear scaling relation, geometric similitude is demonstrated when b equals zero, provided the units of Y_1 and Y_2 are identical. That is, $Y_1 = mY_2$ and therefore $m = Y_1/Y_2$, which shows that the ratio of Y_1 to Y_2 is constant for each increment of Y_2. Thus an anisometric relation is indicated when the value of b differs significantly from zero. That is, even though the ratio $Y_1{:}Y_2$ is constant, for each increment of Y_2, a constant $Y_1{:}Y_2$ is added to an initial value of Y_1 when $Y_2 = 0$. This can result from size-dependent variations in either geometry or shape.

A curious footnote to our discussion of eqs. (1.1) and (1.2) is that both may yield high and very similar correlation coefficients when applied to the same data set, particularly when Y_1 and Y_2 each span one order of magnitude. There are a variety of statistical techniques that can be used to resolve which equation is the more appropriate (see appendix).

Many of the foregoing principles of scaling analysis are best illustrated with real data. Figure 1.4 shows an empirically determined nonlinear anisometric relationship. It is a log-log plot of measurements of cell surface area S (in units of m^2) versus cell volume V (in units of m^3) from twenty-one *Chara* internodal cells. Qualitatively, figure 1.4 shows us that cell surface area variations attend changes in volume. Least squares regression of the data reveals that eqs. (1.1) and (1.2) both yield nearly equal coefficients of determination. Specifically, the power function regression formula ($S = 234\ V^{0.821}$) obtains $r^2 = 0.98$, while the simple linear regression formula ($S = 0.00002 + 4118\ V$) gives $r^2 = 0.97$. Accordingly, these equations have nearly equivalent predictive value. But the data for S as well as V span three orders of magnitude, suggesting that the power function regression formula likely best describes the scaling of S with respect to V. The LS scaling exponent for the relation between S and V is $\alpha_{LS} \pm SE = 0.821 \pm 0.02$. Thus $\alpha_{RMA} = \alpha_{LS}/r = 0.821/0.99 \approx 0.83\ (\pm\ 0.02)$. The lower and upper 95% confidence intervals for α_{RMA} are 0.79 and 0.88, respectively, so the scaling of S with respect to V is anisometric. This is graphically illustrated by the solid diagonal line in figure 1.4, which shows the hypothetical isometric relationship between S and V ($\alpha \approx 0.667$), calculated from the magnitudes of S to V for the smallest of the twenty-one cells. Real data points diverge farther from the diagonal line as V increases in magnitude, meaning that these cells evince size-dependent variations in either geometry or shape. Since all internode cells have the same geometry (a terete cylinder), it seems reasonable to assume that alterations in cell shape account for the anisometry of S. Indeed, the internode cell of *Chara* typically elongates while altering its circumference little, and therefore the departure from geometric similitude is a consequence of size-dependent variations in cell shape rather than geometry. Note, however, that the sample size is very small (only twenty-one cells were measured), and therefore we cannot claim that the trend shown in figure 1.4 will hold for all *Chara* cells.

Figure 1.5 shows another scaling relationship. The Y_2 variable once

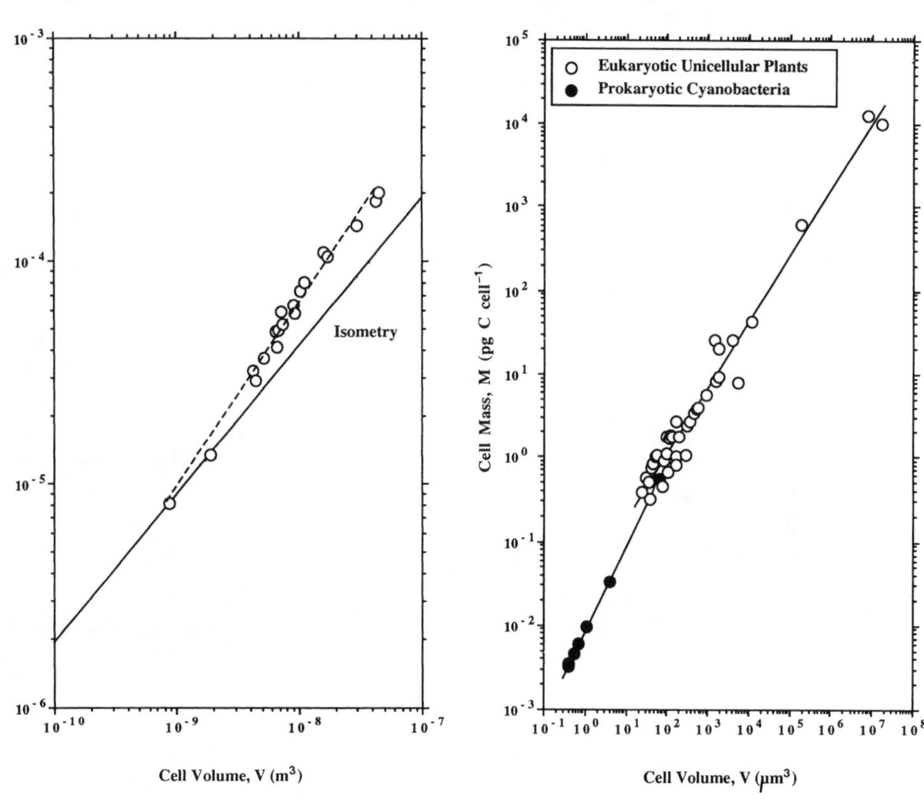

Figure 1.4. Log-log (base 10) plot of cell surface area S versus volume V of the internodal cells of the green alga *Chara*. Solid diagonal line denotes the hypothetical relation resulting from an isometric relation between S and V; dashed diagonal line is the regression curve for the data obtained from LS regression analysis.

Figure 1.5. Log-log (base 10) plot of cell mass M versus volume V for eukaryotic unicellular plants (algae) and prokaryotic cyanobacteria. Solid diagonal lines denote the regression curves obtained for these data from LS regression analysis.

again is volume (now measured in units of cubic micrometers per cell, μm^3), but the Y_1 variable now is cell mass M (measured in units of picograms of carbon per cell, pg C $cell^{-1}$). Values for V and M are from forty-five unicellular eukaryotic plant species and seven prokaryotic species (data from Shuter 1978). The range of each variable is quite large: 10^{-1} $\mu m^3 \leq V \leq 10^8$ μm^3 and 10^{-3} pg C $cell^{-1} \leq M \leq 10^4$ pg C $cell^{-1}$, unlike the limited data set for *Chara* internode cells shown in figure 1.4. Least squares regression of the data from both prokaryotes and eukaryotes yields the power function equation $M = 0.016\ V^{0.865}$ ($r^2 = 0.97$, $N = 52$; $\alpha_{RMA} = 0.88 \pm 0.03$). The 95% confidence intervals for α_{RMA} are 0.94–0.82, demonstrating an anisometric relationship between M and V. However, the RMA scaling exponent 0.88 was computed for the pooled data from organisms that differ in many meaningful ways. Indeed, the scaling exponents of M versus V for prokaryotes and eukaryotes are statistically different at the 1% level. Regression of the data from prokaryotes shows that $M = 0.008\ V^{0.999}$ ($r^2 = 0.99$, $N = 7$; $\alpha_{RMA} \approx 1.0 \pm 0.001$), indicating that M isometrically scales to V, in contrast to the regression formula for eukaryotes, $M = 0.003\ V^{0.794}$ ($r^2 = 0.96$, $N = 45$; $\alpha_{RMA} = 0.81 \pm 0.03$), which reflects an anisometric relation between M and V.

The difference between these two scaling relations is important because it shows that the taxonomic composition of a particular set of data can influence our perception about the scaling relation between Y_1 and Y_2. This is evident from a number of studies. For example, Richard Strathmann (1967) provides scaling equations for estimating the organic carbon content of unicellular marine plants (phytoplankton) from cell or plasma volume. But these empirically determined equations have scaling exponents that vary significantly depending on the taxonomic affiliation of the plants examined: diatoms attain $\alpha \approx 0.76$ ($N = 96$), while other phytoplanktonic species ($N = 13$) scale M with respect to V such that $\alpha \approx 0.87$. Thus we see that the numerical value of the scaling exponent is influenced by the taxonomic composition of the data set. For lack of a better phrase, scaling exponents evince "phyletic dependency." This is not the result of differences in the sample size, nor is it the consequence of data spanning different orders of magnitude in organismic size, although these features do influence α. Rather, it occurs because data from related taxa that share characteristics as a consequence of common ancestry are not independent data points.

A basic assumption of any statistical analysis is that each datum is an independent observation made on the smallest sampling unit. When this assumption is violated a number of problems result, such as "data point inflation," which can dramatically alter the numerical values of the regression parameters α and β. The consequences of phyletic dependency bear on any attempt to test the predictions of an analytical scaling hypothesis derived from first principles. Consider that the analytical approach to scaling predicts a slope based on some physical principle. As such, this approach expects that a broad intertaxonomic comparison will conform to the predicted scaling relation. But the approach also permits scaling exponents to vary numerically both intraspecifically and even among individual closely related genera or families. When testing the predictions of an analytically derived scaling relation, we require very broad intertaxonomic comparisons in which each datum represents a "phyletically" independent observation. Thus, in the absence of a phylogenetic hypothesis (e.g., cladistic suppositions concerning the evolutionary relations among the taxa from which data are gathered), the predictions of an analytical derivation for a scaling relation are difficult to test unambiguously. A variety of statistical techniques may be used to partially circumvent the problems caused by our lack of information on phylogenetic relations. One of these is called nested analysis of variance. This technique is discussed and illustrated with a data set in the appendix (sec. A.6).

Perhaps the broadest intertaxonomic comparison that can be drawn is between plants and animals. These two groups of organisms taxonomically diverged from one another early in the history of life, and therefore any significant convergence in the scaling of their biological features would be strong circumstantial evidence for adaptive evolution. Yet even here we must be extremely cautious. Figure 1.6 illustrates a broad interspecific comparison between plants and animals intended to test the null hypothesis that body mass scales as the cube power of body length. The proportionality $M \propto L^3$ follows mathematically from geometric similitude ($D \propto L^{1.0}$) provided the bulk density ρ of organisms essentially is size independent ($\rho \approx$ a constant). Figure 1.6 is a log-log plot of "fresh weight" body mass (in kg) versus length (in m). These data are from a total of sixty-six unicellular and multicellular plant species and sixty-seven unicellular and multicellular animal species. Virtually every major plant and animal clade is portrayed, and no family is represented by more than a few species. Note the size ranges. Body mass spans twenty-two orders of magnitude; body length spans eight orders of magnitude. These ranges are inclusive of all eukaryotic organisms. Also, the values of M and L are based on direct measurements. As noted, this is important because the magnitude of an "empirically" determined scaling exponent is meaningless whenever Y_1 is estimated from mensuration formulas incorporating Y_2 (or the reverse). Be that as it may, LS regression of the pooled data shows that $M = 1.29\ L^{3.02}$ ($r^2 = 0.91$, $N = 133$; $\alpha_{RMA} = 3.17 \pm 0.08$), indicating that both plant and animal body mass scales roughly as the cube power of body length. The same conclusion is reached when the data from plants and animals are regressed separately—$M_{plants} = 0.12\ L^{2.86}$($r^2$ = 0.94, N = 66; α_{RMA} = 2.95 ± 0.09) and $M_{animals} = 7.51\ L^{2.79}$ (r^2 = 0.97, N = 67; α_{RMA} = 2.84 ± 0.09). Thus it appears that the null hypothesis of geometric similtude is supported in terms of data plotted in figure 1.7.

But do all the organisms considered by this data set share the same geometry and shape? The answer is no. The smaller plants and animals in the data set have a spheroidal geometry (spherical, prolate, and oblate geometries). The largest organisms attain a cylindrical geometry. Like-

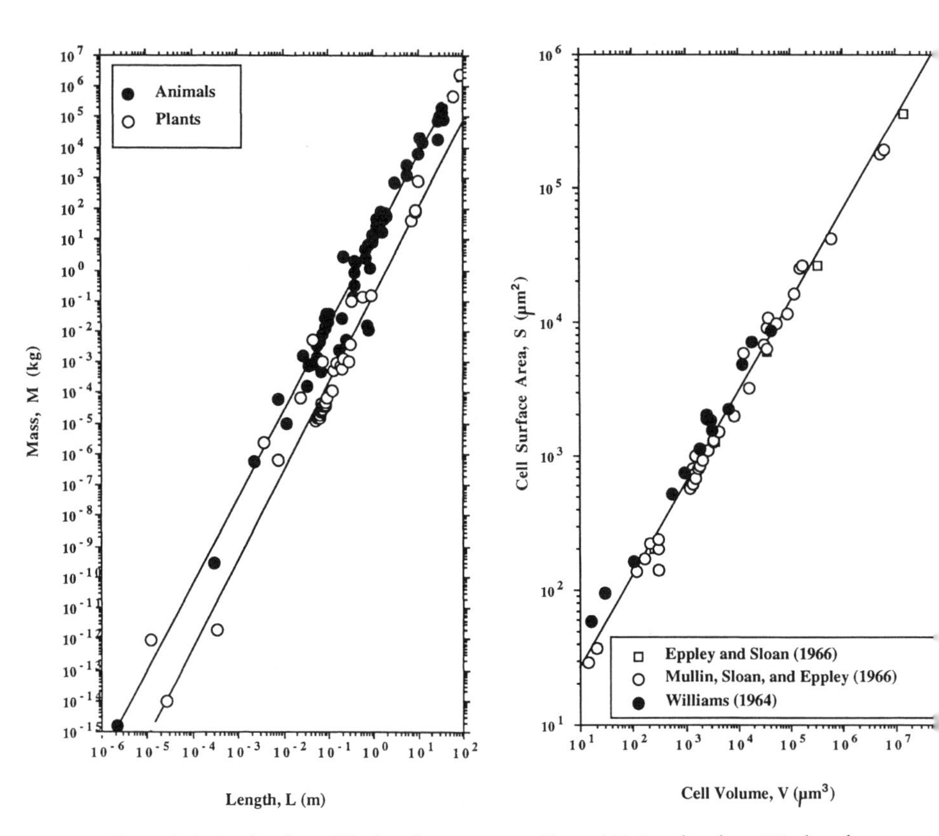

Figure 1.6. Log-log (base 10) plot of body mass M versus length L for unicellular and multicellular plants and animals. Solid diagonal lines denote the regression curves obtained for these data from LS regression analysis.

Figure 1.7. Log-log (base 10) plot of cell surface area S versus volume V for unicellular eukaryotic plants (algae). Data are from three separate reports (see inset). Solid diagonal line denotes the regression curve obtained for the pooled data from LS regression analysis.

wise, within each geometric category, shape varies as a function of body length. We shall see this in operation many times (see fig. 1.8, p. 28), but for now it is sufficient to say that body mass scales to length in a manner predicted by geometric similitude but for very different reasons than adduced by the null hypothesis.

It is worth noting that in the previous examples size was measured in different units as well as in the same units, but in different multiples or submultiples of the same unit (e.g., volume measured as m^3 and μm^3; see figs. 1.4–1.5). Donhoffer (1986) argues that the scaling coefficient α depends on the units used for Y_2. This is not true. If $Y_1 = k\tilde{Y}_2$, where k converts $\tilde{Y}_2$ into the units of Y_2, then inserting $Y_1 = k\tilde{Y}_2$ into eq. (1.1) gives $Y_1 = \beta(k\tilde{Y}_2)^\alpha = \beta k^\alpha \tilde{Y}_2^\alpha$, and therefore α remains unaltered (the lengths of the sides of cubes with equal density are proportional to the cube root of the mass of these cubes regardless of the units in which length or mass is measured).

The final point made in this section is that the relation between α and β for two or more scaling relationships often is determined by the intersection point of the regression curves. When the intersection occurs at $Y_2 = 1$, the values of β are equal to one another regardless of the values of α. When the intersection occurs at $Y_2 > 1$, β varies inversely with α. When the intersection occurs at $Y_2 < 1$, β varies directly with α. From time to time workers have noted that there exists an apparent relation between the values of α and β for a collection of scaling equations approximately expressed by a decreasing exponential function of the form $\beta = ke^{-r\alpha}$, where k is a proportionality constant. If this function is valid, then the scaling equations being compared constitute a family of straight lines (when plotted on log-log paper) intersecting at a common point whose coordinates are (α, log k). In most cases, however, no common point of intersection is seen, regardless of the apparent correlation between α and β for the set of scaling equations considered. The reason, which is particularly instructive, is that many regression curves may be equally effective in their ability to describe the relation between Y_1 and Y_2 (within the limits of the probable error of the observed data points). Hence a particular regression equation cannot be assumed to be valid unless external empirical or theoretical evidence is brought to bear. Hyman Lumer (1936) made this point vividly clear. His contribution to the field of scaling is important but frequently ignored.

1.6 Dimensional Analysis

Empirical scaling analysis is merely regression analysis in one guise or another. By contrast, analytical scaling relies on dimensional analysis (also called similarity analysis; see Günther 1975; Peters 1989). This is an important tool that requires attention before we proceed with the general topic of plant-scaling relations. In this section, therefore, I will review dimensional analysis. This tool will be illustrated by constructing a hypothetical relation between plant cell mass and the rate of cell growth, which will be tested empirically in the following section.

As noted, every quantity in science is associated with an assigned but arbitrary traditional unit based on some external standard. Confusion caused by the use of different standards and multiple units has resulted in the development of the International System of Units, which has seven fundamental quantities, each measured in terms of standard unit. In dimensional analysis, each of these quantities is assigned a symbol. These quantities and their symbols in dimensional analysis are summarized in table 1.1. For example, the basic quantity of mass is symbolized as M and is measured in the base unit of the kilogram, of which the picogram (10^{-12} kg) is a submultiple unit. Other derived physical quantities are expressed in terms of combinations of the seven base quantities. Seventeen of these derived quantities have special names and units. Dimensional analysis is used to show the relation of a derived quantity to the fundamental quantities by manipulating these symbols. Consider the ratio of cell volume to cell mass. Volume is a derived physical quantity from the quantity length, whose unit is the cubic meter. As shown in table 1.1, the symbol for length is L and therefore the dimensional expression for volume is L^3. Weight is a force whose derived unit is the newton. Any force is given by the product of mass M and acceleration a. Acceleration is velocity v over time T, or vT^{-1}. Since v is change of distance L over time T, or LT^{-1}, dimensional analysis shows that $F = Ma = M(vT^{-1}) = M(LT^{-1})T^{-1} = MLT^{-2}$, and that the ratio of F to V has the dimensions $MLT^{-2}L^{-3} = ML^{-2}T^{-2}$. By contrast, the ratio of M to V, which is density, has the dimensions of ML^{-3}.

Since operations involving powers are used in dimensional analyses, the following summary of the most commonly encountered is useful:

$$Y^0 = 1 \quad Y^aY^b = Y^{a+b} \quad \frac{Y^a}{Y^b} = Y^{a-b}$$

$$(Y^a)^b = Y^{ab} \quad (Y^a)^{1/b} = Y^{a/b} \quad \left(\frac{Y_1}{Y_2}\right)^a = \left(\frac{Y_1^a}{Y_2^a}\right)$$

It is very important to remember, however, that these mathematical operations are entirely inappropriate when dealing with empirically (statistically) determined scaling relations, because error terms for Y are not symmetric about the line that provides the best fit in regression analysis.

By way of illustration, consider the LS regression of cell mass versus cell volume shown in figure 1.5, which yields $M = 0.003\ V^{0.794}$ for eukaryotes. Mathematically, this equation obtains $V \propto M^{1.259}$. When cell volume actually is regressed against cell mass, however, LS regression yields $V \propto M^{1.213}$. The difference in α may appear small, but for large cell masses our estimate of V will be way off the mark.

Biological variables, like physical variables, can be expressed in terms of fundamental and derived quantities. Dimensional analyses of biological variables therefore is possible. Typically, however, some of the fundamental quantities are either ignored or presumed constant, for example, temperature K, light intensity I, and electric current E. Indeed, in most case biological variables of a physiological nature, denoted here as G, are expressed only in terms of mass M, length L, and time T. That is, G is a function of M, L, and T:

$$G = f(M, L, T) \propto M^a L^b T^c, \tag{1.3a}$$

where the exponents, a, b, and c are real numbers. It is often very useful to reduce the number of dimensions in eq. (1.3a). This is easily accomplished provided a fixed relationship, a proportionality, exists between pairs of dimensions. Density ρ, which is the quotient of M and V, illustrates this point. If we presume that the density ρ of the cytosol of a cell is constant ($\rho = M/V$ = a constant), then $\rho = ML^{-3}$ = a constant. If we further assume that the cell increases in size but retains exactly the same geometry and shape, then elementary geometry shows that $V \propto L^3$. Since $M \propto V$, it mathematically follows that $L \propto M^{1/3}$. Recall that this is the null hypothesis called geometric similitude. When eq. (1.3a), $L^b \propto M^{b/3}$, is cast in terms of the null hypothesis, it follows that

$$G \propto M^{a+b/3}\, T^c. \tag{1.3b}$$

For most cells, we might assume that the variable G likely depends on the velocity with which the cell exchanges mass with its external environment. From prior dimensional analysis, we know that velocity has dimensions of LT^{-1}. Assuming that velocity is a constant, it follows that $T \propto L$. Therefore, since $L \propto M^{1/3}$ and $T^c \propto M^{c/3}$, eq. (1.3b) becomes

$$G \propto M^{a\,+\,b/3\,+\,c/3}. \tag{1.3c}$$

Our example of a dimensional analysis comes full circle when the dimensions of G are specified, because the magnitudes of the real numbers a, b, and c in eq. (1.3c) depend on both the Y_1 that is being measured and the units elected to express these measurements. For example, if G represents cell growth, which can be measured in a variety of ways, then when G is measured in terms of mass produced per unit time, it has the dimen-

sions of MT^{-1} (see tale 1.1). Accordingly, the real numbers *a, b,* and *c* obtain the values 1, 0, and −1. Inserting these values into eq. (1.3c), we find $G \propto M^{1+0/3-1/3}$ or $G \propto M^{0.666}$. Measuring G in terms of production rate (which has units of ML^2T^{-3}) gives the same scaling exponent: $G \propto M^{1+2/3-3/3}$ or $G \propto M^{0.666}$. If, however, growth is measured in terms of mass produced per unit time, then G has dimensions of MLT^{-3} and $G \propto M^{1+1/3-3/3}$ of $G \propto M^{0.333}$. Obviously, therefore, the numerical value of the scaling exponent α in the proportionality $G \propto M^{\alpha}$ depends on the magnitudes of the real numbers *a, b,* and *c,* which in turn depend on the way G is measured. This general point is true for any dimensional analysis.

1.7 The 2/3-Power Law

The hypothetical relationship between G and M given by eq. (1.3c) is an example of how an argument of similitude is constructed. As noted, arguments of similitude are common in the literature dealing with scaling, and therefore we need to be aware of their logical structure, merits, and problems. A feature of all arguments of similitude is that the value of the scaling exponent α is influenced by the stipulations of similitude among the objects treated in the dimensional analysis regardless of the dimensions of the independent variable. For example, in eq. (1.3c), the scaling factor 1/3 is the consequence of four stipulations: cells differing in size have the same geometry, shape, density, and rate of mass exchange. As a result, length and time scale to the cube root of mass regardless of whether mass is measured in picograms or kilograms. It is important to note two points: the dimensional analysis expressed by eq. (1.3c) *requires* morphological and physiological similitude among cells differing in size, and it can be logically valid but nonetheless biologically incorrect.

There are two extreme ways of looking at any argument of similitude. We can assert that its predicted consequences will hold true whenever similitude occurs. Alternatively, we can assert that its predictions are always true because the similitudes the argument is based on are always true. Obviously, every argument of similitude is subject to falsification. In terms of eq. (1.3c), cell shape, density, and rates of mass exchange can be measured as cells grow in size. If these properties are size independent (i.e., similitude is observed), then the empirical relation between G and M will determine whether cells obtain the predictions mathematically contained in eq. (1.3c). If cells do not, then the argument of similitude is shown not to be true even though the stipulations that length and time scale to $M^{1/3}$ are empirically justified. Logically, an assertion of universality is falsified by finding one exception to the proposed rule. Biologically, however, this is not the case. Because of the tremendous diversity of living things, biologists tend to gravitate toward statements of relative frequency rather than universality. Most arguments of similitude therefore assert a "high" frequency of occurrence, permitting "some" exceptions. To test an argument of biological similitude, very large sets of data are required. Also, judgment must be exercised as to what constitutes sufficient data to warrant the claim that something is generally true. And as previously illustrated, a numerical correspondence between a predicted and an observed scaling exponent cannot be taken as prima facie evidence that all the assumptions underlying an argument of similitude are valid.

Perhaps the best-known similarity argument is called the "surface area law," sometimes called the "2/3-power law": $S \propto V^{2/3}$. As we have seen, this relation follows from geometric similitude. The empirical determination of a 2/3-power scaling exponent for surface area, however, cannot be taken as prima facie evidence that organisms differing in size are similar in geometry and shape. Prothero (1986, 275, fig. 5) has shown that objects differing in size can comply with the 2/3-power rule even when drawn from different geometric classes (solid polygons, spheres, spheroids, cylinders, etc.) Thus a condition of "pseudosimilarity" obtains even though geometric similitude does not occur. Although the point is often not appreciated, objects differing in shape but with the same geometry also can scale *S* according to the 2/3-power relation. And when geometry and shape change as a function of size, the scaling of *S* can take on a variety of scaling exponents (Niklas 1994). These points are easily illustrated with data from unicellular plants. Figure 1.7 plots *S* versus *V* for sixty-one unicellular plant species (data from Williams 1964; Eppley and Sloan 1966; and Mullin, Sloan, and Eppley 1966). Least squares linear regression of these data yields $S = 6.01\ V^{0.682}$($r^2 = 0.98$; $\alpha_{RMA} = 0.69 \pm 0.01$). Based on its 95% confidence intervals, $\alpha_{RMA} \approx 0.69 \pm 0.01$ does not differ statistically from the prediction for geometric similitude. Nonetheless, the plants considered in this data set differ in geometry and, within each geometric class, also differ in shape. Much as with the data plotted in figure 1.6, the smallest plants plotted in figure 1.7 have a spheroidal (= spherical, or prolate or oblate) geometry, while the largest plants typically have a terete cylindrical geometry. True, this size-correlated trend in geometry results in a 2/3-power scaling relation, but the basic assumption that geometry and shape are conserved features among these biological "objects" is magnificently violated.

This result may be shown by comparing the *S* and *V* of computer-simulated and real unicellular plants. Figure 1.8 shows the scaling of *S* for a collection of twenty-six computer-simulated "plants" (differing in geometry and shape) as well as the scaling of *S* observed for the sixty-one unicellular plants plotted in figure 1.7. Least squares regression of S versus V for the simulated "plants" shows that $S \propto V^{0.692}$ ($r^2 \approx 1.0$; $\alpha_{RMA} \approx 0.69 \pm 0.002$), whereas the scaling of S for the real unicellular plants is $S = 6.01\ V^{0.682}$ ($r^2 = 0.98$; $\alpha_{RMA} = 0.69 \pm 0.012$). The two scaling exponents are statistically indistinguishable. Thus the 2/3-power rule adequately *describes* the scaling of S for unicellular plants, but the null hypothesis of geometric similitude must be rejected based on what we know about the morphology of these plants.

The shape of plant cells is remarkably versatile even within a single category of geometry. This versatility can confer numerous advantages. Figure 1.9A, for example, plots diameter versus length for twenty-eight tracheary elements isolated from a small section of *Quercus* wood. Although all these cells were derived from essentially identical meristematic cells within the vascular cambium, inspection of figure 1.9A shows that the aspect ratios of mature vessel members and xylem fibers differ by an order of magnitude. Vessel members are short and wide; fibers are long and narrow. The ability to change shape is important to the function of

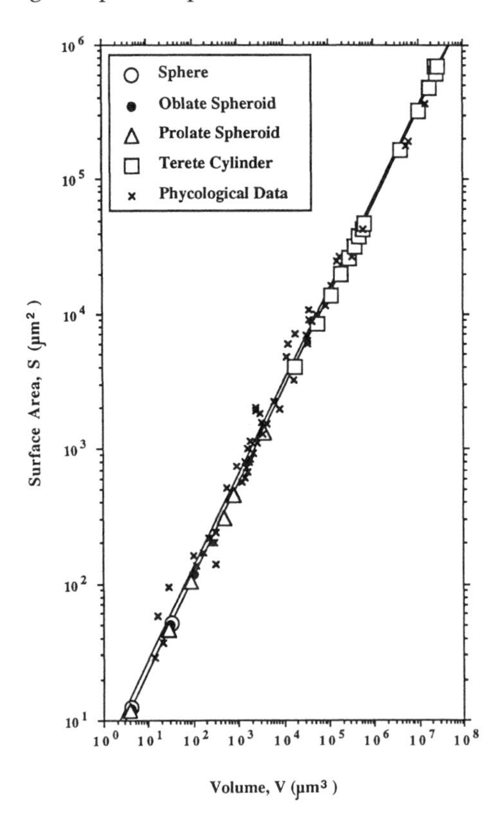

Figure 1.8. Log-log (base 10) plot of cell surface area S versus volume V for unicellular eukaryotic plants (see fig. 1.7) compared with the log-log (base 10) plot of S versus V computed for four computer-simulated geometries differing in shape. Solid diagonal lines denote the regression curves obtained for these data from LS regression analysis.

these cells, since the vessels' large bore permits the rapid transit of water. The ability of a tubular cell to conduct water is related to the fourth power of cell radius R. Thus the functional divergence among mature tracheary elements can be emphasized by plotting R^4 versus L, as shown in figure 1.9B (see Zimmerman 1983).

Intraspecific size-dependent variations in the shape of the entire plant body are biologically important also. For example, Hunt and Nobel (1987) report that the scaling exponent between S and V for the two succulent species *Agave deserti* and *Ferocactus acanthodes* equals 0.75 ($\alpha \approx 3/4$). Hunt and Nobel (1987) also report that the seedlings of either species have a greater root length and smaller shoot surface area per volume than adults do. They speculate that this would increase the ability to take up water, would reduce transpiration relative to adult plants, and

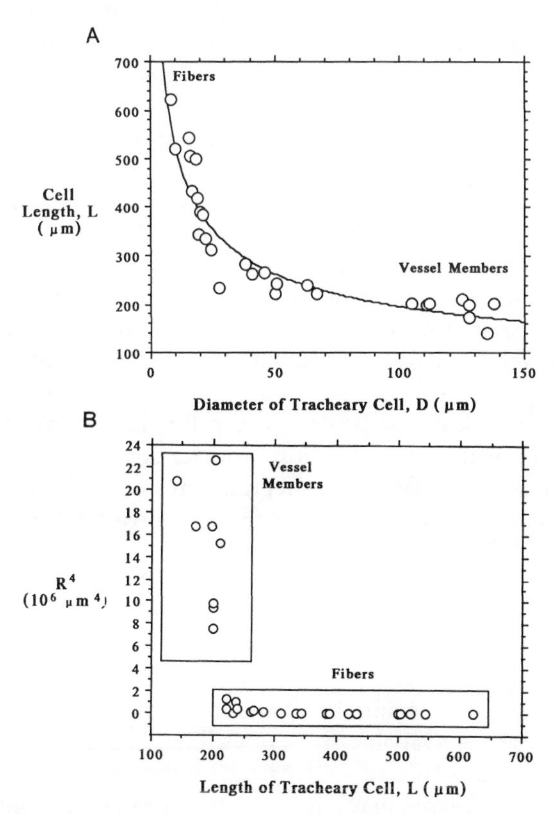

Figure 1.9. Relations between cell dimensions of tracheary elements (individual vessel members and fibers) isolated from a macerated preparation of wood. (A) Cell length L plotted as a function of cell diameter D. (B) Fourth power of cell radius R plotted as a function of vessel member and fiber length L.

therefore could be significant to seedling establishment under field conditions. The way the anisometric scaling of S is achieved is interesting. The mature forms of *Agave* and *Ferocactus* are morphologically very different. *Agave* has large foliage leaves arranged in a rosette. *Ferocactus* has spines on a columnar succulent stem. In *A. deserti,* the length of individual leaves increases faster than either leaf width or thickness. In *F. acanthodes,* the entire stem alters its morphology by changing from a juvenile globular shape to a mature columnar shape. Thus, although the numerical values of the scaling exponents for these two species are equivalent, they reflect very different ways geometry and shape are altered.

Returning to arguments of similitude, we have seen that some unicellular plants appear to "obey" the 2/3-power law in the relation between external surface area and volume, even though the assumptions of geometric similitude do not stand up to empirical scrutiny. The assumptions of the 2/3-power law fail in another important respect, and in this context its failure may be universally true. The law presumes that the bulk density ρ of plants differing in absolute size is constant. If this presumption is true, then $M \propto V$, and therefore $S \propto M^{2/3}$ mathematically follows. Note, however, that $S \propto V^{2/3}$ can be true but $M \propto V$ can be false, and therefore $S \propto M^{2/3}$ need not follow. The scaling of M with respect to V is easily determined. Figure 1.10 plots cell mass M versus volume V for thirty-seven species of unicellular plants (data from Mullin, Sloan, and Eppley 1966). These data are based on direct measurements of M and V. Least squares regression shows that $M \propto V^{0.757}$ ($r^2 = 0.96$; $\alpha_{RMA} = 0.77 \pm 0.03$), indicating that cell mass scales roughly as the 3/4 power of volume rather than isometrically. Why is this so? One possibility is that cell density decreases interspecifically with increasing cell volume. Unfortunately, authors rarely measure ρ directly; most estimate this variable from the quotient of M and V. It is tempting to regress real values of ρ against cell size (measured as either M or V) to determine the scaling exponent for cell bulk density. When this is done for the data plotted in figure 1.10, the relation $\rho \propto M^{-0.27}$ is obtained. It is obvious, however, that this scaling relation is based on an illegitimate practice that yields an autocorrelation between the "independent" and "dependent" variables (M/V versus M or V).

Nonetheless, circumstantial evidence for a size-dependent reduction in cell bulk density is available from other quarters. Figure 1.11, for example, plots cell density ρ versus length L for twenty-one *Chara* internode cells, from which we see that ρ initially decreases rapidly with increasing L and then appears to level off. This trend likely results from the fact that larger *Chara* internode cells have disproportionately larger vacuoles. Additionally, cell wall thickness is a conserved feature among cells differing in size. The density of liquid contained in the vacuole is roughly 1,010 kg m^{-3}, while the density of the cell wall is roughly 1,350 kg m^{-3} (see Wayne and Staves 1991). With increasing cell size, the volume fraction of the vacuole increases while that of the cell wall declines. Both could account for the gradual reduction of bulk density with increasing *Chara* cell size. The same phenomenology may apply to multicellular plants. At the whole plant level, fluid-filled spaces (collectively called the apoplast) typically form within metaphytes as they grow in overall size. It must be kept in mind that the relation between *Chara* cell density and size has no direct bearing on the interspecific relation between these two variables. By the same token, no data were examined to test the notion

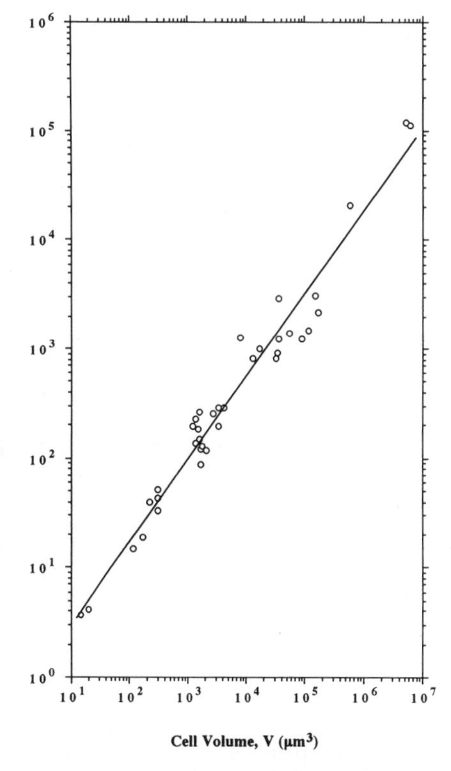

Figure 1.10. Log-log (base 10) plot of cell mass *M* versus volume *V* for eukaryotic unicellular plants (algae). Solid diagonal line denotes the regression curve obtained for the pooled data from LS regression analysis. Data from various sources (see text).

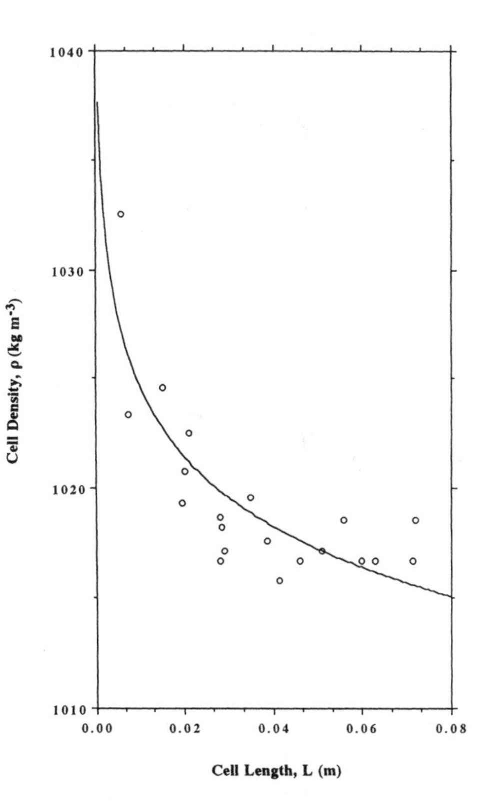

Figure 1.11. Plot of cell density ρ versus length *L* for internodal cells of the green alga *Chara*. Curved line denotes the LS regression curve for log-transformed (base 10) data.

that the bulk density of metaphytes declines with increasing plant size. Clearly, however, there is no a priori reason to assume that the bulk density of unicellular or multicellular plants is size independent. Therefore we cannot presume that $\rho \approx$ a constant.

For this reason we should not *expect* plant metabolic or growth rates to scale as $M^{0.666}$. Indeed, the available data show that they do not. Figure 1.12, for example, plots the rate of growth G versus mass M for fifty-three species of unicellular plants. These data apply to species grown under optimal conditions of light intensity and temperature. Cell mass is expressed in units of picograms of carbon per cell (pg C cell^{-1}) because this permits an unambiguous comparison among cells that may have different volume fractions of vacuoles or cell walls. Least squares regression of these data shows that $G \propto M^{0.787}$ ($r^2 = 0.92; \alpha_{RMA} = 0.82 \pm 0.03$). Analysis of covariance shows that the scaling exponent for the relation between G and M differs significantly from that predicted by the 2/3-power law. Three additional things about figure 1.12 are worth noting. First, although the absolute growth rate of larger plants is greater than that of smaller plants, the exponent 0.85 shows that, on the average, rela-

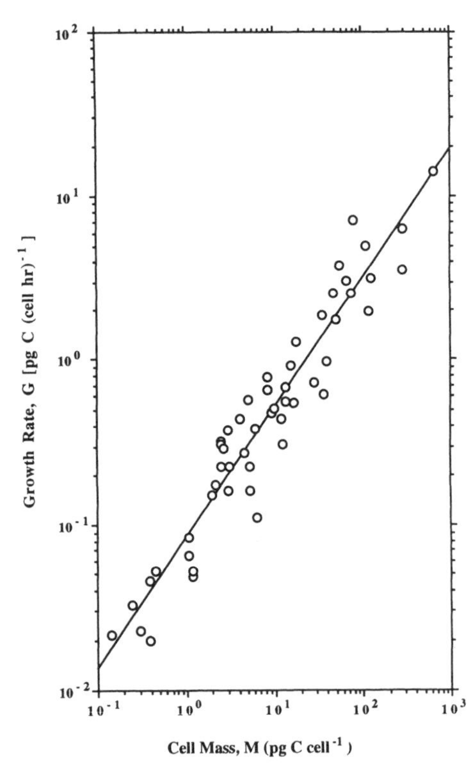

Figure 1.12. Log-log (base 10) plot of growth rate G versus cell mass M for unicellular prokaryotic plants (algae). Data from various sources (see text). Solid diagonal line denotes the regression curve obtained for these data from LS regression analysis.

tive growth rate declines with increasing cell size. That is, larger plants grow proportionally more slowly than their smaller counterparts. Second, the relation between G and M evinces a high correlation because of autocorrelation resulting from the fact that G has the units pg C $(\text{cell hr})^{-1}$ while M has the units pg C cell^{-1}. The effect of autocorrelation is quickly removed by plotting the mass-specific growth rate G/M versus M. Note that G/M, which is the number of cell divisions per hour, has units of hr^{-1} and that $G/M = \beta M^{\alpha-1}$, where α is the scaling exponent for the relation between G and M. Having said all this, regression of G/M for the data plotted in figure 1.12 shows that $G/M \propto M^{-0.213}$ ($r^2 = 0.45$; $\alpha_{RMA} = -0.32 \pm 0.03$). Although the correlation between these two variables has declined appreciably, it is nonetheless significant at the 1% level.

The scaling of G shown in figure 1.12 is consistent with previous research. In an important and extremely thoughtful study of the relation of G to M, Karl Banse (1976) found that the scaling exponent varied between 0.75 and 0.94 for unicellular plant species. Banse attributed this variation to differences in the environmental conditions attending growth and concluded that as growth conditions deteriorate the numerical value of the scaling exponent will increase. Under optimal growth conditions, however, α_{RMA} will more or less equal 0.75. More recent studies also show that the scaling exponent of G may differ among plant lineages even when measured under optimal growth conditions (see Schlesinger, Molot, and Shuter 1981). Both taxonomic affiliation and growth conditions are likely contributing factors. Nonetheless, for each set of data, Banse found that the rate of growth declined as cell mass increased, although gross photosynthesis and metabolic processes underlying growth were size dependent. Also, Banse found that the respiration rate R declines with cell mass for some marine phytoplankton species (Eppley and Sloan 1965), such that $R = 0.017\ M^{0.90}$ ($r^2 = 0.96$, $N = 8$; $\alpha_{RMA} = 0.92$). The scaling of growth for these same species is given by $G = 0.056\ M^{0.94}$ ($r^2 = 0.91$; $\alpha_{RMA} = 0.99$). Note that the quotient of the rates of growth and respiration G/R yields $M^{0.07}$, indicating that G/R is size (mass) dependent. Also, note that net growth is roughly three times the rate of respiration ($\beta_G/\beta_R = 0.056/0.017 = 3.29 \approx 3$).

As a purely conjectural footnote, it is interesting that when cell size is correlated with basal metabolic rate such that the relative rate of growth declines with increasing cell size, a multicellular organism composed of many small cells should, on the average, grow relatively faster than a multicellular organism of comparable size composed of larger cells, because the total size of either organism must equal the product of cell size and number. To my knowledge this prediction has not been examined directly, although there is a wealth of data relating cell size and number to overall plant size. Sachs (1893) and one of his students, Amelung (1893), seem to have been the first to speculate that, for any particular tissue type, cell size is likely to be highly conserved because of the size dependency of physiological activity. Both Sachs and Amelung supported this speculation by detailed studies of cell size and number in growing plants. They concluded that larger plants tend to be composed of more cells rather than bigger cells, although both noted good cases where cell size differed considerably between comparable tissues from plants different in absolute size.

1.8 The 3/4-Power Law

It should not escape our attention that the scaling exponent $\alpha_{RMA} = 0.82$ determined for the data plotted in figure 1.12 does not statistically differ significantly from 0.75 for unicellular plants grown under optimal conditions (Banse 1976). Also, this scaling exponent is statistically indistinguishable from that observed for animal growth rates (Fenchel 1974; Peters 1983; Reiss 1989). Figure 1.13 plots animal and plant growth rates

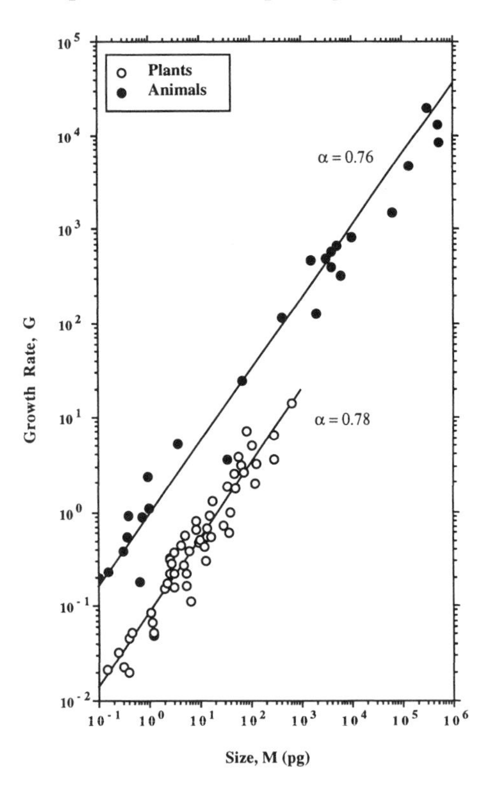

Figure 1.13. Log-log (base 10) plot of growth rate G versus size M for unicellular eukaryotic plants (algae) and unicellular and multicellular aquatic and terrestrial animals (data from Fenchel 1974 and Niklas 1994). Solid diagonal lines denote the regression curve obtained for each data set from LS regression analysis. The slopes of the RMA regression curves (α) obtained for these data sets are given.

versus body size (data from Fenchel 1974 and Niklas 1994, respectively). The difference in the Y_1-intercepts of the two regression curves is attributable to the fact that animal mass was measured in pg (fresh weight) in contrast to unicellular plant mass, which was measured in pg C $cell^{-1}$. Although somewhat distracting, this difference is unimportant because we have already seen that the scaling exponent is indifferent to the way size is measured. Regression analyses show that, for the animal data, $G \propto M^{0.764}$ ($^2 = 0.99$, $N = 35$; $\alpha_{RMA} = 0.77 \pm 0.01$). Recall that regression of the plant data gave $G \propto M^{0.787}$ ($r^2 = 0.92$, $N = 53$; $\alpha_{RMA} = 0.82 \pm 0.03$). Based on their 95% confidence intervals, the scaling exponents for the animal and plant data sets are not statistically different. Likewise, neither differs statistically from $\alpha_{RMA} = 3/4$.

Apparently, despite their numerous physiological differences, plant and animal growth rates scale according to the proportionality $G \propto W^{3/4}$. But why should this be so? Is there a *fundamental* reason why plant and animal growth rates scale roughly as the 3/4 power of body mass, or is this scaling the result of numerous physiological, anatomical, and morphological phenomena operating among taxonomically and ecologically diverse organisms in different ways that statistically yield a 3/4 scaling exponent? Although the "either/or" construction of this question does not imply that the failure to find a robust analytical solution proves the alternative, all attempts to explain the 3/4-power rule based on first principles have been unsuccessful (see LaBarbera 1986, 1989). For example, McMahon (1973, 1980) presumes elastic similtude among all manner of organisms, from which it is argued that metabolic rates will rise as $M^{3/4}$ because the cross-sectional areas of structural members must scale as the 3/4 power of the weight they support. Unfortunately, this explanation has no bearing on the biology of aquatic plants and animals, which essentially operate in a weightless environment yet scale their metabolic and growth rates as a 3/4 power of body mass. Blum (1977) suggests that metabolic rates depend on the functional surface area of organisms and that surface area exists in time as well as in the three spatial dimensions. The surface area of a hypervolume rises as a function of $V^{1-1/n}$, where n is the number of dimensions. If $n = 4$, then $S \propto V^{3/4}$ and, presuming $\rho \approx$ a constant, $G \propto M^{3/4}$ mathematically follows. Although it is reasonable to assume that the surface areas of organisms exist in "time" as well as "space," we have seen that the isometric relation between body volume and mass may be a dubious presumption. Alternatively, Economos (1979) has proposed gravitational loadings, while Gray (1981) argues that local variations in body temperature and their effects on metabolism could account for the $M^{3/4}$ relationship. The mathematics and logic of some of these hypotheses are not easily followed, and in any circumstances some of their predictions flout empirical observations. For example, the gravitational loading hypothesis of Economos predicts that, in a nearly weightless environment, metabolic rates should scale as $M^{2/3}$. Yet the data for unicellular plants do not support this prediction (see fig. 1.13).

Although we cannot explain the 3/4 scaling exponent, a number of reasons can be advanced to explain why the relative growth rate qualitatively declines with increasing organismic size. For example, size-dependent variations in the chemical composition of cells, tissues, or entire organisms may account for $G \propto M^{\alpha>1.0}$. This is a reasonable possibility, since smaller prokaryotic and eukaryotic species tend to have higher RNA and enzyme concentrations, whereas the concentrations of many cellular constituents are proportional to $M^{0.75}$ for animals (Munro 1969; King and Packard 1975). For plants, size-dependent variations in some chemical components occur, although other chemical constituents evince no size dependency (see Parsons, Stephens, and Strickland 1961). Figure 1.14, for example, plots the mass m of chlorophyll a per cell (i.e., the bulk density of this photosynthetic pigment) versus the mass of forty-seven unicellular species. Regression of these data yields the formula $m =$

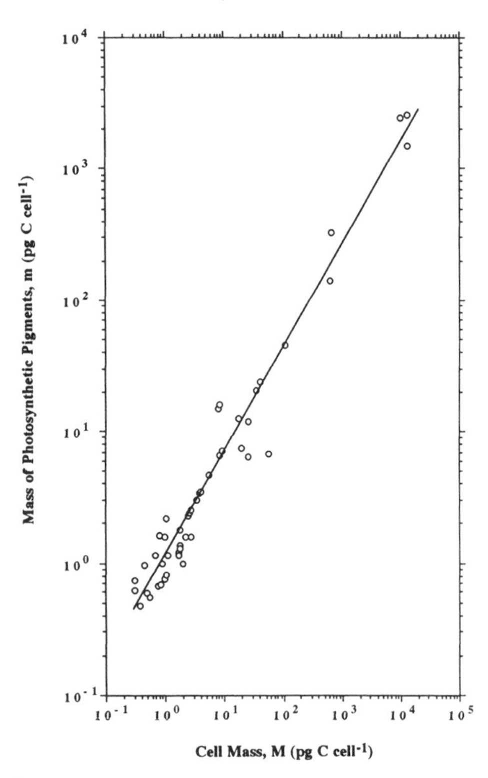

Figure 1.14. Log-log (base 10) plot of mass of photosynthetic pigments per cell versus cell mass M for unicellular eukaryotic plants (algae). Solid diagonal line denotes the regression curve obtained from LS regression analysis.

$0.009\ M^{0.740}$ ($r^2 = 0.86$; $\alpha_{RMA} = 0.80 \pm 0.04$). Figure 1.15 shows that the relation between G and m is nearly isometric. That is, $m \propto G^{0.891}$ ($r^2 = 0.95$; $\alpha_{RMA} = 0.91 \pm 0.03$). Similar "dilutions" of protoplasmic constituents are known. For example, the nitrogen and phosphorus subsistence quotas of unicellular plants are size dependent and decline with increasing cell volume (see Shuter 1978). Nitrogen and phosphorus are structural and metabolic components of plant cells. A size-dependent decline in the requirements for these and other nutrients suggests that structural or metabolic components decrease as cell size increases. True, correlation does not prove causation, but it is difficult to imagine that plant metabolic rates and therefore growth rates do not scale to the quantity of light-harvesting pigments or other chemical constituents. Nonetheless, the only meaningful statement that can be made given our general ignorance of this matter is that similitude arguments like the 2/3- and 3/4-power laws are not intrinsic to the empirical study of scaling. We can describe size-dependent relationships without attempting to uncover their underlying mechanisms. Although this may superficially appear less intellectually satisfying, we must recognize that an empirical approach to scaling is

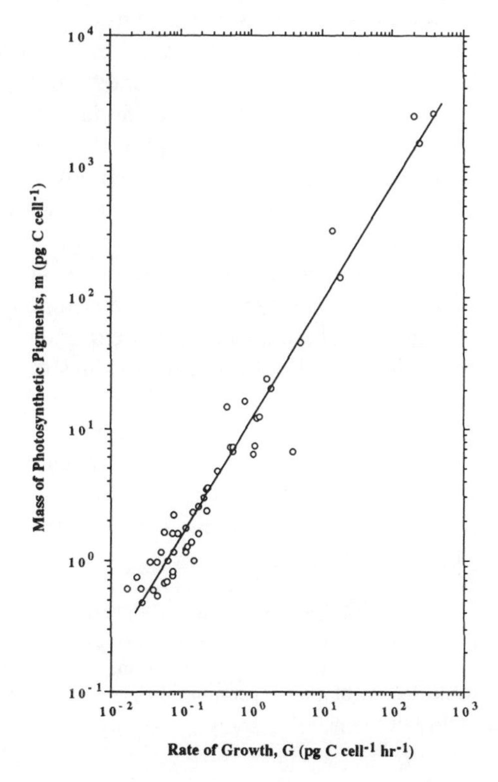

Figure 1.15. Log-log (base 10) plot of mass of photosynthetic pigments per cell versus growth rate G for unicellular eukaryotic plants (algae). Solid diagonal line denotes the regression curve obtained from LS regression analysis.

both unavoidable when dealing with growth and essential to the theorist who attempts to explain the causality of a correlation.

1.9 Growth and Time

In this section, we will consider plant growth measured as a function of time rather than size. To the extent that growth in size is measurable, it is amenable to mathematical description relating an ascending series of quantitative measurements of size, symbolized here by Q, to intervening periods of time t. When a plot of Q versus t is determined, it typically takes the form of a logistic growth curve (also referred to as the sigmoid curve of growth; see Pearl 1925). Figure 1.16A illustrates this curve based on daily measurements of the diameter of a cucurbit fruit taken from a study by Edmund Sinnott (1960, 13, table 2–1). Arbitrary units are used for Q and t to emphasize that the shape of the logistic growth curve fits data from a variety of sources, even the growth of the United States population as measured in the decennial censuses from 1790 to 1910 (but not beyond 1910).

The logistic growth curve consists of two phases, an exponential and an asymptotic phase (fig. 1.16A). The former is easily identified when $\log_{10} Q$ is plotted against t, as shown in figure 1.16B. The resulting semilogarithmic plot is linear for the data corresponding to the exponential phase of growth. The asymptotic phase is identified graphically by the deviation of data points from the predicted linear regression curve of $\log_{10} Q$ versus t. Since it is often useful to know the scaling of growth of the exponential phase and the asymptotic phase before considering the regression curve for the entire logistic growth curve, the equations for each of the two phases typically are determined by regression analysis.

The general equation for any exponential function is $Y = ab^x$, a and b are constants whose numerical values must be determined by regression analysis. Denoting the initial value of Q as Q_0 and substituting Q for Y, the exponential equation takes the form $Q = Q_0 b^t$. For those who know calculus the law that the rate of increase at any stage is proportional to the size already attained is described by $dQ/dt = rQ$, where r is the relative rate of increase. Thus, $\log Q = \log Q_0 + rt \log e$ or $Q = Q_0 b^t = Q_0 e^{rt}$, where e is the base of the natural logarithms ($e = 2.71828$).[1]

1. The exponential phase of growth in the logistic growth curve follows the compound interest law (a phrase first used in this context by Lord Kelvin), wherein the initial size of the system (cell, tissue, organ, organism, or population) is the analogue to the *principle*, while the rate of growth per unit time per unit amount of growing material is the analogue of the rate of interest.

The asymptotic phase of the logistic growth curve is described by the general function $Q = a - b\,e^{-rt}$. Standard regression analysis techniques provide the numerical expression of this function for a particular set of data (see Snedecor and Cochran 1980, 409–12). For the data comprising the asymptotic phase in figure 1.16A, calculations yield the regression formula $Q = 48.45 - 4{,}277\,e^{-0.48t}$, which provides a reasonable approximation, as shown in figure 1.16C.

The entire logistic growth curve is mathematically described by the following equation:

$$Q = \frac{a}{1 + be^{-rt}}, \qquad (1.4) \quad \text{(logistic curve equation)}$$

where a is the value of Q when $t \to \infty$ and $a/(1 + b)$ is the value of Q when $t = 0$. The inflection point of the logistic growth curve (i.e., where

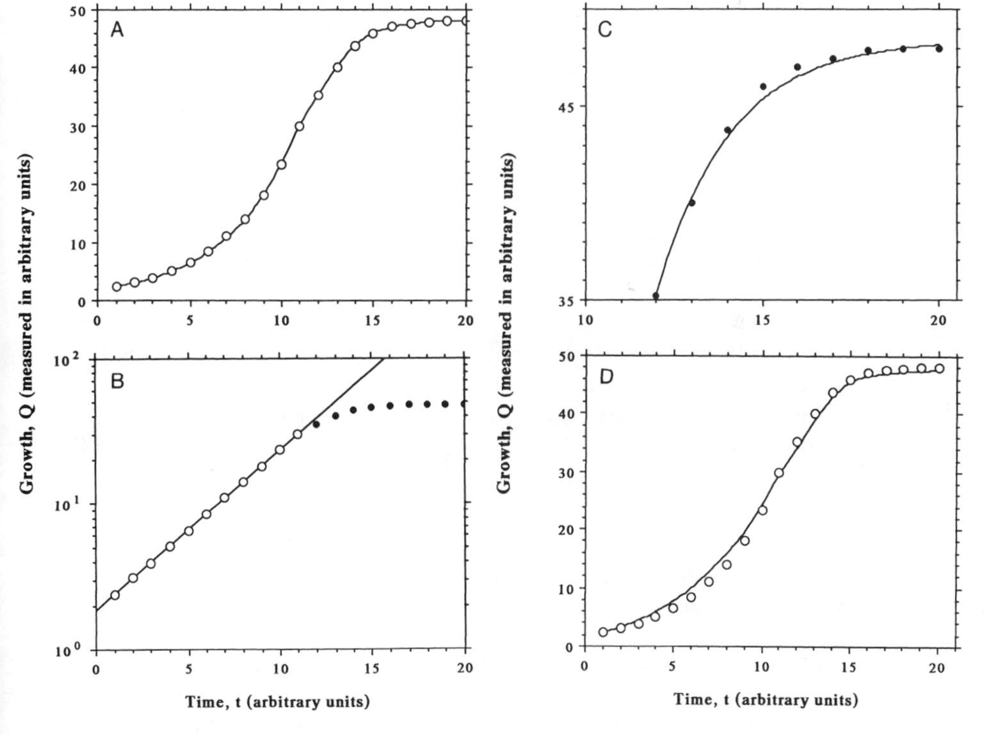

Figure 1.16. Relation between growth Q and time t. (A) Sigmoid plot of Q versus t. (B–C) Log-log (base 10) plots of G versus t. (D) Estimated (sigmoid) regression curve for data points shown in A.

the exponential phase ends and the asymptotic phase begins) has the coordinates $Q = a/2$ and $t = (\log b)/r$. Figure 1.16D shows that the predicted logistic curve complies reasonable well with the data (Sinnott 1960, 13, table 2-1).

It must be recognized that the mathematical description of a growth curve only describes the growth that has occurred. It does not predict growth per se unless the mathematical parameters in the logistic growth curve are given biological meaning. There have been numerous attempts to do so. Most mathematically describe the scaling of growth in terms of the balance between the rate at which metabolites are synthesized (anabolism) and the rate at which metabolites are consumed (catabolism). It is instructive to review these attempts.

The similarity between growth and the chemical process of monomolecular autocatalysis, wherein the concentration of a substrate influences the rate of the chemical process, led Robertson (1923) to draw an analogy between autocatalysis and the growth of a whole organism. In terms of an autocatalytic process, when the amount of a substrate is high, the chemical process will exhibit an acceleration phase. When the concentration of the substrate decreases, it will decelerate. When the substrate becomes exhausted, it will cease. The equation of autocatalysis is

$$(1.5) \qquad \log_{10} \left(\frac{Q_t}{Q_\infty - Q_t}\right) = k\,(t - t_{1/2}), \qquad \text{(autocatalytic equation)}$$

where Q_t is size (the amount grown) at time t, Q_∞ is the final size, $t_{1/2}$ is the time it takes growth to achieve half the final size, and k is the specific growth rate. Equation (1.5) predicts that the rate of growth is determined by the amount of growth that will occur. That is, final size Q_∞ is predetermined upon the initiation of growth. The growth curves for a number of organisms conform very nicely to Robertson's equation (see Reed 1927). That is, the rate of growth constantly decreases. Organisms that exhibit this property can be identified because the curve of $\log Q$ versus t is initially convex. An interesting example is provided by Steenbergh and Lowe (1977), who measured the height h of an individual saguaro cactus over a period of thirty-three years. Figure 1.17 shows that the change in height per year decreases with time for this particular plant (data from Steenbergh and Lowe 1977, 142–43, table 28). Although the growth in height is not determinate in a strict sense, it does decelerate with time, and this could confer a mechanical benefit. The saguaro cactus has little or no capacity to increase the girth of its vertical stems; its slenderness ratio therefore increases with age. As a consequence, the columnar stems of this species become increasingly susceptible to mechanical failure. If the rate of growth decelerates, however, then mechanical failure can be delayed until very late in life.

Not all plants, however, exhibit a deceleration in growth with time. Further, many biological structures can grow at the same rate for different durations and therefore achieve their mature sizes at different times. According to eq. (1.5), the interval between $t_{1/2}$ and t_{∞} should be the same, which is evidently not always the case.

Ludwig von Bertalanffy (1952, 1960) theorized that growth will occur whenever the anabolic rate exceeds the rate of catabolism. When the ratio of these two processes approaches and then drops below unity, growth will decrease and eventually cease. An avid proponent of this view, Bertalanffy provided it with mathematical formality by expressing the rate of anabolism in terms of the rate r at which mass is created per unit area and the rate of catabolism as the product of some proportionality factor k (the coefficient of catabolism) and the bulk of living matter Q within the organism. The change in Q per change in time (which is the rate of

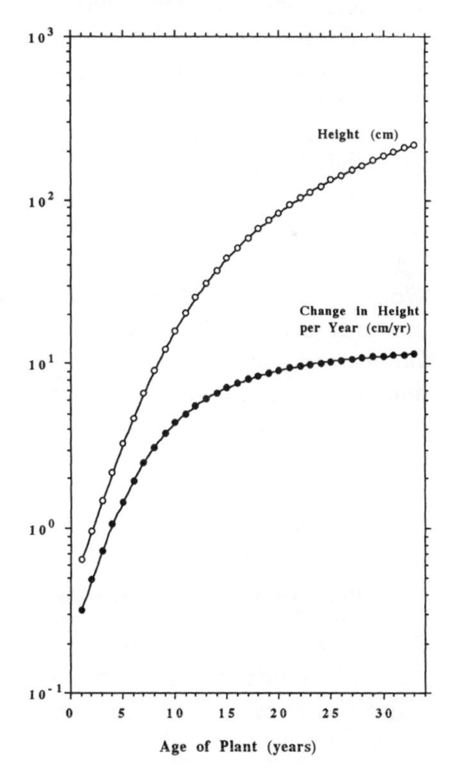

Figure 1.17. Plot of log (base 10) height H and change in H per year versus plant age for a specimen of the saguaro cactus (data from Steenbergh and Lowe 1977, 142–43, table 28). Data points are connected by interpolation.

growth, dQ/dt), therefore, is predicted to equal the difference between anabolism and catabolism:

$$\frac{dQ}{dt} = rS - kQ. \qquad \text{(Bertalanffy equation)} \tag{1.6}$$

Since S has the units of a linear dimension squared L^2, and since Q is a function of volume, which has units of L^3, Beverton and Holt (1957) used eq. (1.6) to determine that the length L or the mass M of an organism measured at any time t should be given by the equations

$$L_t = L_\infty - (L_\infty - L_0)e^{-kt} \qquad \text{(Beverton-Holt equations)} \tag{1.7a}$$

$$M_t = [M_\infty^{1/3} - (M_\infty^{1/3} - M_0^{1/3})e^{-kt}]^{1/3}, \tag{1.7b}$$

where L_t = length at time t, L_∞ = length at time ∞, L_0 = length at time zero, M_∞ = at time ∞, and M_0 = mass at time zero. These equations successfully describe the growth curves of a variety of marine fishes. Likewise, Bertalanffy's equation predicts the growth of some plants under natural conditions (Blackman 1961).

Finally, Richards (1959) derived an equation in which the value of the constant relation between size Q and the metabolic rate is given by nQ^m, where n and m are appropriate constants. If catabolism is denoted by kQ, then the rate of growth is given by the difference between anabolism and catabolism:

$$\frac{dQ}{dt} = nQ^m - kQ. \qquad \text{(Richards equation)} \tag{1.8}$$

When $m = 2$, eq. (1.8) has the characteristics of the autocatalytic equation of Robertson (eq. 1.5). When $m = 1$, eq. (1.8) has the features of the Gompertz equation, that is $Q = ae^{-b\exp(-rt)}$. The Gompertz equation is a double integral equation so flexible that it can fit any growth curve. Since Richard's equation encompasses the full range of eqs. (1.5) to (1.7), it has the maximum flexibility in describing virtually any observed growth curve in terms of the metabolic rates of anabolism and catabolism, although this is hardly surprising, since it contains the most variables.

1.10 Photosynthesis and Two Physical Laws

Every treatment of the scaling of plant growth ultimately gravitates toward a treatment of the size-dependent variations of photosynthesis, because this metabolic process provides the primary metabolites used to construct new cells and tissues. In this section the task is to review the basic requirements for photosynthesis as well as to see how two comparatively simple physical laws, Fick's law of diffusion and Bouguer's law of light attenuation, reveal the influence of size and shape on the acquisition of these requirements.

As shown by its equation,

$$(1.9) \qquad CO_2 + H_2O \underset{\text{chloro}}{\overset{\text{light}}{\rightarrow}} (CH_2O)_n + O_2, \qquad \text{(photosynthesis equation)}$$

photosynthesis requires the interception of light by chlorophyll-bearing tissues and the absorption of inorganic carbon (in the form of carbon dioxide or bicarbonate ions) and water from the physical surroundings. The interception of sunlight is quantitatively dependent on the surface area of photosynthetic tissues projected toward the source of light energy. The interception of light depends on a number of variables, among which the surface area projected toward light is important. The projected surface area of any object depends on geometry, shape, and absolute size. Recalling that geometry and shape are not synonymous terms (see table 1.2), we see that different geometries can have different capacities to intercept light, and the capacity of each geometry can change as the shape is altered. Figure 1.18 plots the photosynthetic capacity $\hat{\imath}$ against the aspect ratio $\mathfrak{R}$ (ratio of diameter to length) for a terete cylinder, an oblate spheroid, and a prolate spheroid.[2] From inspection of these plots, we see that (1) the oblate spheroid has a higher photosynthetic capacity than either the prolate spheroid or the terete cylinder, (2) its capacity to intercept light increases as it *flattens* (as its aspect ratio declines), and (3) the photosynthetic capacities of both the terete cylinder and the prolate spheroid decline as the aspect ratio decreases. Based on figure 1.18, it should come as no surprise that most photosynthetic organs are geometrically approximated by oblate spheroids, and that a cylindrical stem is a fairly reasonable alternative geometry. Based on the calculations giving rise to figure 1.18, a short and stubby green stem can elongate by a factor of 10^5 and still experience less than a 16% decline in its photosynthetic capacity. The absolute size of photosynthetic organs must also be taken into account, since the surface area of any geometry or shape is size dependent as well as shape and geometry dependent.

Regardless of the photosynthetic capacity of a structure, eq. (1.9) indicates that the photosynthetic rate depends on the rates at which water

2. Values of $\hat{\imath}$ are calculated from the area under a graph of projected surface area versus incident light angle (which varies from 0° to 180°) for each aspect ratio. Photosynthetic capacity is a dimensionless variable when incident light angles are expressed in radians. For further details, see Niklas and Kerchner (1984).

and carbon dioxide are supplied. The absorption of CO_2 and H_2O by most plant tissues results from passive diffusion and occurs over the surfaces of cells exposed to these two substances. The ability of a plant to photosynthesize therefore depends on the magnitude of its surface areas, which, as we have already seen, depends on size, shape, and geometry. The formation of large surface areas is limited on land because water is rapidly lost from tissues in contact with a drier external atmosphere. Calculations show that, on the average, roughly 139 mol of H_2O will be lost for every mole of CO_2 that is fixed by photosynthesis. Much of the surface area through which terrestrial plants exchange CO_2 with the external atmosphere is internalized within photosynthetic organs. The invaginated but topologically external exchange surface of the aerenchymatous tissue within a typical dicot leaf is well suited to conserve H_2O while permitting the exchange of CO_2. Restrictions on the elaboration of surface area are relaxed for aquatic submerged plants. Many aquatic species have a one-cell-thick (monostromatic) tissue construction.

The quantitative influence of surface area on the absorption of CO_2 or H_2O is shown by Fick's law of diffusion. This transport equation shows

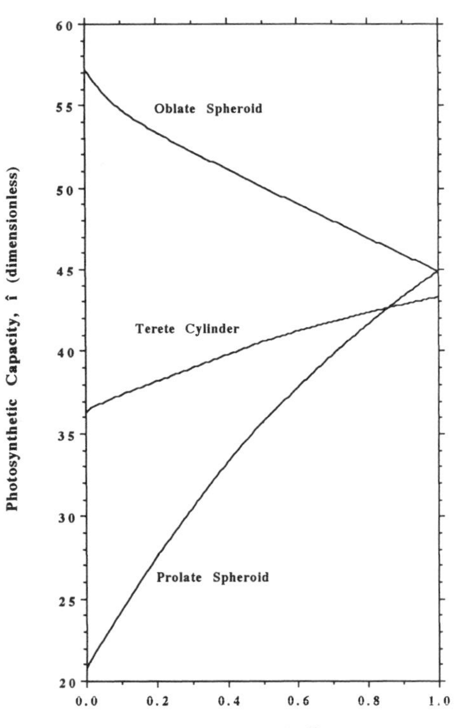

Figure 1.18. Photosynthetic capacity $\hat{\imath}$ of three geometries plotted as a function of the ratio of diameter to length. Data from Niklas and Kerchner 1984.

that the amount of CO_2, H_2O, or any other substance that passively diffuses through a unit surface area per unit time (called the flux density J) is equal to the product of the ability of the substance to diffuse through the surface per unit time (called the molecular diffusivity D) and the change in the density of the substance per change in distance (the concentration gradient $d\rho/dx$):

$$J = -D\frac{d\rho}{dx}. \tag{1.10}$$ (Fick's law of diffusion)

The negative sign placed before D indicates that the net direction of molecular diffusion is in the direction of the lower concentration, that is, the flux is in the negative direction when the gradient is positive.

Based on an argument of similitude, it is easy to show that the requirement for diffusion to adequately supply CO_2 is met when the metabolic need of CO_2 per unit time scales to body mass M with an exponent of ≤ 0.33 (Alexander 1971, 1979). This can be shown by considering a spherical organism with unit radius R. Although any geometry will suffice, a sphere considerably simplifies the mathematics. We begin by noting that the volume V between two concentric *shells* (an inner shell with radius x and the outer one with radius δx) is given by the formula $4\pi[(x + \delta x)^3 - x^3]/3 \cong 4\pi x^2 \delta x$. If the requirement of the organism for a gas is kM^a per unit time, where k is some constant, then the requirement per unit V must be proportional to kM^{a-1}, and the requirement of V for the gas per unit time must be proportional to $4\pi x^2 \delta x k M^{a-1}$. The rate at which the gas is supplied (voume of gas per unit time) is given by eq. (1.10), which in terms of surface area S can be restated as $J = -SPd\rho/d(R - x)$, where S is the surface area through which diffusion occurs, P is the permeability coefficient, ρ is the partial pressure of CO_2, and $(R - x)$ is the distance through which the gas must diffuse. Since the external surface area of the sphere equals $4\pi x^2$, we see that $d\rho/d(R - x) \propto kM^{\alpha-1} x/3P$. Taking ρ_e as the partial pressure of the gas outside the organism and ρ_0 as the partial pressure of the gas at the center of the organism, we find that $\rho_e - \rho_0 \propto (kM^{\alpha-1}/3P) \int_0^R x\delta x$, or $\rho_e - \rho_0 \propto kM^{\alpha-1/3}/6P$, since $R^2 \propto M^{2/3}$. Since the supply of gas must meet metabolic need and since ρ_0 cannot be negative, $\alpha \leq 1/3$. In this regard it is curious that the allometric exponent 0.33 occurs in the Beverton-Holt equation (eq. 1.7b), which works for fishes that rely on a circulatory system to distribute oxygen absorbed passively through gills as well as for plants whose metabolic needs for CO_2 per unit time are met by passive diffusion.

For many practical cases, the integrated form of eq. (1.10) is desirable, since it is difficult to actually measure concentrations of a molecular species over small distances. This is easily accomplished provided the concentration of a molecule can be measured at the surface through which diffusion occurs and at some distance in the ambient fluid. The difference between the two measurements ΔC divided by the distance x between the two measurements (which reflects a resistance to diffusion) yields an expression for flux density in a mathematical form analogous to Ohm's law:

$$J = D\frac{\Delta C}{x} = D\frac{\Delta \rho}{x}, \tag{1.11}$$

where the concentration gradient ΔC of CO_2, H_2O, or the like is expressed in terms of the difference in its density $\Delta\rho$. For convenience, the negative sign showing that diffusion is in the direction of the lower concentration has been eliminated.

With the aid of Fick's law and the "rules" of elementary geometry, the consequences of size and shape on photosynthesis can be examined. Since plants have little control over the physical constants D and ρ, all things being equal (light intensity, ambient CO^2 concentration, temperature, etc.), the objective is to reduce the magnitude of x whenever the rate of diffusion needs to be increased. This leads to a large $S{:}V$. For example, the time t required for the concentration of a nonelectrolyte j, like CO_2, initially absent from a tissue, to reach one-half the concentration of the external ambient concentration of j is given by the equation (Nobel 1983, 32):

$$t = \frac{V}{P_j S}\ln\left[\frac{(c_o - c_i)_{t=0}}{(c_o - c_i)_{t=1/2}}\right], \tag{1.12a}$$

where c_o is the external ambient concentration of j, c_i is the internal concentration of j, P_j is the permeability coefficient of j, $(c_o - c_i)_{t=0}$ is the initial difference in the external and internal concentrations of j (at time zero), and $(c_o - c_i)_{t=1/2}$ is the difference in the external and internal concentrations of j when $c_i = c_o/2$ (i.e., t is the time required for the internal concentration of j to equal one-half the ambient concentration of j)[3]. Since $\ln[(c_o - c_i)_{t=0}/(c_o - c_i)_{t=1/2}] = \ln[(c_o - 0)_{t=0}/(c_o - c_o/2)_{t=1/2}] = \ln 2 = 0.693$, eq. (1.12a) takes a comparatively simple form:

3. Values of P_j for small nonelectrolytes typically range between 10^{-6} and 10^{-5} m s^{-1} for most plant cells. However, water can have a permeability coefficient of 10^{-4} m s^{-1} for charophycean algae (Nobel 1983, 30). Gimmler et al. (1990) report that the permeability coefficient (conductivity) of algal plasma membranes for CO_2 measured in the dark varies between 0.1×10^{-6} and 9×10^{-6} m s^{-1}. A value of $Pj = 10^{-5}$, therefore, is not unreasonable for illustration.

$$t = 0.693 \frac{V}{P_j S}, \qquad (1.12b)$$

which shows that t is inversely proportional to the ratio of $S{:}V$. Recall that S decreases relative to V when morphologically identical objects increase in size. Thus morphological isometry is not benefical in considering the passive diffusion of substances into cells or organisms. Figure 1.19 shows the consequences of morphological isometry and anisometry on t. In this figure, eq. (1.12b) (where $P_j = 10^{-5}$) was used to calculate values of t based on measured values of S for twenty-one *Chara* internode cells. The contrasting relationship expected if these cells maintained the same shape (constant cell aspect ratio $\Re = R{:}L$) is also shown. Figure 1.19 demonstrates that t increases linearly with respect to the actual surface areas measured for internode cells: $t = 8.0 + 3.7\,S$ ($r^2 = 0.82$). By contrast, t would dramatically increase nonlinearly if the shape of all internode cells remained constant: $t = 26\,S^{0.5}$. The morphological anisometry of the population of *Chara* internode cells therefore is theo-

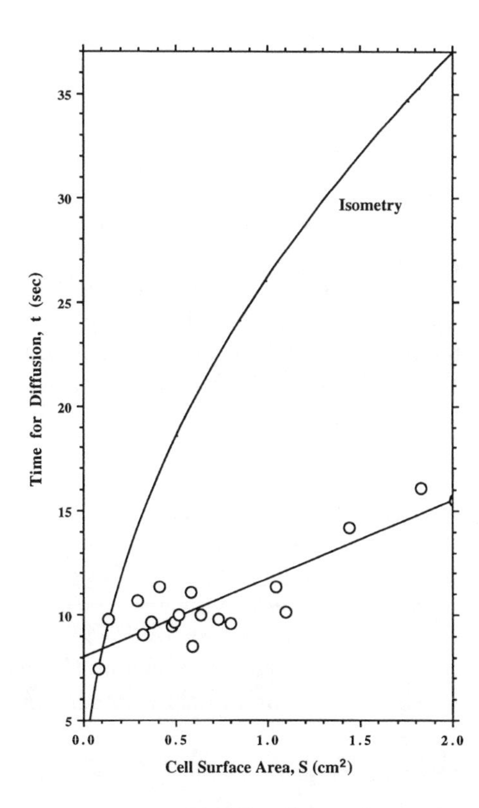

Figure 1.19. Estimated time t required for the internal concentration of a substance j to equal one-half the ambient concentration of j (see eq. 1.12b) for internodal cells of the green alga *Chara* plotted as a function of cell surface area S. Solid diagonal line is the regression curve obtained for the data from LS regression analysis; solid curved line is the hypothetical relation between t and S predicted assuming an isometric relation between S and cell volume V.

retically beneficial in terms of the diffusion of metabolites like CO_2. Conversely, eq. (1.12b) shows that isometry is less desirable in this context.

Whether unicellular or multicellular, every plant consists of a comparatively thin mantle of photosynthetic material draped over a nonphotosynthetic core. The individual chloroplasts within the cytosol of unicellular plants typically are sandwiched between the cell wall and a central vacuole or core of nonphotosynthetic material. And analogously, the photosynthetic cells in terrestrial metaphytes typically lie just beneath the epidermis. Fick's law shows that this arrangement is functionally adaptive in permitting the passive diffusion of CO_2. However, regardless of how much the external fluid in which a plant is submerged is brought into proximity or direct contact with photosynthetic cells, the thickness of photosynthetic rind within a plant is limited by the amount of light that reaches chloroplasts at sufficient intensity to drive photosynthesis. A typical chloroplast absorbs roughly 70% of the red and blue wavelengths of visible light it intercepts. Since chloroplasts shade one another, the efficiency of a population of chloroplasts decreases above an optimal concentration.

Perhaps the most interesting expressions of a "photosynthetic mantle" are seen among desert plants like the saguaro cactus. The outer portions of the stems of this species consist of an epidermis E and hypodermis H, collectively measuring roughly 1 mm in thickness (a total of ten layers of cells), through which long substomatal canals percolate. Beneath the E-H, a chlorenchymatous layer Ch extends to the depth of light extinction. The E-H-Ch rind remains alive and functional for most of the 200-year life span of individual plants. The E-H layers have been shown to protect the Ch from ultraviolet alpha and beta radiation, permit the absorption of much of the incident photosynthetically active PAR light, and reduce the heat load on the plant (Darling 1989). That is, the E-H is essentially opaque to UV radiation, yet it transmits roughly 86% of PAR, while the high reflectance of the Ch in the near-infrared reduces the absorption of incident radiation to 40%. Nonphotosynthetic storage parenchyma occurs beneath the E-H-Ch layers. At the very least, therefore, a treatment of the scaling of photosynthesis for this species must deal with the volume of a plant as a heterogeneous physiological compartment consisting of an outer E-H-Ch rind and a nonphotosynthetic core. It is noteworthy that, although the absolute size of a plant can increase, the thickness of its photosynthetic rind is limited by the amount of light reaching chloroplasts regardless of how much carbon dioxide is supplied.

The attenuation of light through a medium is given by Bouguer's law:

$$I = I_o e^{-\varepsilon_\lambda t}, \qquad (1.13) \quad \text{(Bouguer's law)}$$

where I is the attenuated flux density of light through a medium, I_o is the unattenuated flux density at the surface of the medium (the ambient light intensity), ε_λ is an extinction coefficient of the medium, and t is distance within the medium at which I is measured. Bouguer's law indicates that the decay of light intensity through a medium is exponential. Accordingly, if 50% of the available light energy is absorbed by the first 1 cm of a tissue, it is weakened yet another 50% by the second 1 cm. Because of this attenuation, PAR may eventually reach a level of intensity where no net photosynthesis can occur. This level is called the *light compensation point,* denoted as I_c. With the aid of Bouguer's law, the limiting condition on the magnitude of rind thickness $\tilde{t}$ set by I_c can be estimated:[4]

$$\tilde{t} = \frac{\log(I_o/I_c)}{\varepsilon_\lambda c V f}, \tag{1.14}$$

where c is the concentration of chlorophyll molecules in an average chloroplast and Vf is the volume fraction of chloroplasts. Equation (1.14) presumes that the scattering and reflection of light are negligible. This is a gross oversimplification, particularly for tissues with gas-filled intercellular spaces, whose surfaces can reflect and scatter light. Nonetheless, eq. (1.14) provides a measure of the sensitivity of $\tilde{t}$ to the values of I_o, c, Vf, and ε_λ. The following sensitivity analysis is instructive even if purely hypothetical.

The magnitude of I_o at the earth's surface varies with atmospheric conditions. For a sunny day with little or no cloud cover, the magnitude of I_o can reach 2,000 μmol PAR m^{-2} s^{-1}. At 20°C and an ambient atmospheric concentration of 350 ppm CO_2 (roughly 0.0125 mol CO_2 m^{-3}), the light compensation point for C_3 plants ranges between 6 and 10 μmol PAR m^{-2} s^{-1}; that of C_4 plants ranges between 4 and 8 μmol PAR m^{-2} s^{-1} (Nobel 1983, 438). For illustration, 8 μmol PAR m^{-2} s^{-1} can be taken as a crude average value for I_c. If so, then the numerator in eq. (1.14) becomes $\log_{10}[(2 \times 10^3$ μmol PAR m^{-2} $s^{-1})/(8$ μmol PAR m^{-2} $s^{-1})] = 2.398$ or roughly 2.4. The absorption coefficient of chlorophyll in the red and blue bands ε_λ is roughly 10^4 m^2 mol^{-1}, while the concentration of chlorophyll molecules c in foliage leaves typically equals 30 mol chlorophyll m^{-3}. The value of Vf varies among species. For most leaves, however, chloroplasts occupy roughly 3% of the total leaf volume. Based on these values, the denominator in eq. (1.14) equals (10^4 m^2 mol^{-1})(30 mol chlorophyll m^{-3}) (1/30) or 10^4 m^{-1}. Therefore $\tilde{t}$ is estimated to equal 2.4×10^{-4} m or

4. Bouguer's law applies only to an optically transparent medium and therefore cannot be used to calculate the actual rind thickness of a plant. Equation (1.14) is presented for purely didactic reasons.

24 mm. This value varies, however, as a function of *Vf*. If *Vf* is 50%, as it is for some small unicellular algae like *Chlamydomonas*, whose single chloroplast can occupy up to half of the total cell volume, then $\tilde{t}$ equals $(2.4)/(10^4\ m^2\ mol^{-1})(30\ mol\ chlorophyll\ m^{-3})(1/2)$ or 0.016 mm. That is, an overall cell diameter of 0.032 mm is predicted (the average diameter of a typical *Chlamydomonas* cell equals 0.035 mm). A decrease in the ambient light intensity or an increase in the light compensation point will affect the estimate of $\tilde{t}$ similarly. For example, if I_o = 500 μmol PAR $m^{-2}\ s^{-1}$ and I_c = 8 μmol PAR $m^{-2}\ s^{-1}$, then $\log_{10}[(500\ \mu mol\ PAR\ m^{-2}\ s^{-1})/(10\ \mu mol\ PAR\ m^{-2}\ s^{-1})] = 1.699 \approx 1.7$. Assuming once again that $Vf = 50\%$, $t = (1.7)/(10^4\ m^2\ mol^{-1})(30\ mol\ chlorophyll\ m^{-3})\ (1/2)$ or 0.011 mm. These naive calculations reveal that a comparatively high concentration of chloroplasts in the photosynthetic mantle of plants is not a particularly good *design* when ambient light intensities are comparatively low.

Conversely, even if light can penetrate deeply into a tissue, how fully carbon dioxide can diffuse into it limits the magnitude of $\tilde{t}$. This is shown by the following equation:

$$\tilde{t} = \left(\frac{J_{CO_2}}{P_{chloro}}\right)\left(\frac{Af}{Vf}\right), \tag{1.15}$$

where J_{CO_2} is the flux density of CO_2 into a plant (which has units of mol $CO_2\ m^{-2}\ s^{-1}$), P_{chloro} is the photosynthetic rate per unit volume of chloroplasts (which has units of mol $CO_2\ m^{-3}\ s^{-1}$), and *Af* is the dimensionless ratio of the area of the cell walls exposed to intercellular fluid (air or water) to the total external surface area of a photosynthetic organ. Once again, the dimensionless ratio *Vf* denotes the volume fraction of the photosynthetic tissue occupied by chloroplasts. The value of J_{CO_2} varies depending on a number of factors such as the thickness of the unstirred layer of air blanketing the surface of an organ and ambient temperature and light intensity (Nobel 1983). For illustration, a value of 8×10^{-5} mol $CO_2\ m^{-2}\ s^{-1}$ provides a crude approximation for the magnitude of J_{CO_2} for an average terrestrial plant (Raven, Beardall, and Griffiths 1982; Raven 1985, 275). Likewise, 2.5 mol $CO_2\ m^{-3}\ s^{-1}$ can be taken for P_{chloro} under optimal conditions, althought this value is contingent on a variety of assumptions and empirical data reveal that P_{chloro} can vary widely even for a particular species, particularly since CO_2 in the chloroplast generally limits photosynthesis. Be that as it may, based on presumed values for J_{CO_2} and P_{chloro}, the first parenthetical term in eq. (1.15) becomes $(J_{CO_2}/P_{chloro}) = (8 \times 10^{-5}\ mol\ CO_2\ m^{-2}\ s^{-1}/(2.5\ mol\ CO_2\ m^{-3}\ s^{-1}) = 3.2 \times 10^{-6}$ m.

The second parenthetical term (Af/Vf) is measurable directly for a particular organ, but it varies widely among organs even on the same plant (e.g., sun and shade leaves). Af can range between 5 and 20 for mesophytes and can drop below unity for bryophytes and algae, whose photosynthetic organs lack intercellular spaces. As anticipated, the magnitude of $\tilde{t}$ is extremely sensitive to values of Af (our measure of how far a photosynthetic organ is profused by intercellular spaces) as well as Vf (our measure of the relative *concentration* of chloroplasts in the photosynthetic tissue). If $Af = 1$ (no intercellular spaces) and $Vf = 1/30$ (a typical value for foliage leaves, then t equals $(3.2 \times 10^{-6}$ m)(30) or 0.096 mm, which is three orders of magnitude less than predicted based on the attenuation of light through the chlorophyllous tissue of a typical foliage leaf (24 mm). For a mesophyte leaf with a very spongy mesophyll, Af can equal 20. Thus $(Af/Vf) = 600$. For these values, t equals $(3.2 \times 10^{-6}$ m)(600) or 1.9 mm. This is roughly 8% of the value of $\tilde{t}$ predicted from light attenuation.

Table 1.3 gives the formulas for the ratio $S{:}V$ in terms of $\tilde{t}$ for some simple geometries that serve as reasonable analogues to plant organs or entire plants. In conjunction with eqs. (1.4) and (1.5), these formulas can

Table 1.3. Mensuration Equations for Ratios of Surface Area S to Volume V and Aspect Ratios $\Re$ of Geometries with Rind Thickness $\tilde{t}$

	$S{:}V$	
Sphere	$\dfrac{3\Re^2}{\tilde{t}\,[3\Re(\Re - \tilde{t}) + \tilde{t}^2]}$	where $\Re = 1$
Cylinder	$\dfrac{2\Re L}{\tilde{t}(2\Re L - \tilde{t})}$	where $\Re = R/L$
Cone	$\dfrac{3\Re^2 L^2(\Re^2 + 1)^{1/2}}{\tilde{t}[3\Re L(\Re L - \tilde{t}) + \tilde{t}^2]}$	where $\Re = R_1/L$
Frustum of cone	$\dfrac{(\Re + 1)L^2\Re_0[1 + \Re^2\Re_0^2 - 2\Re\Re_0^2 + \Re_0^2]^{1/2}}{\tilde{t}[\Re_0 L(\Re + \tilde{t}) - \tilde{t}]}$	where $\Re = R_1/R_2$ $\Re_0 = R_2/L$
Prolate spheroid	$\dfrac{3a^2}{2\tilde{t}}\left\{\dfrac{2\Re + \dfrac{\sin^{-1}e}{e}}{4a^2\Re(\Re + 1) - a\tilde{t}\,(4\Re + 1) + \tilde{t}^2}\right\}$	where $\Re = b/2a$
Oblate spheroid	$\dfrac{3a^2}{2\tilde{t}}\left\{\dfrac{1 + \dfrac{2\Re^2}{e}\ln\dfrac{(1 + e)}{(1 - e)}}{a(4a\Re - \tilde{t}) + (2a\Re - \tilde{t})^2}\right\}$	$e = (1 - 4\Re^2)^{1/2}$

Note: a, b = major and minor semiaxes of elliptical cross section; e = eccentricity of ellipse $= (a^2 - b^2)^{1/2}/a$; L = length; R unit radius; R_1 = bottom radius; R_2 = top radius; $\log_e$ = natural logarithm.

be used to estimate the influence of size, shape, and geometry on photosynthesis. Although purely pedagogical, table 1.3 and eqs. (1.4) and (1.5) quickly reveal that some plant geometries are impractical because the volume fraction of photosynthetic cells rapidly declines as a function of absolute size. The most convenient geometry with which to illustrate this is the sphere whose shape and absolute size are defined by a single parameter, the unit radius *R*. As intuition freely reveals, a solid sphere cannot change external shape, since its aspect ratio is always equal to unity ($\Re = R/R = 1$). The application of eqs. (1.14) and (1.15) also shows that a spherical plant should be determinate in growth because, as size increases indefinitely, the volume fraction of photosynthetic material steadily declines once $R > \tilde{t}$. True, spherical plants exist, although they are comparatively rare. The multicellular volvocalean algae are essentially spherical; they also are determinate in growth (see chap. 2). But the larger volvocaleans, like *Volvox* are hollow, unlike their solid smaller counterparts. As a natural experiment in the relation between size and shape, the biology of spherical plants is consistent with the expectation that growth in size must be truncated or shape must change as a function of absolute size. Indeed, the morphogenesis of larger volvocalean algae involves a transition from a solid to a hollow sphere.

Other plant shapes offer more versatile solutions and are comparatively more common. The terete cylinder can grow indefinitely in length (but not in girth) without alteration in *S:V* or $\tilde{t}$. This is a preposterous mechanical design on land; however, the yearly decrease in the rate of vertical growth previously noted for the saguaro cactus (see fig. 1.17) is one developmental solution to the limitations imposed by mechanical considerations. Another is to terminate vertical growth before mechanical failure ensues. Many vertically growing stems that lack secondary tissues are determinate in vertical growth. Their aspect (slenderness) ratio steadily declines, but its attenuation falls short of the critical value that leads to mechanical failure. Most of these determinate stems are flower stalks, which are expendable once their reproductive function is fulfilled. It is also noteworthy that most of these stalks are hollow and therefore have a fairly uniform wall thickness through which light can penetrate and external gases can diffuse.

The frustum of a cone, a geometric analogue to a tree trunk, is another excellent geometric design whenever a plant organ functionally persists over many years. As it grows, its aspect ratio *R:L* increases as a result of the seasonal deposition of internal growth layers of wood, a tissue noted for its capacity to sustain extremely high compression stresses. Although the ratio *S:V* of a tree trunk decreases as the girth of the trunk increases, this is of little physiological consequence because the only living tissues (vascular cambium, phloem, and phellogen) within a trunk are sandwiched between the dead internal wood (secondary xylem) and the bark. Table 1.3 shows that as more and more growth layers of wood are deposited, the *S:V* of these living layers of tissue (which are configured into a hollow conical frustum) can remain the same or even increase. At the same time, the distance of these living tissues from the external atmosphere (which provides oxygen) remains relatively constant.

The aspect ratios of other plant geometries can be altered by development in similarly functionally useful ways (table 1.3). Consider prolate and oblate spheroids, the three-dimensional solids that result from the rotation of an ellipse around either its major or its minor axis. The former rotation yields the cigar-shaped prolate spheroid; the latter gives rise to thalloid-shaped oblate spheroid. Many small unicellular aquatic plants are essentially prolate spheroids (e.g., *Chlamydomonas*), as often are the individual cells of filamentous multicellular plants. Provided its aspect ratio is held constant, a prolate cell can grow indefinitely without altering its *S:V* (see tables 1.2 and 1.3). Likewise, *S:V* can be increased by holding the minor axis constant and increasing the major axis (i.e., elongation). Siphonous and filamentous algae take advantage of this simple geometric fact. They grow in length and retain a more or less constant girth. By the same token, an oblate spheroid can grow in size and increase *S:V* by flattening, as do many foliage leaves. And as revealed by figure 1.18, the oblate spheroid is superb at light interception.

1.11 Organs versus Organisms

As noted earlier (see sec. 1.3), most plants continue to grow indefinitely by periodically producing additional parts (leaves, stem internodes, etc.) that in most cases are determinate in growth. Exceptions occur naturally. The foliage leaves of the climbing fern (*Lygodium*) are indeterminate in growth. Their petioles mechanically function as aerial stems. Their leaflets (pinnae), however, are determinate in growth, as are the preponderance of the foliage leaves of other vascular plants. Likewise, stem internodes for most plants are determinate in vertical growth, although the production of secondary tissues results in indeterminate growth in stem girth. The juxtaposition of the determinate growth of plant organs with the indeterminate growth of the individual plant raises a number of issues that are discussed in this section.

Let me begin by noting that the ontogeny of the individual may be insulated from the direct influence of the physical environment. During its formative stages, some or all of the metabolic requirements for the growth of the juvenile can be assumed by the parent. The yolk sac and complex feeding behaviors of oviparous and viviparous vertebrates show that the parent can metabolically or physically intervene to temper the direct consequence of the environment on its juveniles. Among species of embryophytes (plants with archegonia), the embryo is protected and metabolically provided for by the gametophytic generation in which it is contained. In seed plants, the gametophyte is likewise nurtured by the previous sporophytic generation. These reproductive systems permit latitude in the products of ontogeny provided the end result, the mature individual, is ultimately capable of survival and reproduction. For organisms characterized by indeterminate growth and the production of numerous organs, each juvenile organ is physiologically and mechanically sustained by older organs. A juvenile leaf metabolically imports photosynthates from mature exporting foliage leaves. An analogous situation in the animal kingdom occurs when juvenile *Hydra* bud from an adult. Since the sporophytes of many species continue to increase in size indefinitely throughout their lifetime, as do the adults of many species of animals, it is evident that the mature portions of these organisms metabolically assist the growth and development of juvenile organs.

The scaling relation between the size of organs and the size of the individual is clear for animals but not for plants. Large animals tend to have proportionately large body parts and vice versa. Likewise, the size of an adult animal can influence either the number or the size of juveniles produced. For example, body-cavity volume in salamanders correlates well with total egg volume (Kaplan and Salthe 1979); birth weight, litter weight, and gestation time are positively correlated with the adult size of eutherian mammals (Miller 1981); and frog body size correlates with total egg volume and individual egg size (Berven 1982). By contrast, the empirical data for plants are problematic. Intraspecific comparisons, such as those of Turrell (1961) for *Citrus sinensis* and Sinnott (1921) for *Phaseolus vulgaris*, typically show little or no correlation between organ and organismic size. However, our concern is with interspecific comparisons. In this regard Primack (1987) reports positive correlations between plant height and leaf length for several herbaceous genera. Likewise, Thompson and Rabinowitz (1989) found positive and significant Spearman-Rank correlations ($0.38 \leq r_s \geq 0.68$; $0.05 \leq p \geq 0.001$) between the size of the individual plant and the size of propagules for species within the Caryophyllaceae ($N = 41$), Brassicaceae ($N = 47$), Fabaceae ($N = 118$), and Poaceae ($N = 81$). Yet these authors report no statistically significant correlations for species in the Apiaceae ($N = 33$), Asteraceae ($N = 87$), Lamiaceae ($N = 38$), and Scrophulariaceae ($N = 37$). Additionally, they report no meaningful correlations between seed size and the height of trees. As a consequence, Thompson and Rabinowitz conclude that the size of seeds is not likely to be physically restricted by the size of the parent plant. Although this supposition may be in jeopardy for mechanical reasons (e.g., Peters et al. 1988 show that fruit mass is disproportionately large compared with the mass of small-diameter branches), Thompson and Rabinowitz (1989) insightfully propose alternative but not mutually exclusive hypotheses to explain their findings: that seed size may differ interspecifically as a function of the successional habitat occupied by species, increasing as a habitat matures; and that seed size and plant height may be related to interspecific seed-dispersal mechanisms, becoming better or more highly correlated in species lacking specialized dispersal mechanisms. Circumstantial evidence in support of these hypotheses varies among the families they examined.

The equivocation in the literature regarding whether bigger plants produce bigger organs is in part a consequence of failure to define precisely what is meant by the individual plant. An individual organ or even the entire aerial shoot of an herbaceous plant is determinate and represents a modular unit of the individual plant. The size, shape, and geometry of a modular unit are influenced by the environment attending the growth and development of the unit. By contrast, the total size of an herbaceous plant and the height of a tree are nonmodular characters. Both are legitimate albeit incomplete measures of the size of the individual plant. Obviously larger plants can produce more organs, and they may produce larger organs. For example, Kang and Primack (1991) report significant correlations between the number of flowers per plant and plant size measured in terms of lateral spread for two populations of the celandine poppy *Chelidonium majus* (Papaveraceae). They also report that within species larger plants produce larger flowers. Nonetheless, it is evident that the size of the mature plant organs produced by the same individual can differ dramatically. Figure 1.20 illustrates this point for leaves by plotting the surface areas of leaf laminae against the mass of laminae for randomly selected leaves taken from an orange tree and a maple tree. These data are for fully mature leaves. Likewise, the size, morphology, or anatomy of an organ type also can vary as a function of genotype regardless of environmental similitude. Consider a trait such as plant height, resulting from five genes, each with three alleles. In these circumstances the possible number of only the homozygous combinations would equal 3^5, or 243. Assuming that one allele is dominant for each gene, a minimum number of 243 phenotypic classes would be observed.

The variation in leaf size shown in figure 1.20 is probably due to variations in the sizes of the apical meristems from which leaves arise. It is well known that the size of an apical meristem alters over the ontogeny of the shoot and in response to changes in the physical environment. Likewise, recent studies suggest that plants are not capable of producing uniform seed crops in the face of variations in the availability of resources from one growing season to another (Herrera, 1991; Winn 1991). In the context of scaling, this was first pointed out by Edmund Sinnott (1921, 400), who concluded that "the size of the plant body is not the direct causative factor in determing the size of leaves, fruits or seeds which it produces. . . . the size of any given organ depends rather upon the size of the growing point (apical meristem) out of which it has been developed." Sinnott regressed the average dry mass of the vegetative shoot of the bean variety Red Kidney against the dry mass of stems (Sinnott 1921, table 2). This regression reflects Sinnott's view that the shoot is the *individual,* although it is evident that each plant consists of roots as well as shoots. Be that as it may, the problem of autocorrelation also exists, since a vegetative shoot is the sum of its leaves and stems. Fortunately, Sinnott's data permit us to subtract leaf mass m from total shoot mass such that we can regress the dry mass m of leaves against that of stems M without fear of

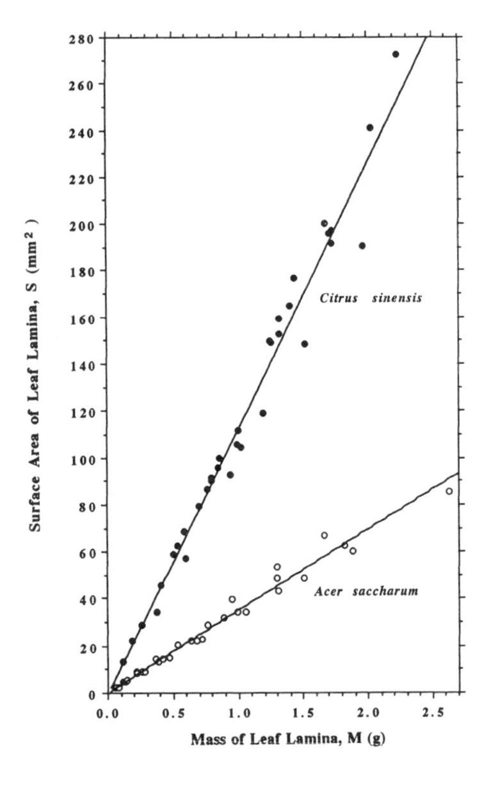

Figure 1.20. Surface areas S of leaf laminae plotted against the mass M of the laminae of randomly selected leaves from an orange tree (*Citrus sinensis*) and a maple tree (*Acer saccharum*). Solid diagonal lines denote regression curves obtained from LS regression analysis.

autocorrelation. Figure 1.21 plots m versus M based on the original data. The scaling relationship is $m = 0.13\ M^{0.47}$ ($r^2 = 0.76$, $N = 344$; $\alpha_{RMA} = 0.54 \pm 0.05$), indicating that leaf mass scales roughly as the square root of stem mass. A more desirable protocol for determining the scaling of leaf size to stem size is to regress the mass of each leaf aganst the mass of its subtending internode. These data, however, cannot be retrieved from the original publication.

Sinnott (1921) further distinguished between immature nonflowering plants (harvested daily) and mature fruit-bearing plants (simultaneously harvested at the end of the growing season). Regression of the data for mature plants shows that no correlation exists between m and M ($r^2 = 0.021$, $N = 98$), in contrast to the data from immature plants, which give $r^2 = 0.76$ ($N = 246$). Based on these relations, Sinnot inferred that the apical meristems of immature plants increased in size until flowering ensued, thereby accounting for a correlation between the size of juvenile individuals and their organs. By contrast, the apical meristems of the ma-

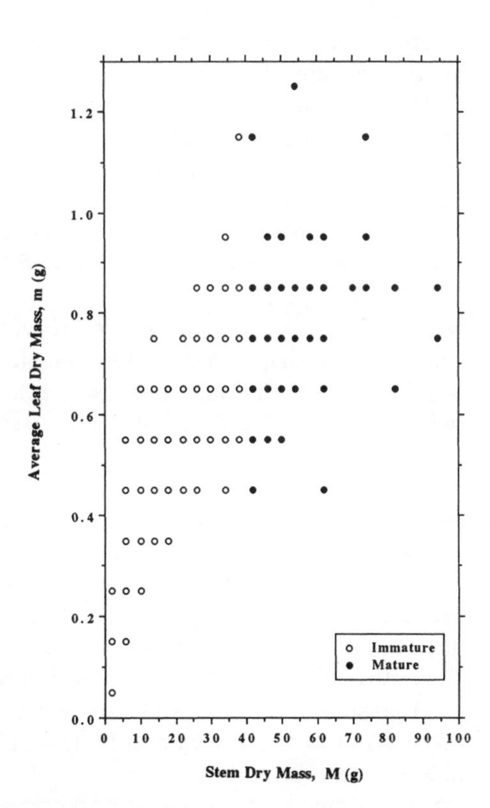

Figure 1.21. Average leaf dry mass m plotted as a function of stem dry mass M for immature and mature specimens of the bean variety Red Kidney. Data from Sinnott 1921.

ture plants were assumed to have reached their maximum size, although the variation in the size of apical meristems within the bean population was not measured.

1.12 Summary

Operational definitions for growth and development were presented. Growth was defined as any permanent increase in size; development was defined as any orderly changes in extenal shape (morphogenesis) or internal structure (anatomy), which typically attend growth in size. These definitions were used to explore the notion that plant size, shape, and structure are under separate but integrated genetic control, permitting plants potentially independent routes for expressing organic form. The functional significance of multicellularity and the modular construction of metaphytes were briefly discussed. Scaling was next shown to empirically describe variations in the form and metabolic rates of unicellular and multicellular plant species differing in size. Empirically observed morphological and metabolic differences among plants were seen to be highly size dependent.

This was followed by a treatment of dimensional analyses. Based on arguments of similitude, dimensional analyses was used to derive the 2/3- and 3/4-power laws, which attempt to explain a priori size-dependent variations in shape and metabolic rates. To test these laws, the shapes of unicellular plants were quantified and expressed in terms of surface area S and volume V. Empirical data showed that S scales to $V^{0.67}$. Although the scaling exponent of this relation complied with the 2/3-power law, which follows from geometric similitude, the assumption that organismic geometry and shape are size independent was shown to be incorrect.

Likewise, the presumption that cell-density similitude also exists (which predicts that metabolic rates associated with V and dependent upon S must scale to the 2/3 power of biomass M) was shown to be incorrect. The rate of growth G of unicellular and multicellular plants and animals was shown to scale roughly as the 3/4 power of body mass. This scaling relation may be the consequence of changes in cellular constituents attending growth in size. Among unicellular plants, the concentration m of chlorophyll a per cell scaled as $M^{0.74}$ ($\alpha_{RMA} = 0.80$), while a nearly isometric relation exists between m and $G(\alpha_{RMA} = 0.91)$.

Photosynthesis was then discussed in terms of the ability of light and external substances (e.g., CO_2) to penetrate photosynthetic tissues based on Fick's law of diffusion and Bouguer's law of the attenuation of light intensity. The plant body was conceived of as a rind of photosynthetic material draped over a nonphotosynthetic core. Different plant geometries were shown to have very different capacities to deal with the metabolic consequences of this rind-core construction.

Finally, the relation between the size of organs and that of individual plants was considered. The available evidence suggested that intra- and interspecific scalings of organ size to the size of the individual plant likely vary as a consequence of variations in the physical environment attending the growth and development of organs.

2 Aquatic Plants

2.1 Introduction

This chapter has three objectives that, taken as a package, reflect what may be called an engineering approach to the scaling of aquatic plants. The first objective is to describe the physical environment in which plants live. Even a qualitative description of the physical properties of water exposes the biological advantages and disadvantages of living in this fluid medium. Water is roughly one thousand times denser than air and, unlike air, has a pronounced capacity to absorb and scatter light. As a consequence, most aquatic plants are weightless, requiring little or no mechanically specialized supportive tissue, and the growth of most is limited by the availability of light. Consequently they tend to invest virtually all their energy into biomass whose surface area is maximized to enhance light harvesting. I said in chapter 1 that plants with large surface areas relative to their volume have an advantage in photosynthesis. Large ratios of surface area to volume reduce the time required for substances to passively diffuse into and out of cells as well as increasing the capacity of photosynthetic tissues to intercept sunlight. Large surface areas that aid metabolism in general and photosynthesis in particular are permissible provided access to water is not limited. The morphological elaboration of surface areas, which enhances rates of passive diffusion and the ability to harvest the available light energy, also provides virtually every cell within even comparatively large aquatic metaphytes with ready access to an unlimited supply of water and other nutrients.

In the terrestrial ecosystem, a disproportionate increase in surface area at the expense of volume is not mechanically advantageous, owing in part to the influence of gravity. By contrast, in aquatic ecosystems, where the effects of gravity on plants are greatly reduced by virtue of the negligible difference between the bulk density of plant tissues and the density of liquid water, the principal mechanical support system in aquatic metaphytes is the cell wall infrastructure of tissues.[1] Obviously there are excep-

1. The average densities of plant cell sap, endoplasm, and cell wall are 1,010, 1,015, and 1,360 kg m^{-3}, although significant intra- and interspecific variation occurs depending upon physiological condition. The average density of a parenchyma cell is 1,035 kg m^{-3}; the average density of parenchyma tissue (with intercellular airspaces) is 950 kg m^{-3}. The density of pure water at 20°C is 998.4 kg m^{-3}. Roughly, therefore, the ratio of the density of an average parenchyma cell to the density of water is 1.04; the ratio of the density of parenchyma tissue to the density of water is 0.95.

tions to this generalization. The medullary tissues in some genera of intertidal algae (e.g., *Fucus* and *Postelsia*) function mechanically much like the secondary xylem in the woody stems of terrestrial plants—that is, new tissues are added every growing season. The appearance of these specialized tissues in marine macrophytes demonstrates that the forces generated by wave action and by gravity place analogous mechanical demands on aquatic and terrestrial plants. In general, however, both the absolute volume and the volume fraction of mechanically supportive tissues in aquatic plants are greatly reduced compared with terrestrial plants, in large part because aquatic plants are not subjected to the compressive stresses induced by the effect of gravity on their biomass. As a consequence, the volume fraction of photosynthetic tissues is proportionally larger for aquatic than for terrestrial metaphytes.

The second objective of this chapter is to introduce physical laws that quantitatively describe the transfer of mass or energy between a plant and its external fluid environment. The exchange of mass and energy is essential to photosynthesis and therefore to plant growth. Photosynthesis depends on the transfer of mass (e.g., carbon dioxide and water dissolved in the ambient fluid environment) into the plant. Likewise, energy in the form of light, which is absorbed by chlorophyll, is used to dissociate water molecules and subsequently synthesize carbohydrates from carbon dioxide. The physical laws that describe the transfer of mass or energy are expressed in terms of transport equations that are described in this chapter. These equations have comparatively simple and analogous mathematical forms. Each describes a flux density in terms of a concentration gradient, a resistance, and a proportionality factor. Each of the transport equations refers to the physical variables that define the behavior of mass or energy in a fluid medium. They also refer either directly or indirectly to the biological variables of organic size and shape. Consequently the transport equations provide a direct mathematical link between the description of how the physical environment affects plant mechanics and physiology and how plant size and morphology influence the operation of the physical environment.

The complexity of this reciprocity can be greatly reduced with the aid of dimensionless groupings of variables that figure in the transport equations. Using these groupings, the behavior of one physical system can be correlated with the behavior of another. This is a powerful tool. For example, the rate of nutrient absorption by unicellular algae is extremely difficult to measure directly. This process, however, is physically analogous to the transfer of heat between water and metal pellets of varying size, geometry, and shape. Since the dissipation of heat is comparatively easy to measure, with dimensionless groupings of variables we can use rates of heat dissipation to estimate the transfer of nutrients between plants and their fluid environment.

The third objective of this chapter is to show that the physical environment and the laws that describe its behavior do not operate on a passive, totally submissive organism. By its growth in size and its potential to alter shape and structure, an organism can influence and even alter the extent to which the physical environment affects the rates of mass and energy transfer. The persistent operation of physical laws on highly adaptive living things is the syzygy—the conjunction through opposition—that defines biology.

2.2 Absorption Coefficients of Water and Plant Cells

The scaling of aquatic plants is influenced by the fact that water is a significant obstacle to the penetration of light. This and the following section discuss why this is so. The aquatic light field decreases in intensity and alters in spectral properties as a function of how deeply light penetrates the water column. Figure 2.1, for example, which plots the light absorp-

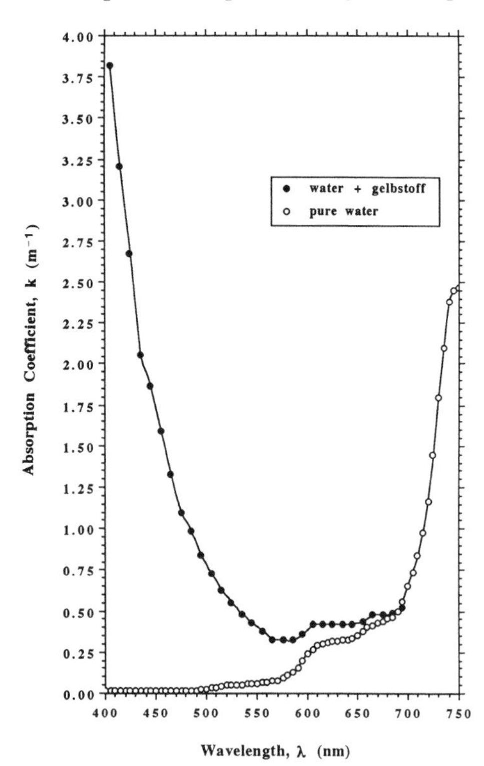

Figure 2.1. Light absorption coefficient k of pure water and water plus gelbstoff plotted as a function of wavelength λ in the 440 to 750 nm range. Data from Smith and Baker 1981 and Kirk 1975, 34, table 6, respectively.

tion coefficient k (in units of m^{-1}) of pure water as a function of wavelength λ in the 440 to 750 nm range (data from Smith and Baker 1981), shows that values of λ increase as a nonlinear function of increasing values of λ. The absorption of wavelengths below 550 nm is small but nonetheless important to plant physiology in determining the penetration of short wavelengths. Equally important is that ultraviolet light inhibits photosynthesis and therefore reduces the rate of growth. For longer wavelengths, water absorbs weakly, as in the blue and green regions of the visible spectrum. Figure 2.1 shows that water absorbs more as the wavelength increases above 550 nm. Two shoulders occur in the visible spectrum, one at $\approx$ 604 nm and another (weaker) one at $\approx$ 514 nm (these two shoulders correspond to the fifth and the sixth harmonics of the O–H stretch vibration of liquid water; Tam and Patel 1979). The fundamental harmonic is roughly 3 μm. Pure water significantly obstructs the penetration of light in the red region of the spectrum and therefore is an excellent infrared absorber. An absorption peak occurs at $\approx$ 745 (fig. 2.1) and, another at $\approx$ 960 (not shown). These peaks respectively correspond to the fourth and third harmonics of the O–H stretch vibration. Thus water is a blue liquid, as is apparent when large bodies of water are viewed from above or from below at some depth.

The absorption coefficients of pure water are not particularly instructive when considering the attenuation of light in most natural ecosystems, because water (fresh and salt) typically contains substantial quantities of dissolved and suspended materials, which absorb and scatter light and therefore further attenuate its intensity. In large part, the dissolved materials in natural water columns come from the decomposition of plant tissues. These dissolved materials are referred to generally as humic substances (Schnitzer 1978; Kirk 1983, 51–60). They can accumulate in open bodies of water or in soils where they can drain into lakes or oceans. Significant qualitative and quantitative differences in the water-soluble humic substances are reported for different bodies of water (see Kirk 1983, 57–59, table 3.2) and therefore there exists no standard, *typical* absorption spectrum for naturally occurring water. Nevertheless, most water-soluble humic substances are yellow (German *Gelbstoff,* literally "yellow substance") and therefore typically absorb blue wavelengths of light. Thomas Kirk (1983, 56) has proposed a method to standardize the concentration of gelbstoff according to the absorption coefficient at 440 nm, symbolized by g_{440}. In general g_{440} is low for oceanic waters, moderate for coastal and estuarine waters, and comparatively high for inland waters. Figure 2.1, for example, plots the absorption spectrum of water (over the wavelength range of 355 to 695 nm) containing gelbstoff, for which g_{440} = 2.00 (data from Kirk 1975, 34, table 6). As can be seen, gelbstoff also significantly absorbs short wavelengths and ultraviolet radiation. The attenuation of light through a water column in which gelbstoff is dissolved can be substantial. Figure 2.2, for example, plots the spectral distribution of light quanta flux (in units of $cm^{-2}s^{-1}nm^{-1}$) incident just below the water surface (i.e., refraction has been allowed for) based on spectral data reported by Kirk (1975, 35, table 7). The total flux of light radiation in the 300 to 700 band is 99.41×10^{15} quanta $cm^{-2}s^{-1}$.

By far the most significant attenuation of light in the water column is the result of the absorption of light by suspensions of phytoplankton. In highly productive waters, concentrations of phytoplankton may be high enough to limit plant growth. This is the consequence of self-shading at the level of the population. For any wavelength of light, the absorption coefficient is the sum of the absorption coefficients of all the light-absorbing dissolved and suspended constituents in the water column. The inanimate suspended material, consisting of particulates collectively called tripton, typically does not absorb light strongly, although it can

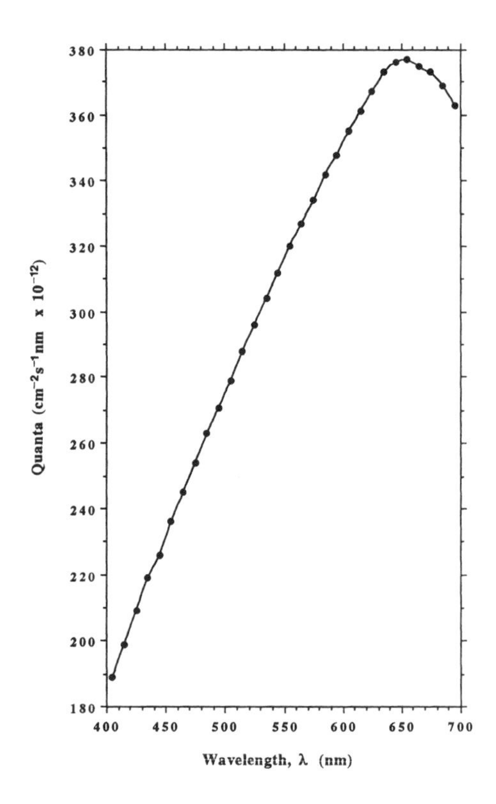

Figure 2.2. The spectral distribution of light quanta flux incident just below the water surface (refraction has been allowed for) plotted as a function of wavelength λ in the 400 to 700 nm range. Based on spectral data reported by Kirk (1975, 35, table 7).

significantly scatter light. By contrast, phytoplanktonic species strongly absorb light owing to the presence of photosynthetic pigments within cells. The in vitro absorption spectrum of an extracted pigment is easily determined spectrophotometrically. The absorption coefficient for each wavelength measured for the solvent-pigment is divided by the pigment concentration (expressed in appropriate units, like mg m^{-3}). This quotient is called the specific absorption coefficient α. It has units of m^2 mg^{-1}. When α is reported in terms of its natural logarithm, it is called the natural logarithm specific absorption coefficient, denoted as γ (i.e., $\gamma = 2.303 \alpha$). Values of γ depend on the concentrations of pigments in cells. For many species of green algae (i.e., Chlorophyta), the intracellular chlorophyll *a* concentration roughly equals 2.9×10^6 mg m^{-3}, although from broad interspecific comparisons we know that the concentration of chlorophyll *a* decreases with increasing cell volume (see fig. 1.14). The average chlorophyll concentration per green algal cell can equal 4.0×10^6 mg m^{-3}, presuming that the average ratio of chlorophyll *a* to chlorophyll *b* equals 2.7.

Values of γ determined from in vitro solutions of pigments are not particularly useful in practical terms when dealing in natural ecosystems. The absorption spectrum of a living plant cell is the result of the absorption spectra of the mixture of pigments within the cell (chlorophyll *a,* and *b*, carotenoids, etc.). Additionally, the absorption spectrum of a pigment bound to a chloroplast membrane differs from that of an extended pigment dissolved in a solvent. For example, the chlorophyll red peak in ether is 660 nm. In acetone the peak is shifted to 663 nm (French 1960, 254, table 1). Unfortunately, the in vivo absorption peaks and specific absorption coefficients of pigments cannot be estimated with the utmost confidence by the method of proportion from data for extracted pigments, although this is frequently the practice. Figure 2.3, for example, plots values of γ determined for a living suspension of the unicellular green alga *Chlamydomonas reinhardtii* (data kindly provided by Thomas Owens, Section of Plant Biology, Cornell University). For comparison, estimated values of γ are coplotted for the pigment mixture of the unicellular green alga *Chlorella pyrenoidosa* (see Kirk 1975, 23, table 1, based on data from Emerson and Lewis 1943).[2] Although the two graphs are in qualitative agreement, quantitative differences between the data for *C. reinhardtii* and *C. pyrenoidosa* are significant and highlight the difficult-

2. *Chlorella* has much the same complement of light-harvesting pigments as *C. reinhardtii* (i.e., chlorophylls *a* and *b*, carotenoids), although quantitative differences exist in the concentrations of these pigments.

ies and potential errors in estimating γ from the absorption spectra of pigments extracted from living cells.

2.3 The Package Effect and the Scaling of Cell Geometry and Shape

Thus far we have seen how the absorption properties of water and dissolved and suspended materials influence the light field. We now examine how the interception of light by unicellular plants is influenced by cell geometry and shape, pigment concentration per cell, and number of cells per volume of suspension. From this treatment, light interception is shown to be a scaling phenomenon (i.e., a size-dependent variable).

Light interception depends on cell size because of the *package effect*, so named because equivalent amounts of pigments contained in discrete packages or units (chloroplasts, individual cells, or colonies of cells) are less effective at harvesting light than pigments suspended in solution. The consequences that compartmentalizing photosynthetic pigments into cells has on the attenuation of light through a suspension of unicellular organ-

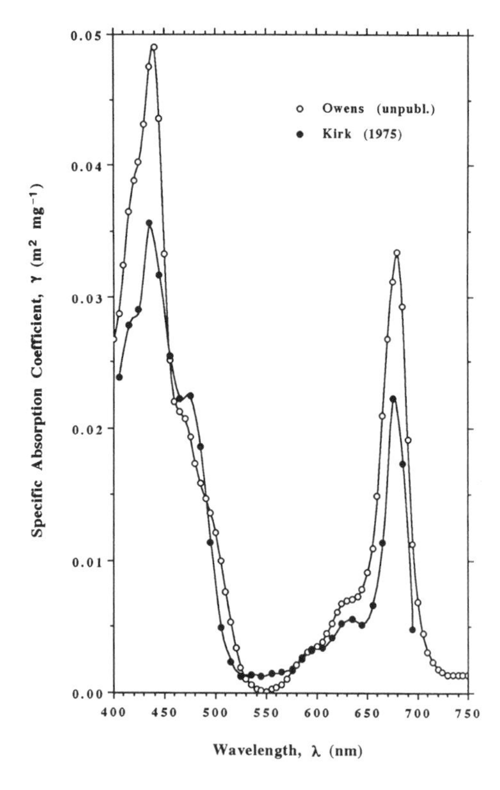

Figure 2.3. Specific absorption coefficient γ determined for a living suspension of the unicellular green alga *Chlamydomonas reinhardtii* plotted against wavelength λ in the 400 to 750 nm range (data provided by Thomas Owens, Cornell University) and estimated values of γ for the pigment mixture of the unicellular green alga *Chlorella pyrenoidosa* plotted against λ in the 400 to 690 nm range (based on data from Emerson and Lewis 1943; see Kirk 1975, 23, table 1).

isms have been modeled by Kirk (1975, 1976, 1983, 203–9; see also Morel and Bricaud 1981), based on the equation

$$2.1 \qquad I = I_0 e^{-(k + n\bar{A}\bar{a})t \operatorname{cosec}\beta},$$

where I is the downward flux of radiation per unit area on a horizontal plane at depth t meters, I_0 is the flux of radiation on a horizontal plane just beneath the water surface, k is the total absorption coefficient of the liquid (water plus gelbstoff), n is the number of cells per cubic meter, $\bar{A}$ is average projected cell area in the suspension, $\bar{a}$ is the average proportion of the total light incident on a cell that is absorbed, and β is the angle of incident light to the horizontal. When the direction of incident light is normal to the horizontal plane just beneath the water, $\operatorname{cosec}\beta = 1.0$, and therefore eq. (2.1) takes the formal $I = I_0 e^{-(k+n\bar{A}\bar{a})t}$, where $n\bar{A}\bar{a}$ is the average absorption cross section of the cell suspension. Since the average absorption cross section has units of surface area, it provides a convenient measurement of the effect of packaging pigments into cells differing in size. Before we examine this, however, it is worth noting the eq. (2.1) is an expression of Bouguer's law (see eq. 1.13). Thus, modeling the package effect for algal suspensions must presume very dilute concentrations of cells (the absorbance of the suspension is assumed to equal the sum of the absorbance of all the individual cells). Essentially, the suspension is considered optically transparent.

Regardless of this practical limitation on the applicability of models based on Bouguer's law to natural ecosystems, eq. (2.1) is pedagogically useful because it shows that the average absorption cross section decreases as a function of cell size. For this comparatively simple cell geometry, Duysens (1956) has shown that the fraction of intercepted light per cell is given by the formula

$$(2.2) \qquad a = 1 - \frac{2[1 - (1+\gamma CD)e^{-\gamma CD}]}{(\gamma CD)^2},$$

where γ is the natural logarithm specific absorption coefficient and C is the intracellular pigment concentration. Figure 2.4 plots the absorption cross section for four hypothetical algal suspensions composed of cells differing in diameter (D = 4, 8, 16, and 64 μm) against wavelength. These plots are based on eq. (2.2). Each plot was obtained from the stipulation that all four algal suspensions have equivalent total fresh-weight biomass (the total cellular volume V_T equals $1 \times 10^{-6} m^3$). Each also assumes an equivalent intracellular pigment concentration ($C = 4 \times 10^6$ mg m^{-3}). Figure 2.4 shows that the average absorption cross section decreases as cell size increases, as predicted by the package effect. A comparison among the four plots reveals that the algal suspension for which

$D = 64$ μm is the least efficient at harvesting incident light, while the suspension consisting of cells with $D = 8$ μm is the most efficient of the four. Note that the differences among the four plots are more pronounced at the absorption peaks than within the troughs. Thus the size dependency of $n\bar{A}\bar{a}$ is most readily seen for maxima in the absorption spectra of the complement of pigments characterizing a suspension.

Figure 2.5 plots the average absorption cross section (calculated for 675 nm) against D. A decrease in $(n\bar{A}\bar{a})_{675}$ with cell size is readily apparent. This was first noted by Kirk (1983, fig. 9.2), to whom credit is given for this analysis format. A few additional points are worth mentioning. Although the values used for V_T, C, and γ are not unrealistic for many

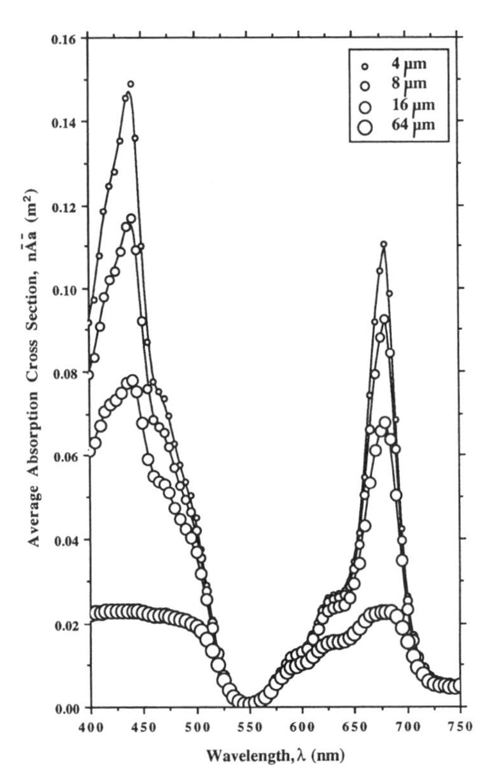

Figure 2.4. The absorption cross section $n\bar{A}\bar{a}$ (computed from eq. 2.2) for four hypothetical algal suspensions composed of cells differing in diameter D (4, 8, 16, and 64 μm) against wavelength λ in the 400 to 750 nm range.

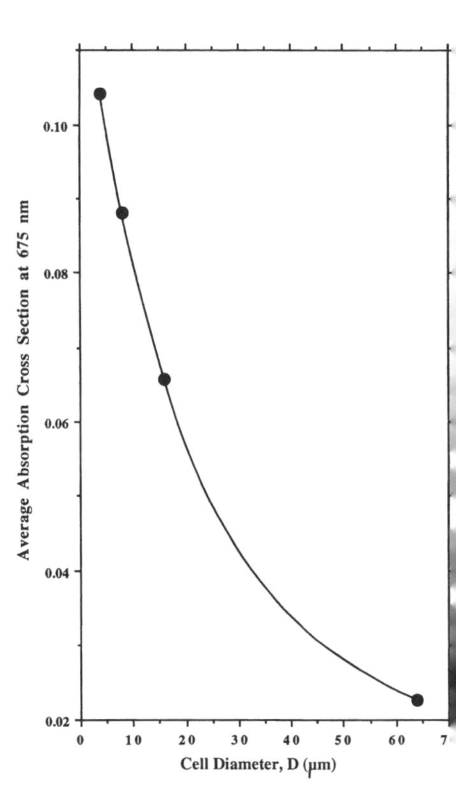

Figure 2.5. The average absorption cross section (calculated for 675 nm) against cell diameter D. Data from figure 2.4.

natural aquatic systems, they are not appropriate for all algal suspensions. Accordingly, although the qualitative phenomenology of the package effect shown in figures 2.4 and 2.5 is obtained under natural conditions, it cannot be used to predict qualitative effects. Also, since cell size D differs among the four suspensions but V_T does not, the number of cells N in suspension also differs ($N = 2.99 \times 10^{10}$, 3.73×10^{9}, 4.66×10^{8}, 7.28×10^{8} cells for $D = 4$, 8, 16, and 64 μm, respectively). Therefore the relations shown in these figures are influenced by cell number (population density) as well as cell size (scaling).

The package effect implies that when the number of phytoplankton cells per unit volume of suspension increases, the average absorption cross section will decrease regardless of the average cell size or intracellular pigment concentration. Thus the efficiency of light interception is increased by keeping the number of individual unicellular organisms within a population low as well as by keeping cell size small.

Cell shape also plays an important role in the capacity of a suspension of phytoplankton to harvest light. Thomas Kirk (1976) evaluated the consequences of cell shape on the attenuation of light in algal suspensions by deriving equations for the average absorption cross section of cellular suspensions with prolate, oblate, and cylindrical geometries. Figure 2.6 shows that, for these geometries, the average projected area $\overline{A}$ and the average proportion of incident light absorbed by cells $\overline{a}$ are comparatively intricate functions of the angle θ subtended between the light beam penetrating each cell and any specified plane within the cell. Referring to the notation in figure 2.6 (see Kirk 1976, figs. 1–2), for a suspension of randomly oriented cells with a terete cylindrical geometry, the value of $\overline{A\overline{a}}$ is given by the formula

$$\text{(2.3)} \quad \overline{A\overline{a}} = \int_0^{\pi/2} DL\cos^2\theta\left[1 - \frac{2}{D}\int_0^{D/2} e^{-\gamma CD\left(1-\frac{4Z^2}{D^2}\right)^{1/2}} dZ\right] d\theta. \quad \text{(cylinder)}$$

For suspensions of randomly oriented prolate or oblate spheroids, the following formula applies:

$$\text{(2.4)} \quad \overline{A\overline{a}} = \int_0^{\pi/2} \pi lb\cos\theta\left[1 - \frac{4}{\pi lb}\int_0^{L}\int_0^{S/2} e^{-\gamma CQ\left(1-\frac{4Z^2}{S^2}\right)^{1/2}} dX dZ\right] d\theta, \quad \text{(spheroids)}$$

where b is the semiminor axis of the spheroid.

With the aid of eqs. (2.3) and (2.4), Kirk showed that cells with extended shapes (long cylinders, long prolate spheroids, or extremely flattened oblate spheroids) capture nearly the same amount of light as a greater number of smaller spherically packaged cells. This conclusion presumes that suspensions have equivalent total algal volumes V_T. This is easily illustrated by a few simple graphs. Figure 2.7, for example, plots average absorption cross sections versus wavelength for two suspensions of spherical cells differing in size (D = 4 and 64 μm) as well as suspensions of prolate spheroidal cells (semimajor × semiminor axes = 250 × 50 μm), and terete cylindrical cells (length × diameter = 3,500 × 10 μm). These plots were obtained for $V_T = 1 \times 10^{-6}$ m³. Therefore the numbers of cells differ in the four hypothetical algal suspensions. Specifically, for spherical cells with D = 4 and 64 μm, $N \approx 2.99 \times 10^{10}$ and 7.28×10^6 cells. For prolate spheroids, $N \approx 3.82 \times 10^5$ cells and, for terete cylinders, $N \approx 3.64 \times 10^6$ cells. Inspection of figure 2.7 shows that

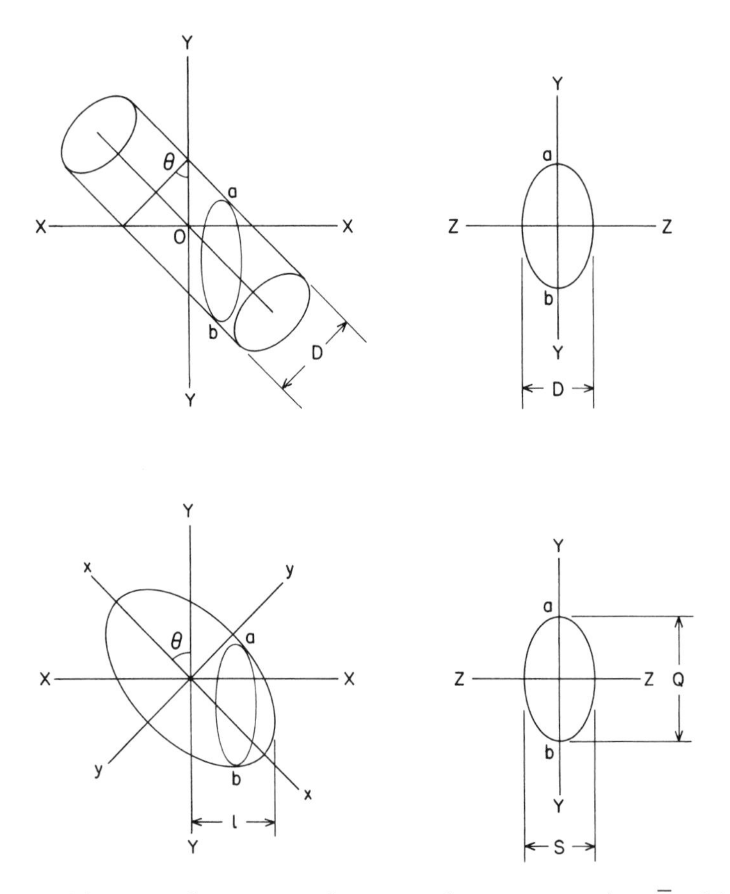

Figure 2.6. Schematics and parameters used to compute the average projected area $\bar{A}$ and the average proportion of incident light absorbed by cells $\bar{a}$ as a function of the angle θ subtended between the light beam penetrating a cylindrical or spheroidal cell (top and bottom geometries, respectively) and any specified plane (shown in diagrams to the right of each figure) within the cell. Redrawn after Kirk 1976, figs. 1–2.

differences in cell geometry can influence the efficiency of harvesting light in the water column. For example, approximately 3.64×10^6 cylindrical cells are roughly 50% as efficient as $\approx 2.99 \times 10^{10}$ small spherical cells at gathering light. This is somewhat surprising, since the cell number in these two suspensions differs by four orders of magnitude. The difference in the efficiency of light interception for these populations of cells is a consequence of the way randomly oriented cells differing in geometry project their surface areas toward the incident light.

Thus far we have assumed that intracellular pigment concentration is independent of cell size. In chapter 1 this assumption was shown to be incorrect in terms of broad interspecific comparisons among unicellular algae (see fig. 1.13). Potentially, size-dependent variation in the concentration of photosynthetic pigments per cell has repercussions on the light harvesting of individual cells. In theory, a reduction in pigment concentration per cell could yield a concomitant reduction in intra- and intercellular self-shading yet decrease the efficiency of intracellular absorption. This possibility has been explored by Susana Agustí (1991), who investigated the scaling of light absorption (and scattering) by individual phyto-

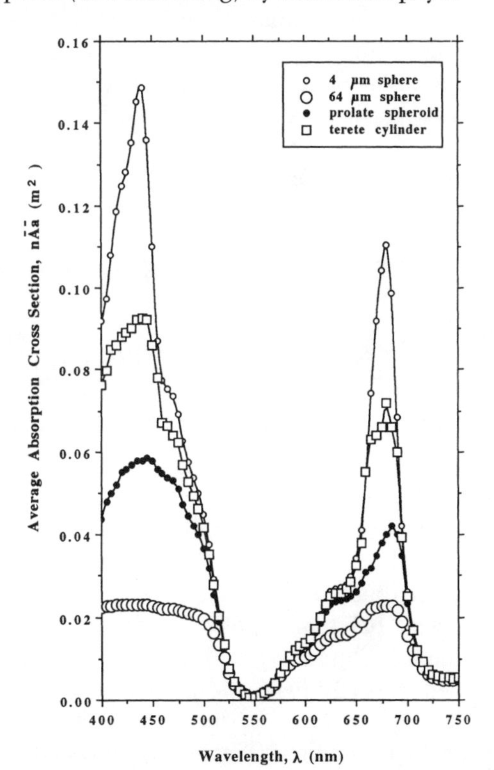

Figure 2.7. Average absorption cross sections $n\bar{A}\bar{a}$ plotted as functions of wavelength λ (over the range of 400 to 750 nm) for two suspensions of spherical cells differing in size (D = 4 and 64 μm) and suspensions of prolate (semimajor x semiminor axes = 250 × 50 μm) and terete cylindrical cells (length × diameter = 3,500 × 10 μm).

plankton cells from twenty-eight species. Her data show that light absorption scales to the cross-sectional area of an individual cell and that individual cells absorb less light, relative to their volume, as they increase in size. This scaling relation between the light absorption of individual cells and intracellular chlorophyll *a* concentration suggests that higher pigment concentrations in cells grown under low light conditions produce a proportional increase in the amount of light absorbed per cell. Thus, the results reported by Agustí (1991) allow for a simple description of algal photoadaptation to low irradiance that typically involves an increase in pigment content per cell.

Unlike the effect of cell size and geometry, the consequence of cell shape on light harvesting has not been modeled because of numerous practical and theoretical difficulties. Recall that geometry and shape are not the same thing. For example, terete cylinders or prolate spheroids differing in their aspect ratios may or may not have the same shape as defined by dimensionless ratios constructed from their surface areas and volumes. The effect of changes in the aspect ratio of spheroidal and cylinder geometries on light interception was treated in chapter 1 (see fig. 1.18). Similar analyses are required to determine the role of shape on the harvesting of light by phytoplankton. Since for very weakly absorbing randomly oriented packages the product of the average projected area and average package absorptance $\overline{A}\overline{a}$, regardless of shape, equals γCV, where V is the volume of the package (Kirk 1976, 345), shape is irrelevant for packages that essentially are optically transparent. This is not true, however, for pigmented cells that are optically translucent to opaque. For example, when eqs. (2.3) and (2.4) are cast in terms of dimensionless aspect ratios $\Re$ (see table 1.2), we find that, for a terete cylinder,

$$\overline{A}\overline{a} = \int_0^{\pi/2} 2\Re L^2 \cos^2\theta \left[1 - \frac{2}{2\Re L}\int_0^{\Re L/2} e^{-2_{\pi} C\Re L\left(1-\frac{2Z^2}{(\Re L)^2}\right)^{1/2}} dZ\right] d\theta.$$

Although rather cumbersome, this formula shows that the average absorption cross section depends on cell shape ($\Re$ = the ratio of radius to length) as well as size (length, L).

The package effect also has not been modeled for metaphytes. Although the pigments in multicellular plants are aggregated in chloroplasts, these packages are nonrandomly distributed in tissues and can reorient or change position within cells in response to changes in light intensity, quality, or direction. For example, the barlike chloroplast of *Mougeotia*, a filamentous green alga, rotates about its length and projects its edge toward low-intensity white light and its broad surface toward incident red light (Wagner, Haupt, and Laux 1972). Light-oriented chloroplast movements also occur in vascular plants (e.g., *Adiantum* protenema; see Yatsuhashi and Wada 1990). Unfortunately, no simple mathematical treatment exists for these and other responses that bear on the light interception of chloroplasts suspended in plant tissues.

Nonetheless, the size and shape of many benthic metaphytes are known to vary with the prevailing light environment, presumably in an adaptive response to maximize light interception. For example, both the total surface area of branches and the number of branch segments on the siphonous green alga *Halimeda tuna* increase with depth over the range of 7 to 16 m (Mariana Colombo et al. 1976). Likewise, the submerged leaves of the angiosperm *Potamogeton* increase in area and decrease in thickness as a function of water depth such that the specific leaf area (cm^2 mg^{-1} dry mass) increases linearly with water depth (Spence, Campbell, and Chrystal 1973). These variations in morphology reflect the capacity of nonvascular and vascular plants to modify their phenotypes in response to the light field. In other cases genetic differences among intermixing populations account for morphological variations along a light gradient. Thus morphologically adaptive responses to ambient light conditions are achieved by genetic diversity within a population as well as by the capacity of an individual genotype to manifest different shapes. Concerning the former, Lipkin (1979) showed that morphological differences among closely spaced individuals of the seagrass *Halophila stipulacea* account for the observation that leaf area increased about 2.5 times with depth from the surface down to about 30 m. Plants also have the capacity to alter their chemistry in response to light conditions. For example, morphologically heterophyllous species of *Potamogeton* differ in flavonoid chemistry. Interspecific phytochemical comparisons show that the submerged leaves of *Potamogeton* species typically lack glycoflavones (chemicals that protect tissues from ultraviolet [UV] radiation), which are present in the floating leaves (exposed to higher UV levels) of other species (Les and Sheridan 1990). A curious footnote to these findings is that they lend credence to the speculation that glycoflavone evolution was an important step in the colonization of the terrestrial landscape (Swain 1975).

I now turn briefly to differences in the pigment composition of plants, which is important to understanding many features associated with aquatic plants. Plant groups differ in the chemical composition of their photosynthetic pigments, and therefore phyletic groups differ in their ability to absorb different wavelengths of photosynthetically active light (PAR). Table 2.1 gives the distribution of chlorophylls and major chloroplast carotenoids in various algal groups. These distributions affect the capacity of plants to harvest light differing in wavelength. Figure 2.8 plots

Table 2.1. Distribution of Chlorophylls and Major Chloroplast Carotenoids among Different Plant Groups

	Chlorophylls				Carotenoids				
Plant Group	*a*	*b*	c_1	c_2	α-Carotene	β-Carotene	Lutein	Zeoxanthin	Neoxanthin
Embryophytes	+	+	–	–	+	+	+	+	+
Chlorophytes	+	+	–	–	+	+	+	+	+
Euglenophytes	+	+	–	–	–	+	–	–	+
Chrysophytes	+	–	+	+	+[a]	+	–	–	–
Pyrrophytes	+	–	– (+)	+	–	+	–	–	–
Phaeophytes	+	–	+	+	–	+	–	–	–
Cryptophytes	+	–	–	+	+	+	–	–	–
Rhodophytes	+	–	–	–	+	+	+	+	–
Cyanophytes	+	–	–	–	–	+	–	+	–

Source: Data from Kirk and Tilney-Bassett 1978.

[a]Found in Chrysophyceae but not in other families.

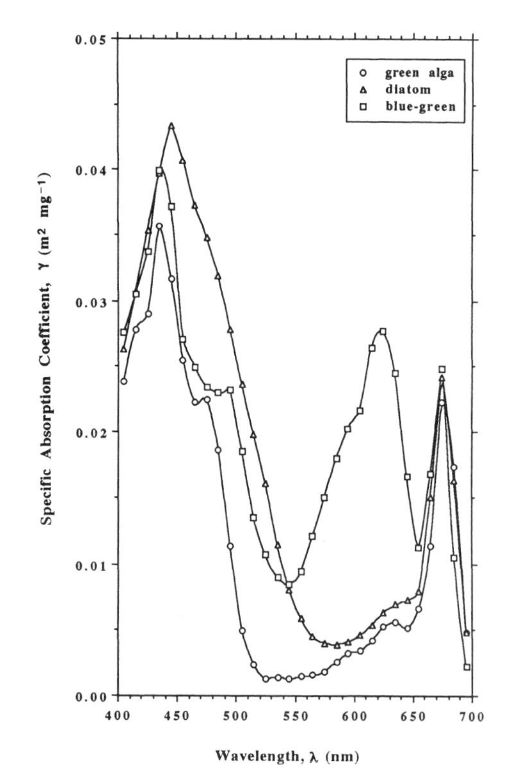

Figure 2.8. The natural logarithm specific absorption coefficient γ of a *typical* green alga, a diatom, and a blue-green bacterium plotted against wavelength λ over the range of 400 to 700 nm. Data from a variety of sources.

the natural logarithm specific absorption coefficients of a *typical* green alga, a diatom, and a blue-green bacterium against wavelength. The latter significantly absorbs in the 550 to 650 nm region, while more subtle but nonetheless important differences exist between the spectra of the two eukaryotes. It is reasonable to suppose, therefore, that some photosynthetic organisms are better equipped to exploit particular zones within an attenuated light field and that the pigments characteristic of each aquatic plant group can play an important role in determining ecological niche. The presence or absence of glycoflavones in the leaves of species of *Potamogeton* is but one example, while at a broader level of comparison qualitative and quantitative differences in photosynthetic pigments among species are correlated with discernibly predictable floristic changes along water depth gradients.

Two theories have been proposed to account for this effect. The chromatic adaptation theory of Engelmann (1883) asserts that the spectral qualities of light determine algal zonation. The attenuation theory of Berthold (1882) and Oltmanns (1892) argues that variation of light intensity with depth dictates the distribution of different algal groups. Since the spectral quality and the intensity of light vary simultaneously as a function of depth in the water column, these theories must be considered complementary, although they have needlessly been placed in opposition. Even within a particular species, the ratios of characteristic light-harvesting pigments can be altered as a function of water depth. Thus chromatic adaptation operates even within the context of a phylogenetically well defined pigment composition. Since the proportion of different pigments can change within an individual plant in accordance with the prevailing environmental conditions attending its growth and development, and since light intensity and spectral quality change in the water column, physiological homogeneity among all individuals within a population should never be assumed.

2.4 The Transport Equations and Dimensionless Groupings of Variables

In addition to attenuating light, water can exert substantial forces on objects obstructing its flow. The physical properties of water also influence the supply of dissolved metabolites to living cells as well as the extent to which submerged organisms can excrete materials. Thus the physical properties of water have profound effects on the growth of aquatic plants. To understand these effects we need to examine three physical laws, summarized mathematically by three transport equations. With the aid of these equations, we also will be able to quantify how modifications in plant size, shape, and geometry can mitigate or utilize the physical effects of water on growth.

The three transport equations that will occupy our attention are mathematical expressions of fundamental physical laws: Newton's law of viscosity, Fick's law of diffusion, and Fourier's law of heat transfer. Newton's law of viscosity states that the shearing force per unit area (called the shear stress, τ)exerted by a moving fluid (water or air) equals the product of the fluid's dynamic viscosity μ and the change in the speed of fluid flow per change in distance (the velocity gradient) du/dx:

(2.5) $$\tau = \mu \frac{du}{dx}.$$ (Newton's law of viscosity)

A simple comparison between water and air illustrates how the physical properties of a fluid influence the magnitude of shear stresses. Referring to table 2.2, we see that the dynamic viscosity of fresh water at 20°C is 1×10^{-3} kg m^{-1} s^{-1}. In contrast, the dymanic viscosity of air at the same temperature equals 18×10^{-6} kg m^{-1} s$^{-1.}$ At 20°C, therefore, the viscosity of air is only 1.8% that of fresh water. For this reason, aquatic plants typically sustain much higher shear stresses than their terrestrial counterparts at equivalent fluid flow.

Fick's law of diffusion (see chap. 1) states that the amount of a molecular species diffusing through a unit surface area per unit time (the flux density J) is equal to the product of the molecule's ability to diffuse through the surface per unit time (the molecular diffusivity D) and the change in the density of the molecular species per change in distance (the concentration gradient $d\rho/dx$):

(2.6) $$J = -D\frac{d\rho}{dx}.$$ (Fick's law of diffusion)

Table 2.2. Physical Properties of Pure Water and Air at 20°C

Physical Property	Water	Air
Density, ρ (kg m^{-3})	998.4	1.204
Dynamic viscosity, μ (kg m^{-1} s^{-1})	1.00×10^{-3}	18.1×10^{-6}
Kinematic viscosity, υ (m^2 s^{-1})	1.01×10^{-6}	1.51×10^{-5}
Thermal diffusivity, D_H (m^2 s^{-1})	144.0×10^{-9}	2.15×10^{-5}
Specific heat kJ, c_p (kg^{-1} °C^{-1})	4.19	1.01
Thermal conductivity, k (kJ m^{-1} s^{-1} °C^{-1})	602.0×10^{-6}	26.1×10^{-6}
Diffusion coefficient, D (m^2 s^{-1})		
H_2O	2.4×10^{-9}	2.42×10^{-5}
CO_2	1.8×10^{-9}	1.47×10^{-5}
O_2	2.0×10^{-9}	2.02×10^{-5}
Phosphate	2.0×10^{-9}	1.51×10^{-5}

The minus sign placed before *D* is the conventional way of indicating that net molecular diffusion is in the direction of the lower concentration (i.e., the flux is in the negative direction when the gradient is positive). Once again, a comparison between water and air illustrates the importance of the physical properties of the fluid an organism lives in. For plants, a comparison based on the diffusivity of carbon dioxide is particularly useful. From table 2.2, we see that the molecular diffusivity of CO_2 in air is 1.47×10^{-5} m^2 s^{-1}, while that of CO^2 in water is 1.8×10^{-9} m^2 s^{-1}. The difference in $D_{[CO_2]}$ between air and water therefore is four orders of magnitude. Since CO_2 can diffuse easily through air compared with water, the land plant has an advantage over its aquatic counterpart in the rapid acquisition of inorganic carbon for photosynthesis.

The last transport equation we shall consider, Fourier's law of heat transfer, states that the power per unit area (the heat flux density *H*) equals the product of the rate of heat transfer across a unit area of unit thickness with a unit temperature difference between its surfaces (the thermal conductivity *k*) and the change in temperature per change in distance (the temperature gradient, *dT/dx*):

$$H = -k\frac{dT}{dx}. \qquad (2.7) \quad \text{(Fourier's law of heat transfer)}$$

As in eq. (2.6), the minus sign in Fourier's law indicates that heat transfer is in the direction of the lower temperature. In this regard it is helpful to know that *k* is the reciprocal of the insulation value of the medium (e.g., air or water). Referring to table 2.2, the value of *k* for air is 26.1×10^{-6} kJ m^{-1} s^{-1} °C^{-1}, while that of water is 602×10^{-6} kJ m^{-1} s^{-1} °C^{-1}. These values indicate that the temperature of an aquatic plant will be very close to the ambient water temperature. By contrast, the difference between the temperature of a terrestrial plant and that of the air around it can be significant.

The three transport equations show that the properties of the fluid around a plant influence metabolic processes dependent on passive diffusion and temperature as well as the magnitudes of mechanical forces exerted on a plant obstructing the flow of the fluid. That the size and shape of a plant are equally important is shown by the fact that each equation refers to some measure of "size" (e.g., distance or area) whose magnitude typically changes as a plant or plant organ grows. From this we can see that the mathematics describing the operation of physical laws and the mathematics describing plant growth are interrelated.

The relation between physical laws and plant growth becomes more obvious when the transport laws are expressed as diffusion equations. Table 2.3 shows that, by converting the proportionality factors in the transport equations for Newton's and Fourier's laws into expressions of *diffusivities* and converting the driving forces into concentration gradients, these two laws take the same mathematical form as Fick's law. For Newton's law, the diffusivity of momentum through water or air is measured by the kinematic viscosity υ of the fluid, while the product of a fluid's density ρ and its speed u is called the concentration of momentum ρu. For Fourier's law, the concentration of heat is the product of temperature T and the volumetric heat capacity of the fluid ρc_p (the product of the fluid's density and its specific heat c_p). Thus,

$$\tau = \mu \frac{du}{dx} = \upsilon \frac{d(\rho u)}{dx} \qquad (2.8) \quad \text{(Newton's law restated)}$$

and

$$H = -k \frac{dT}{dx} = -D_H \frac{d(\rho c_p T)}{dx}, \qquad (2.9) \quad \text{(Fourier's law restated)}$$

where D_H is called thermal diffusivity ($D_H = k/\rho c_p$). From eqs. (2.8) and (2.9), the concentration of momentum $d(\rho u)/dx$ and heat $d(\rho c_p T)/dx$ are seen to take the same mathematical form as the concentration of a molecular species $d\rho/dx$ dealt with Fick's law of diffusion. It is also extremely

Table 2.3. Terminology and Units for Proportionality Factors and Driving Forces

	Rate of Transport		Diffusivity Equation	
Law	Proportionality Factor	Driving Force	Proportionality Factor	Driving Force
Newton's law of viscosity	Dynamic viscosity μ (kg m^{-1} s^{-1})	Velocity gradient $\frac{du}{dx}$ (s^{-1})	Kinematic viscosity[a] υ (m^2 s^{-1})	Momentum $\frac{d(\rho u)}{dx}$ (kg m^{-3} s^{-1})
Fick's law of diffusion	Molecular diffusivity D (m^2 s^{-1})	Concentration gradient $\frac{d\rho}{dx}$ (kg m^{-4})	Molecular diffusivity D (m^2 s^{-1})	Concentration gradient $\frac{d\rho}{dx}$ (kg m^{-4})
Fourier's law of heat transfer	Thermal conductivity k (kJ m^{-1} s^{-1} °C^{-1})	Temperature gradient $\frac{dT}{dx}$ (°C m^{-1})	Thermal diffusivity D_H (m^2 s^{-1})	Heat concentration $\frac{d(\rho c_p T)}{dx}$ (kJ m^{-4})

[a]Also known as momentum diffusivity.

convenient that both the proportionality factor ("diffusivity") of momentum and heat assume the same units (m^2 s^{-1}) as molecular diffusivity D when the transport laws are expressed as diffusion equations (see table 2.3).

Unfortunately, it is extremely difficult (or simply impractical) to actually measure concentrations of a molecular species of differences in temperature over very small distances. Likewise, it is difficult to measure concentrations of momentum over very small distances. For these reasons, the integrated forms of the transport equations are required for most practical applications. Since the momentum of a fluid, the concentration of a molecule, and the concentration of heat (temperature) can be measured at the surface of an organism C_s and at some distance within the ambient fluid C_a, the difference between the two measurements ΔC divided by the resistance to the transfer of mass r_J or heat r_H (resistance has the units of s m^{-1}) yields an expression for flux density in a mathematical form analogous to Ohm's law. For example, the macroscopic form of Fick's law is

$$J = \frac{\Delta \rho}{r_j}, \tag{2.10}$$

where the concentration gradient of a molecular species is expressed as the difference in density $\Delta\rho$. Similarly, Fourier's law becomes

$$H = \rho c_p \frac{\Delta T}{r_H}. \tag{2.11}$$

And finally, the macroscopic form of Newton's law of viscosity is given by the equation

$$\tau = \frac{\rho U^2}{2} C_D, \tag{2.12}$$

where C_D is called the drag coefficient, whose meaning and numerical measure will occupy our attention later (see fig. 2.13).

The important point to note here is that the transport equations reveal that the rate of mass or energy transfer between a plant and its fluid medium depends on the rate at which the ambient fluid moves relative to the plant. The term "relative" is required because a fluid can move with respect to a sedentary plant, or a plant can move with respect to a comparatively unmoving fluid—do not forget that many plants are capable of self-propulsion (see secs. 2.6 and 2.7). The significance of fluid motion is easily demonstrated. For example, eq. (2.12) states that the shear stresses developed over the surfaces of a plant increase as the square of the ambient flow speed. Therefore a small increase in fluid speed can be magnified into a large shear force. The influence of the rate of fluid flow on the transport of mass or heat is just as significant, albeit perhaps less obvious until it is recognized that the resistance to transport is a function of the thickness of the blanket of comparatively unmoving fluid that envelops any object submerged in a fluid, which is called the boundary layer. In zero or very slow flow, the thickness of the boundary layer extends far beyond the surfaces of a submerged object. The resistance to the passive diffusion of metabolites or to the transfer of heat therefore increases with decreasing flow speed. In some circumstances a stagnant flow condition can jeopardize growth or even kill an organism. As the ambient rate of flow increases, the boundary layer is progressively thinned, and therefore the resistance to the passive diffusion and heat transfer progressively decreases. At the high end of ambient flow speeds, the exchange of mass or energy is mediated largely by forced convection. The boundary layer becomes extremely thin, and flow near surfaces is turbulent. Once nonlaminar flow occurs, however, the principal assumptions made for molecular and thermal diffusion no longer apply. Although this severely restricts the use of eqs. (2.10) to (2.12), these relations remain relevant provided the resistance to mass and energy transfer can be determined.

The dependency of the rates of physiological processes like respiration and photosynthesis on ambient flow rates has been known for a very long time. William James (1928) compared the metabolism of *Fontinalis antipyretica* growing in still water and in slowly moving water currents and found that the rate of respiration of this aquatic moss increases as the ambient flow speed increases from zero to 6 cm s^{-1}. At higher flow speeds metabolic rates were largely unaffected, as expected. Similar effects of slow water currents on rates of respiration and photosynthesis have been reported for freshwater algae (Schumacher and Whitford 1965) and for aquatic angiosperms (e.g., *Ranunculus pseudofluitans* and *Potamogeton pectinatus;* see Westlake 1967). In passing, it is also worth noting that differences in ambient flow speed are sometimes correlated with morphological variations among individual plants selected from the same population, leading to taxonomic problems when variations in shape are used to distinguish among species within a genus. For example, Parodi and Cáceres (1991) report that the number of branches of the green alga *Cladophora* increases with higher water speeds, thereby casting doubt on the use of ramification frequency as a taxonomic criterion. An interesting example of how flow speeds can influence reproductive success and the survival of juvenile plants, possibly as a consequence of nutrition, is provided by Reed, Neushul, and Ebeling (1991).

In order to deal with the influence of fluid flow on the transport of mass or energy, dimensionless numbers (composed of groupings of variables) are frequently used. Table 2.4 lists some of the more commonly encountered dimensionless numbers, among which the Reynolds number, denoted *Re,* is perhaps the best known. This dimensionless grouping of variables is used to characterize fluid flow regimes. The numerical value of *Re* reflects the relative importance of inertial and viscous forces that develop in a particular fluid flow system. Very low Reynolds numbers indicate that viscous forces dominate flow. In these circumstances, flow around the object is symmetric and laminar, and the surrounding fluid is diffusive in character. When Reynolds numbers are high, however, inertial forces dominate and the flow field around an object becomes increasingly asymmetric, nonlaminar, and more difficult to deal with in terms of diffusion theory. Beyond a certain Reynolds number, flow is turbulent.

The transition between laminar and nonlaminar flow and between nonlaminar and turbulent flow occurs at different *Re* for different geome-

Table 2.4. Seven Dimensionless Ratios for Correlating Mass or Heat Transfer

Reynolds number	$Re = \frac{Ud}{\upsilon}$	Ratio of inertial force to viscous force
Grashof number	$Gr = \frac{gad^3\Delta T}{\upsilon^2}$	Ratio of buoyant force times inertial force to the square of kinematic viscosity
Nusselt number	$Nu = \frac{Ht}{\rho c_p D_H \Delta T}$	Ratio of heat flux to that of the same ΔT for a still fluid layer of thickness t
Schmidt number	$Sc = \frac{\upsilon}{D}$	Ratio of kinematic viscosity to molecular diffusivity
Prandtl number	$Pr = \frac{\upsilon}{D_H}$	Ratio of kinematic viscosity to thermal diffusivity
Sherwood number	$Sh = \frac{Jd}{D\Delta\rho}$	Ratio of mass flux density to that of the same concentration applied to a still layer of fluid of thickness t
Péclet number	$Pe = PrRe = \frac{Ud}{D_H}$	Product of Prandtl and Reynolds numbers

Symbols: U = fluid speed (m s^{-1}); g = gravitational acceleration (9.8 m s^{-2}); a = coefficient of thermal expansion ($^{0}C^{-1}$); d = characteristic dimension (m); υ = kinematic viscosity (m^2 s^{-1}); ΔT = difference in temperature measured at surface T_s and some distance T_f in fluid (^{0}C); H = heat flux density (kJ m^{-2} s^{-1}); t = still fluid layer thickness (m); ρ = density (kg m^{-3}); c_p = specific heat capacity (kJ kg^{-1} $^{0}C^{-1}$); D_H = thermal diffusivity (m^2 s^{-1}); J = mass flux density (kg m^{-2} s^{-1}); D = molecular diffusion coefficient (m^2 s^{-1}); and $\Delta\rho$ = difference in density (concentration) measured at surface ρ_s and some distance ρ_f in fluid (kg m^{-3}).

tries. Nonetheless, the Reynolds number at which one flow regime changes into another has been empirically determined for a variety of very simple geometries like the sphere and the terete cylinder. Although these critical *Re* are often used to treat biological structures, it is important to bear in mind that organic forms rarely have simple geometries. Indeed, most biological structures are geometrically complex and have highly irregular surfaces. Thus the transition from one flow regime to another may not be abrupt, as is often the case for objects with very simple geometries and smooth surfaces. Also, since many biological objects have a composite geometry (e.g., the shoot has platelike leaves and cylindrical internodes), one part of the same organism may experience laminar flow while another part encounters nonlaminar or even turbulent flow. More will be said of this later.

Since the Reynolds number is so important, it is useful to explore its derivation. When an object moves through a fluid, it exerts a force on the fluid in the direction of transit. In turn, the fluid exerts an equal and opposite retarding force. This force is composed of two components, the inertial force component F_I and the viscous force component F_V. For objects that move very fast or that have large mass, F_I dominates. For objects that have low mass or move very slowly, F_V dominates. Obviously, fast and slow, like large and small, are useless expressions until quantified. Considering the inertial force, Isaac Newton showed the $F_I = ma$, where m is the mass of the object and a is its acceleration. The acceleration of a fluid equals the change in its speed U per change in time t. Thus, $F_I = ma = m(du/dt) = d(mu)/dt$, where mu is momentum. The mass flux m/t of a moving volume of fluid equals ρSU, while the momentum flux of the fluid volume mu/t equals ρSU^2, where S is the fluid's surface area projected in the direction of motion. The momentum flux for a fluid can be expressed in terms of the inertial force described by Newton. That is, $mu/t = F_I = \rho SU^2$. By considering the relation between a moving surface and a nonmoving surface, Newton also showed that the viscous force F_V generated between the two equals $\mu SU/d$, where d is the distance between the moving and nonmoving surfaces. Since the viscous force divided by the moving surface S is the shearing stress τ(treated by Newton's law of viscosity; see eq. 2.8), we see that $F_V = \tau S = \mu Su/d = \rho \upsilon Su/d$. In summary,

$$F_I = \rho SU^2 \qquad \text{(inertial force)} \tag{2.13a}$$

$$F_V = \rho \upsilon \frac{SU}{d}. \qquad \text{(viscous force)} \tag{2.13b}$$

The Reynolds number is simply the ratio of the inertial force to the viscous force:

$$Re = \frac{F_I}{F_V} = \frac{\rho S U^2}{\rho \upsilon \dfrac{SU}{d}} = \frac{Ud}{\upsilon}, \tag{2.14}$$

(Reynolds number)

where d now designates the characteristic dimension of the object measured in the axis parallel to the direction of relative fluid flow. The Reynolds number provides the sought-after quantification of slow and fast as well as small and large. It is a truly remarkable dimensionless number.

From the preceding, we see that the Re depends on three variables: the characteristic dimension d, the ambient flow speed U, and the kinematic viscosity υ. High Reynolds numbers therefore occur as a result of either very rapid flow or a very large characteristic dimension. Conversely, a low Re indicates that the flow rate is very slow or that the characteristic dimension is very small. Recall that the characteristic dimension typically is taken as the dimension parallel to the direction of fluid flow (e.g., the diameter of a cylinder whose length is perpendicular to the direction of flow, the length of a flat plate).[3] Since the characteristic dimension is a measure of size, Re depends in part on scaling phenomena. In some circumstances the flow speed also may be subject to scaling. For example, plants that are suspended in the water column either sink or float or swim. The rate at which passively suspended plants move relative to the surrounding fluid depends on a variety of factors, among which size, shape, geometry, and cell density are important. Mobile plants also can alter their speed according to the rate at which cilia beat. Perhaps the most stable parameter in the Reynolds number equation is the kinematic viscosity of a fluid υ, which is constant for a particular fluid at a specified temperature. The magnitude of υ does, however, differ *among* fluids and is influential in determining flow characteristics (imagine swimming through asphalt rather than water).

In the next section, we shall use the Reynolds number to correlate data on heat or mass transfer in systems differing in ambient fluid flow rates U or in the size d of the object obstructing fluid flow. In this regard, note that the variables contained in Re also occur in the other dimensionless numbers listed in table 2.4. Thus all the dimensionless numbers can be mathematically related to Re as well as to each other by means of Re.

3. The purist will note that d is the distance traversed by a moving fluid particle as it moves from the leading edge to the rear of an object obstructing flow. Thus, d is one-half the circumference C of a sphere or terete cylinder. Since $C/2 = \pi D/2 \approx 1.57\ D$, where D is diameter, we see that $d \neq D$. Further, if a terete cylinder is tilted from the vertical, then its projected cross section in the plane normal to the direction of flow becomes elliptical, and therefore d increases to the limiting case where it equals the length of the cylinder. Nonetheless, Re is useful only to distinguish among "orders of magnitude" differences in U or d. Excessive precision, therefore, will yield only illusory differences in Re.

This is extraordinarily useful when dealing with mass transfer, which is extremely difficult to measure empirically in comparison with heat transfer. Fortunately, it is possible to infer mass transfer from measurements of heat transfer for systems having the same Reynolds numbers.

2.5. Fluid Flow and Mass and Heat Transport: Phytoplankton

The usefulness of the transport equations and dimensionless numbers reviewed in section 2.4 is illustrated in this section by considering the transfer of nutrients and heat between aquatic plants and their external fluid medium. In terms of mass transfer, our objective is to find the Sherwood number for a plant. This number is the ratio of the mass flux density of a plant moving relative to its fluid surroundings to what would occur if the same mass concentration gradient was applied to a still layer of fluid whose thickness equals the characteristic dimension of the moving plant (see table 2.4). For heat transfer, the focus is on the Nusselt number, which is the ratio of the heat flux density of a moving plant to that of a still layer of fluid. In either case, we need to determine the Reynolds number for the system in order to correlate mass with heat transfer as well as to relate the behavior of one fluid flow system to that of another.

From the preceding discussion, we know that determining *Re* is an empirical affair. We need to know the size of the organism and the ambient flow rate. In an aquatic habitat, *Re* depends on whether a plant is passively suspended in the water column or actively mobile, or attached to a substrate. For passively suspended phytoplanktonic species, the minimum *Re* (and therefore the minimum Sherwood or Nusselt number) is determined by the settling velocity of the individual plant in the water column. Small plants whose cell bulk density is near that of water will settle either very slowly or not at all. Consequently *Re* will be very low (viscous forces will dominate). Accordingly, the rates of mass or energy transfer for these plants will be comparatively slow. For flagellated plants, the minimum *Re* is determined by the minimum speed at which a plant can move. The Reynolds numbers of these plants will tend to be higher than those of passively suspended plants of comparable size. Therefore the rates of transfer of mass or energy will be comparatively higher than for passively suspended plants. Nonetheless, as noted earlier, orders of magnitude changes in *Re* are required to produce significant changes in the flow regime. Consequently, passively suspended and self-propelled unicellular plants operate metabolically in much the same flow environment. For plants attached to a substrate, *Re* is largely under the control of the physical environment. Turbulent flow conditions can dominate the fluid environments of these plants. Clearly, the physical environment and the nature of the plant have a combined influence on the transfer of mass or energy. To illustrate these points, let us consider biological systems characterized by progressively higher Reynolds numbers, starting with a stagnant fluid medium and ending with plants attached to a substrate over which water moves rapidly.

In a stagnant fluid medium the Reynolds number equals zero, and the problem of the mass transport is analogous to the electrostatic problem of a charged conductor in a charge-free homogeneous dielectric medium. In electrostatics, the quotient of the capacitance and the permittivity of the medium equals the external conductance $\mathfrak{I}$, which is units of length. With the exception of spheres, the local external conductance is not uniform over the surface area. Edges and corners, which have high curvatures, have high local conductances and, by analogy, high diffusive flux densities. Table 2.5 gives the equations for $\mathfrak{I}$ for some common geome-

Table 2.5. External Conductances for Geometries in Zero Flow

Geometry	Conductance $\mathfrak{I}$
Sphere	$4\pi R$
Touching spheres	$8\pi(\ln 2)R$
Intersecting spheres	$4\pi\left[R_1 + R_2 - \frac{R_1 R_2}{(R_1^2 + R_2^2)^{0.50}}\right]$
Oblate spheroid ($b/a < 1$)	$\frac{4\pi a\left[1 - \left(\frac{b}{a}\right)^2\right]^{0.50}}{\cos^{-1}\left(\frac{b}{a}\right)}$
Prolate spheroid ($b/a > 1$)	$\frac{4\pi a\left[\left(\frac{b}{a}\right)^2 - 1\right]^{0.50}}{\ln\left\{\left(\frac{b}{a}\right) + \left[\left(\frac{b}{a}\right)^2 - 1\right]^{0.50}\right\}}$
Cylinder ($0 \le L/D \le 8$)	$\left[8 + 6.95\left(\frac{L}{D}\right)^{0.76}\right]R$
Cube	$2.62\ \pi l$
Thin plate	$\frac{2\pi L}{\ln\left(\frac{L}{W}\right)}$

Source: Clift, Grace, and Weber 1978.

Symbols: R = radius; L = length; D = diameter; l = edge length; w = width; a and b are semimajor and semiminor axes of an oblate spheroid or semiminor and semimajor axes of a prolate spheroid ($b/a < 1$ for an oblate spheroid; $b/a > 1$ for a prolate spheroid).

tries. These equations show that $\mathfrak{F}$ is size dependent, regardless of geometry. It also is clear that external conductance is a shape-dependent variable, since except for spheres and cubes, which cannot change shape, each equation contains two variables denoting the two linear dimensions of a particular geometry. For example, figure 2.9 plots $\mathfrak{F}$ against the unit radius R of a sphere and the radii of touching and orthogonally intersecting spheres. These geometries serve as crude analogues to isolated cells and to cells arranged in filaments, respectively. In each case, $\mathfrak{F}$ is seen to increase monotonically with increasing R. The geometry with the highest conductance for any absolute size is that of touching spheres. By contrast, the isolated sphere has the lowest conductance regardless of absolute size.

With the aid of the equations listed in table 2.5, the Sherwood number for a geometry can be calculated from the following equation:

$$Sh_0 = \frac{\mathfrak{F}d}{S}, \tag{2.15}$$

where d is the characteristic length (e.g., diameter of a sphere, length of the edge of a cube, length of a cylinder) and S is surface area. The sub-

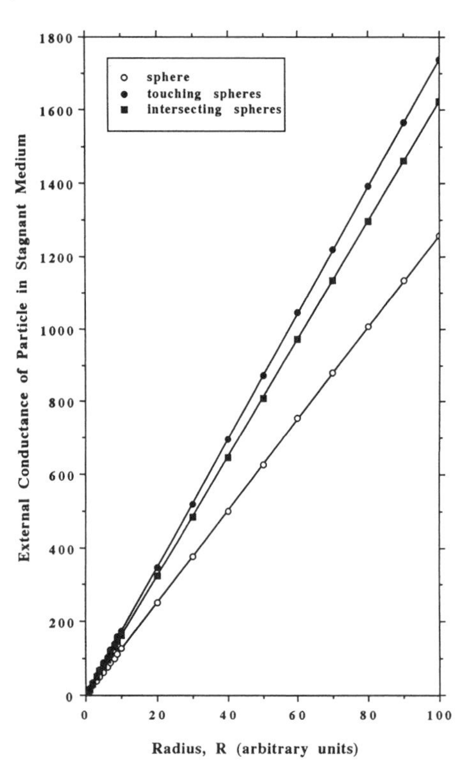

Figure 2.9. Estimated external conductance $\mathfrak{F}$ in a stagnant medium (calculated with the equations provided in table 2.5) plotted as a function of the radius R of a sphere, a string of touching spheres, and a string of intersecting spheres.

script *0* denotes the absence of external flow. For geometries incapable of changing shape, Sh_0 is a constant. That is, the zero-flow Sherwood number is size independent for these geometries. For example, $SH_0 = 2.0$, 1.39, and 1.23 for the sphere, intersecting spheres, and touching spheres, respectively. Likewise, for the cube, which cannot change shape, $Sh_0 =$ 1.37. By contrast, the Sherwood number of geometries capable of changing shape varies as a function of shape. Among these geometries are the terete cylinder and the flat plate. For cylinders with ratios of length to diameter ≈ 0 and 8, Sh_0 equals 1.27 and 2.05, respectively. For thin rectangular plates with aspect ratios of length to width equal to 1.1 and 100, Sh_0 equals 4.66 and 105, respectively. The equations listed in table 2.5 together with eq. (2.15) can be used to explore other geometries for which shape can be altered.

At this juncture it is useful to recall that the time t required for a substance to passively diffuse into a plant cell (see eq. 1.12) depends on the ratio of the surface area to the volume of the cell. Accordingly, we can express the time for diffusion in terms of the zero-flow Sherwood number:

$$t = \frac{V}{P_j}\left(\frac{Sh_0}{\Im d}\right) \ln\left[\frac{(c_0 - c_i)_{t=0}}{(c_0 - c_i)_{t=1/2}}\right] \tag{2.16}$$

Once again, we see that the rate of passive diffusion decreases as size decreases or as surface area increases. The importance of remaining small or changing shape with increasing size whenever mass transfer depends on passive diffusion cannot be overemphasized.

When $Re > 0$, the Sherwood number at nonzero flow is a function of Sh_0, Re, and cell shape and size. This can be illustrated for spheroids experiencing creeping flow ($0 < Re < 1$), where viscous forces dominate. Table 2.6, for convenience, gives the equations for Sh_0 for oblate and prolate spheroids that are variously oriented to the direction of ambient flow. These equations are expressed in terms of the aspect ratio $\Re$ of each spheroid and show that the orientation as well as the shape and geometry of a spheroid is important to mass transfer. The following equation permits Sh to be calculated by the equations for Sh_0:

$$Sh = \left(\frac{Sh_0}{2}\right) + \left[\left(\frac{Sh_0}{2}\right)^3 + K^3 Pe\right]^{0.333}, \tag{2.17}$$

where Pe is the Péclet number (the product of the Prandtl and Reynolds numbers; see table 2.4), and K is the mass thermal factor. Figure 2.10 plots K versus the aspect ration $\Re$ of spheroids (for the sphere, $\Re = 1$). This figure shows that K depends on shape as well as geometry. The maximum K occurs for an oblate spheroid whose aspect ratio equals 0.5 (the ratio of the minor to the major axis equals one-half). *K* decreases for progressively more attenuated (cigar-shaped) prolate spheroids. Indeed, all other things being equal, the Sherwood number is higher for oblate than for prolate spheroids, reaching a maximum for this geometry when $\Re = 0.5$; *Sh* progressively declines for longer and thinner oblate spheroids; and *Sh* will decrease, regardless of geometry, as the Reynolds number declines. Figure 2.11 summarizes these relationships by plotting *Sh* as a function of $\Re$ for three values of *Re*. These plots were obtained by assuming that the direction of ambient flow was parallel to the major axis of oblate as well as prolate spheroids. A reversal of the direction of ambient flow would alter the magnitude of *Sh* but not the general trends shown in figure 2.11. Thus, regardless of orientation, the maximum *Sh* occurs for the oblate spheroid with an aspect ratio equal to 0.5. Interestingly, erythrocytes and many unicellular aquatic plants occupying relatively low *Re* environments tend to be oblate in geometry, with aspect ratios near

Table 2.6. Approximate Equations of Sh_o for Spheroids

Oblate spheroid ($\Re < 1$), flow parallel to major axis ($d = 2b$)

$$Sh_o = \frac{8(1 - \Re^2)^{0.50}}{\left[2 + \frac{\Re^2}{(1 + \Re^2)^{0.50}} \ln \frac{[1 + (1 - \Re^2)^{0.50}]}{[1 - (1 - \Re^2)^{0.50}}\right] \cos^{-1}\Re}$$

Oblate spheroid ($\Re < 1$), flow parallel to minor axis ($d = 2a$)

$$Sh_o = \frac{8(1 - \Re^2)^{0.50}\Re}{\left[2 + \frac{\Re^2}{(1 + \Re^2)^{0.50}} \ln \frac{[1 + (1 - \Re^2)^{0.50}]}{[1 - (1 - \Re^2)^{0.50}}\right] \cos^{-1}\Re}$$

Prolate spheroid ($\Re > 1$), flow parallel to major axis ($d = 2b$)

$$Sh_o = \frac{4(\Re^2 - 1)^{0.50}}{\Re\left[1 + \frac{\sin^{-1}\frac{(\Re^2 - 1)^{0.50}}{\Re}}{(\Re^2 - 1)^{0.50}}\right] \ln[\Re + (\Re^2 - 1)^{0.50}]}$$

Prolate spheroid ($\Re > 1$), flow parallel to minor axis ($d = 2a$)

$$Sh_o = \frac{4(\Re^2 - 1)^{0.50}}{\Re^2\left[1 + \frac{\sin^{-1}\frac{(\Re^2 - 1)^{0.50}}{\Re}}{(\Re^2 - 1)^{0.50}}\right] \ln[\Re + (\Re^2 - 1)^{0.50}]}$$

Symbols: $\Re$ = aspect ratio (ratio of *b* to *a*); *a* = semimajor axis of oblate spheroid or semiminor axis of prolate spheroid; *b* = semiminor axis of oblate spheroid or semimajor axis of prolate spheroid; *d* = characteristic dimension for Reynolds number.

0.5. Be that as it may, regardless of shape or geometry, the absolute value of *Sh* always declines as a function of decreasing *Re*. For example, when $\Re = 0.5$, $Sh = 2.12\ Re^{1/3}$.

Does the Sherwood number influence the overall metabolic rate of an organism? Specifically, does a higher *Sh*, all other things being equal, promote higher growth rates? Circumstantial evidence indicates that the answer is a qualified yes. Figure 2.12 plots the rate of growth *G* versus the Sherwood number for fourteen species of algae with oblate and prolate

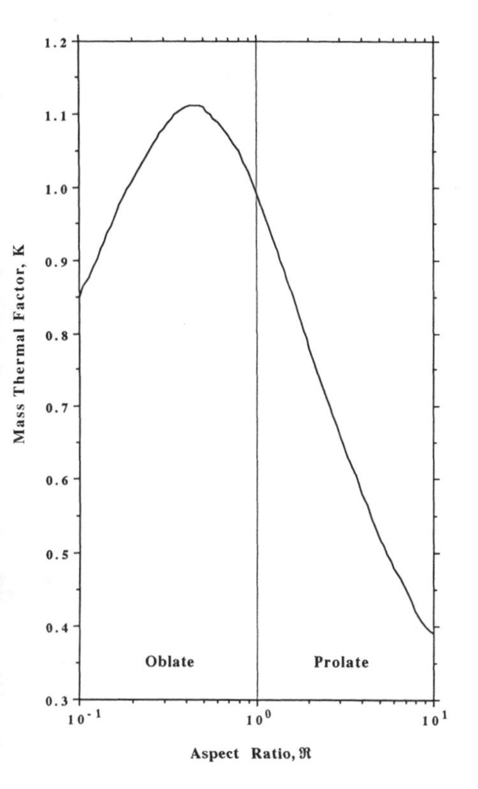

Figure 2.10. The mass thermal factor *K* plotted against the aspect ratio $\Re$ (the ratio of the minor to the major axis) of oblate and prolate spheroids (for the sphere, $\Re = 1$). The maximum *K* occurs for an oblate spheroid with $\Re = 0.5$.

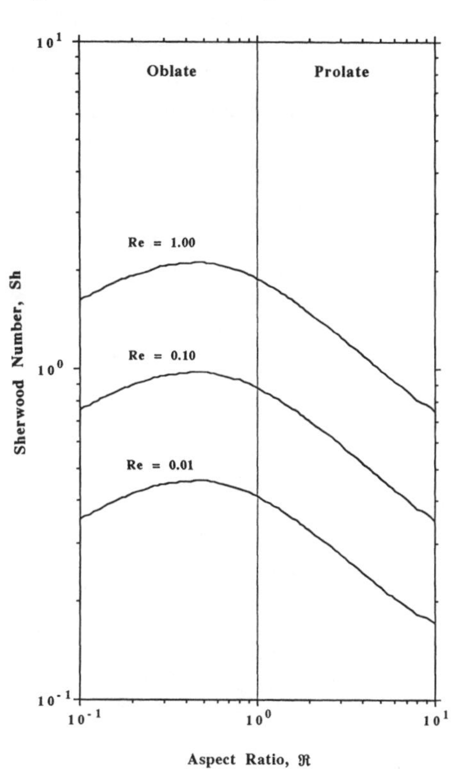

Figure 2.11. Log-log (base 10) plot of the Sherwood number *Sh* versus the aspect ratio $\Re$ (the ratio of the minor to the major axis) of oblate and prolate spheroids (for the sphere, $\Re = 1$) for three different Reynolds numbers ($Re = 0.01$, 0.10, and 1.00). These plots were obtained by assuming that the direction of ambient flow was parallel to the major axis of the oblate or the prolate spheroid.

geometries differing in shape and size. Values of G, which are expressed as number of cell divisions per day, were taken from the primary literature; species were selected from very different algal lineages to reduce the "phyletic dependency" of the scaling exponent (see sec. 1.6). Sh was calculated (from eq. 2.17 and those listed in table 2.6) based on the cell dimensions for each species. Regression of G versus Sh yields $G = 0.26 + 0.08\ Sh$ ($r^2 = 0.94$), indicating that G increases linearly as a function of increasing Sh, although this scaling must be viewed with some skepticism, since it is based on a few organisms. The three highest growth rates, however, are reported for algal species with oblate morphologies, which accords with the prediction that oblate spheroids have higher Sh_0 and K than those of prolate spheroids (see fig. 2.11).

Unfortunately, close-form equations like eq. (2.17) are not available for many geometries other than spheroids. To my knowledge the only other one that has been developed from theory and that has been empirically verified to some degree is for an infinitely long cylinder (Clift, Grace, and Weber 1978):

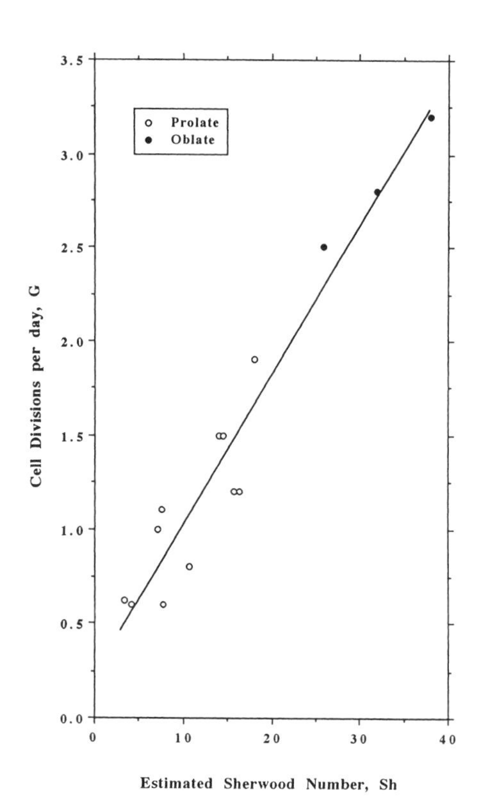

Figure 2.12. Rate of growth G plotted as a function of the estimated Sherwood number for fourteen species of algae with oblate and prolate geometries differing in shape and size. Data from a variety of sources.

(2.18a) $$Sh = \frac{2}{\ln\frac{8}{Pe} - \gamma}\left[1 - \frac{a_3}{\left(\ln\frac{8}{Pe} - \gamma\right)^2}\right]$$ (flow normal to length)

(2.18b) $$Sh = \frac{4}{\ln\left(\frac{16\chi}{DPe}\right) - \gamma},$$ (flow parallel to length)

where γ is Euler's constant (≈ 0.577), a_3 varies depending on the Schmidt number Sc, χ is a shape factor, and D is diameter. Although eq. (2.18) predicts that Sh will increase as a function of increasing Re, it is hardly useful in practical terms for cylindrical plants that have a finite length.

The approach typicaly taken to deal with the mass transfer of nonspheroidal geometries is to assume that the Sherwood number is proportional to the Nusselt number. The logic underlying a physical analogy between mass transfer and heat transfer is quickly seen when the equations for Nu and Sh are compared (see table 2.4). Each of these dimensionless numbers deal with the ratio of a flux density in a moving fluid versus a stagnant fluid. The following equations have been developed for the proportional relationshhip between Sh and Nu based on empirically determined relations between Nu and Re (see Kramer 1946; Kreith 1965; Happel 1973):

(2.19) $$Sh \propto Nu = 2.0 + 1.3\,Pr^{0.15} + 0.66\,Pr^{0.31}Re^{0.5}$$ (sphere)

(2.20) $$Sh \propto Nu = 0.66\,Pr^{0.33}Re^{0.5}$$ (flat disk or plate)

(2.21a) $$Sh \propto Nu = 0.91\,Pr^{0.31}Re^{0.385}$$ (cylinder, $0.1 \leq Re > 50$)

(2.21b) $$Sh \propto Nu = 0.80(Pr\,Re)^{0.31}$$ (cylinder, $Re < < 1$)

(2.21c) $$Sh \propto Nu \approx Re^{0.30}.$$ (cylinder, $Re = 1$)

These equations can be further simplified. As shown in table 2.7, the Prandtl number is a constant whenever υ and fluid temperature are specified. Thus, for a sphere submerged in water at 20°C, Pr=7.03 and eq. (2.19) reduces to $Sh \propto Nu = 3.74 + 1.21\,Re^{0.50}$. Likewise, for a cylinder ($0.1 \leq Re > 50$) in water, $Sh \propto Nu = 1.66\,Re^{0.385}$.

From eqs. (2.19) to (2.21), we expect Sh to increase anisometrically with respect to Re such that a further increase in Re results in a disproportinately smaller increase in Sh. The practical application of these equations, however, requires that the magnitude of Re be known. Obviously these magnitudes can vary, but for passively suspended plants in the water column, the minimum Re occurs when the plant obtains its settling velocity U_T. This occurs when the gravitational force exerted on the passively suspended plant is precisely balanced by the drag force. The equations for the settling velocity of a number of common geometries have been determined. Table 2.8 lists the equations for a few simple geometries that serve as crude albeit practical analogues to the morphologies of phytoplankton. Note that these equations can provide only first-order approximations of U_T. In each case, U_T refers to the characteristic dimension; thus U_T can be expressed in terms of *Re*. Although the equations for U_T are close-form solutions, those for spheres and cylinders require that the magnitude of the drag coefficients C_D for different values of *Re* be known. For a very small sphere, $C_D = 24\ Re^{-1}$. When $Re > 1.0$, however, the values of C_D for

Table 2.7. Prandtl and Schmidt Numbers for Water and Air at 20°C

	Water	Air
Prandtl number	7.01	0.70
Schmidt number		
CO_2	560	1.03
Phosphate	505	1.00
O_2	505	0.75
H_20	421	0.62

Table 2.8. Approximate Equations for Settling Velocity U_T of Simple Geometries That Apply When $Re << 1$

Geometry	U_T
Sphere	$\left[0.75\,\frac{g(\rho_o-\rho_f)d}{C_D\rho_f}\right]^{0.5}$
Flat plate	$\left[0.72\,\frac{g\sin\theta(\rho_o-\rho_f)td^2}{\rho_f\upsilon^2}\right]^{0.66}\left(\frac{\upsilon}{d}\right)$
Flat disk	$\frac{0.09\,g(\rho_o-\rho_f)td}{\rho_f\upsilon}$
Cylinder	$\left[\frac{1.57\,g(\rho_o-\rho_f)d}{C_D\rho_f}\right]^{0.50}$
Spheroids	$0.33\,\frac{g(\rho-\rho)V}{\rho_f\upsilon}\left(\frac{1}{c_1}+\frac{2}{c_2}\right)$

Symbols: $g = 9.8\text{m s}^{-2}$; ρ_o = density of object; ρ_f = density of fluid; d = characteristic dimension; C_D = drag coefficient (see fig. 2.13); θ = angle subtended between plane perpendicular to direction of settling and surface of a flat plate; t = thickness of plate or disk; υ = kinematic viscosity of fluid; V = volume of spheroid (see table 1.2); $c_1 = 1.2\pi a\ (4 + b/a)$ and $c_2 = 1.2\pi a\ (3 + 2b/a)$, where a and b are semimajor and semiminor axes of an oblate spheroid or semiminor and semimajor axis of prolate spheroid (note that $b/a < 1$ for an oblate spheroid; $b/a > 1$ for a prolate spheroid).

spheres and other geometries must be determined empirically. Figure 2.13 plots C_D for a sphere and cylinder as a function of *Re*.

With the aid of the equations and relations thus far developed, we can now speculate on the ability of phytoplankton differing in size and geometry to transfer mass as they passively descend through the water column. We will assume that the settling velocities of these plants are given by the equations listed in table 2.8. For illustration only, we shall consider the flux density of carbon dioxide J_{CO_2}, although we could just as easily consider the absorption of other materials essential to growth, such as phosphate (see table 2.2).

We begin by expressing the Sherwood number in terms of the carbon dioxide flux density, characteristic dimension, diffusion coefficient, and carbon dioxide concentration gradient. For a sphere in water, eq. (2.19) gives

$$\frac{J_{CO_2} d}{D_{CO_2}(\rho_c - \rho_a)} \approx 3.74 + 1.21\, Re^{0.5}, \tag{2.22a}$$

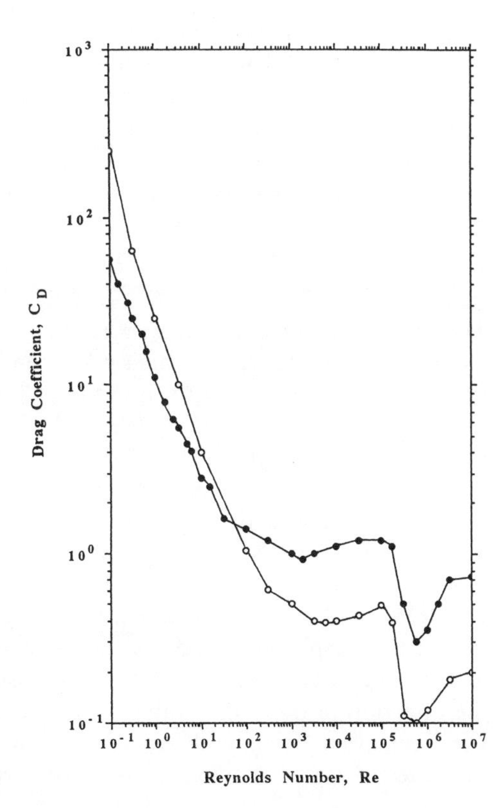

Figure 2.13. Log-log (base 10) plot of the drag coefficient C_D for a sphere and cylinder versus Reynolds number *Re*. Data obtained from unpublished flume and wind tunnel experiments by the author.

where ρ_c is the intracellular CO_2 concentration, ρ_a is the ambient concentration of CO_2 (measured many cell diameters from the surface of the cell), and D_{CO_2} is the diffusion coefficient of CO_2. Since $d = Re\nu/U_T$ and $D_{CO_2}/\nu = Sc$, eq. (2.22a) becomes

$$(2.22b) \qquad J_{CO_2} \approx (3.74 + 1.21\ Re^{0.5})(\rho_c - \rho_s) U_T (ReSc)^{-1},$$

where Sc is the Schmidt number (see table 2.4). For CO_2 dissolved in water, $Sc = 5.6 \times 10^2$ (table 2.7). Inserting the appropriate equation for U_T (see table 2.8) into eq. (2.22b), where d is expressed in terms of Re, we find

$$(2.22c) \quad J_{CO_2} \approx \left[\frac{3.74 + 1.21 Re^{0.5}}{C_D{}^{0.33}\, Re^{0.66}\, Sc}\right] (\rho_c - \rho_a) \left[\frac{4g\nu(\rho_0 - \rho_f)}{\rho_f}\right]^{0.33}.$$

Equation (2.22c) contains three physiological variables: the intracellular CO_2 concentration ρ_c, the ambient CO_2 concentration ρ_a, and cell density ρ_o. Note that the term $(\rho_c - \rho_a)$ indicates whether the flux of CO_2 is into $(\rho_c < \rho_a)$ or out of $(\rho_c > \rho_a)$ the cell. During steady-state photosynthesis in water, $\rho_c = 7$ mmol CO_2 m^{-3} while $\rho_a = 11.7$ mmol CO_2 m^{-3} (see Raven 1985, 275). Thus the flux density of CO_2 is into the cell. The term $(\rho_o - \rho_f)$ indicates whether the cell floats $(\rho_o < \rho_f)$ or sinks $(\rho_o > \rho_f)$. The density of seawater at 20°C is about 1,024 kg m^{-3}. For convenience, we will assume that $\rho_o = 1{,}200$ kg m^{-3} simply to specify that the cell will sink rather than float. Inserting these values into eq. (2.22c), we obtain

$$(2.22d) \quad J_{CO_2} \approx (-1.35\ \mu\text{mol } CO_2 m^{-2} s^{-1}) \left[\frac{3.74 + 1.21\ Re^{0.5}}{C_D{}^{0.33}\, Re^{0.66}}\right].$$

The expression in parentheses has the units of flux density—that is, amount of mass per surface area per second. The expression in brackets (to the far right in eq. 2.22d) is entirely dimensionless, albeit dependent on size, geometry, and shape. It reveals that the flux density of carbon dioxide will anisometrically decrease as Re increases because Re can increase only as a consequence of an increase in cell size, which in turn increases U_T. This is the Scylla and Charybdis—the physical dilemma—that every passively suspended very small plant faces. Figure 2.14 shows this graphically for spheres as well as a few other simple geometries for which U_T can be easily calculated. In this figure, J_{CO_2} is plotted as a function of Re for spheres, thin disks, flat plates, and terete cylinders differing in size. The orientation of the disk was assumed to be perpendicular to the direction of descent; a value of $\theta = 45°$ was used for the flat plate. Different orientations with respect to descent would result in different estimates of J_{CO_2}. Likewise, if the nonspherical geometries were given a

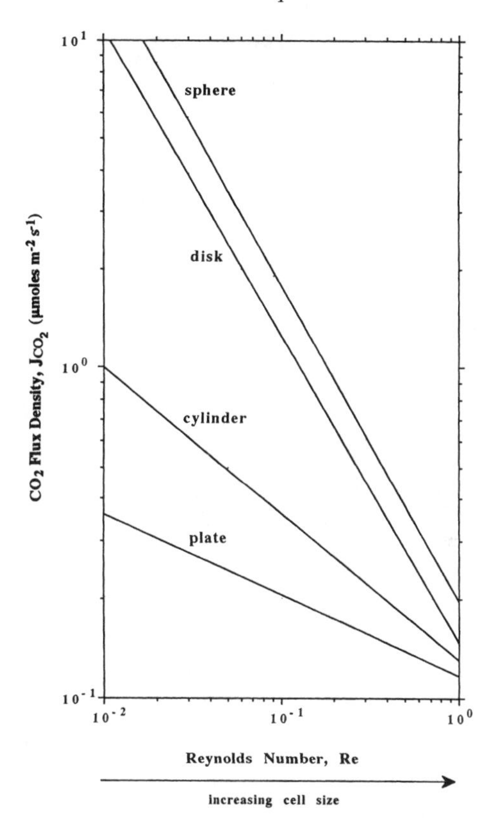

Figure 2.14. Log-log (base 10) plot of flux density of carbon dioxide J_{CO_2} versus Reynolds number *Re* for hypothetical cells differing in geometry.

slight rotation, then each would continue to rotate as it sinks. Since energy is used in the process of rotation, the settling velocity of a rotating object is slower than the same object that does not rotate. Leaving this issue aside, figure 2.14 reveals that the most efficient size and geometry among those considered is an extremely small sphere. The next most efficient is a very small thin disk. The least efficient geometry is a small flat plate. The hypothetical relationships among J_{CO_2}, cell size, and geometry are similar to those presented in a classic paper by Munk and Riley (1952).[4]

Intuitively, we might have expected that the rate of mass exchange would increase as the speed of the ambient fluid increases relative to a sinking plant. This expectation is certainly true for high Reynolds num-

4. Those who refer to this otherwise excellent paper should be warned of typographical errors in eqs. (11) and (18) as well as the error in eq. (22). In eq. (11) substitute *q* for 6; in eq. (18) replace the exponent 0.15 with 0.5; for eq. (22) see the often missed footnote on p. 240.

bers, where the boundary layer enveloping a plant is stripped away by the passage of the nutrient-replenishing ambient fluid. Indeed, at very high *Re,* forced convection occurs. However, at the very low Reynolds numbers that likely characterize the settling velocities of phytoplankton, the increase in the resistance to passive diffusion resulting from an increase in cell size is estimated to far outweigh the consequences of thinning the boundary layer as a cell settles through the water column at a faster rate. Essentially, free-floating plants are at a metabolic advantage when they remain *very* small. By contrast, aquatic plants attached to a substrate can benefit from forced convection in terms of mass exchange provided they get large enough to experience turbulent flow regimes (high *Re*). Consider a flat, platelike organism attached to a substrate. For such an organism, $Sh \propto Re^{0.50}$. Thus $J \propto Re^{0.50}d \propto (U/\nu)^{0.50} d^{1.50}$. The scaling relationship between the flux density and the characteristic dimension indicates that a comparatively small increase in d can cause a disproportionately large increase in J, regardless of the ambient flow speed. In rapidly moving water, the strategy for such an organism is to assume an orientation nearly parallel with the direction of ambient flow, since this maximizes the characteristic dimension and therefore J. This orientation has mechanical as well as physiological benefits, since it reduces the surface area projected toward the oncoming stream of water, thereby reducing the drag force exerted on the plant (however, see fig. 2.23). It is evident that remaining flexible and going with the flow is much more desirable than maintaining a rigid posture.

2.6 Self-Propulsion: The Volvocales

The nonbotanist may think of plants as immobile and subject to mercurial fluid flows. Such a misconception is aggravated by protozoology texts that persist in calling mobile plants "green animals." However, a good number of plant species have the ability to move by means of ciliated cells. These plants can change their speed, and therefore they can change their Reynolds numbers to a limited degree. Mobile plants present a particular challenge when we deal with the transfer of mass or energy. True, the physical laws governing the transfer of mass or energy that were previously developed for passively floating plants or plants attached to a substrate are equally capable of dealing with self-propelled plants. These laws, however, must be reconfigured to allow for the consequences of the motion of cilia, which can stir the boundary layer and therefore reduce resistance to the transfer of mass or energy through external cell surfaces. The fluid mechanics of cilia are left to the next section of this chapter. Here I simply focus on the consequence of geometry and shape on the biology of small aquatic self-propelled plants. I shall use the Volvocales to illustrate a few scaling relations because most of these plants are spheroidal in geometry. Thus we can test some of the predictions of geometric similitude in terms of plants that can move on their own.

Although the Volvocales (an order of freshwater green algae) serve as a model natural taxon with which to explore the consequences of plant mobility on mass or energy transfer, we must be alive to the criticism that the taxa within this order do not yield independent data points when we determine scaling relations. Recall from chapter 1 that the numerical value of the scaling exponent is influenced by the taxonomic composition of the sample of species considered (the "phyletic effect"). Even within a particular suprageneric taxon, overrepresentation of one or more genera within a plant group can bias the scaling exponent. For these reasons we cannot extrapolate the scaling relations determined for the Volvocales to other algae capable of self-propulsion, and we must be fastidiously aware of the taxonomic relations within the Volvocales to avoid stacking the phyletic deck when considering the scaling relations for this particular group of green algae.

Unfortunately, the phyletic relations among genera and species within the Volvocales are far from clear, although the patterns of cell cleavage, cell differentiation, and embryogenesis, which are shared by all the volvocalean algae, support the classic view that the Volvocales constitute a monophyletic group (Bisalputra and Stein 1966; Starr 1984; Kirk 1988). Traditionally, the genera within this order have been arranged in an evolutionary series beginning with comparatively small unicellular plants and ending with the largest genera, which can achieve cell numbers on the order of 10^4. Figure 2.15 diagrams some volvocalean genera that will be repeatedly referred to and that are often arranged in a hypothetical evolutionary sequence. Curiously, purported evolutionary series like that suggested in figure 2.15 have been used, sometimes uncritically, in the context of a variety of evolutionary propositions unrelated to the issue of scaling, as in discussions of the evolution of individuality and multicellularity (see Buss 1987, 72–77). However, evidence from a variety of studies casts doubt on the legitimacy of viewing the smaller genera diagrammed in figure 2.15 as archaic or the larger genera within the Volvocales as more evolutionarily derived. Indeed, although preliminary in nature, the available evidence strongly supports the view that some of the genera may be polyphyletic in origin. For example, cell wall and extracellular matrix chemical analyses indicate that *Volvox carteri* and *Chlamydomonas reinhardtii* are more closely related to one another than *C. reinhardtii* is to other species of *Chlamydomonas* (Adair et al. 1987; Adair, pers. comm., as cited by Kirk 1988, 34).

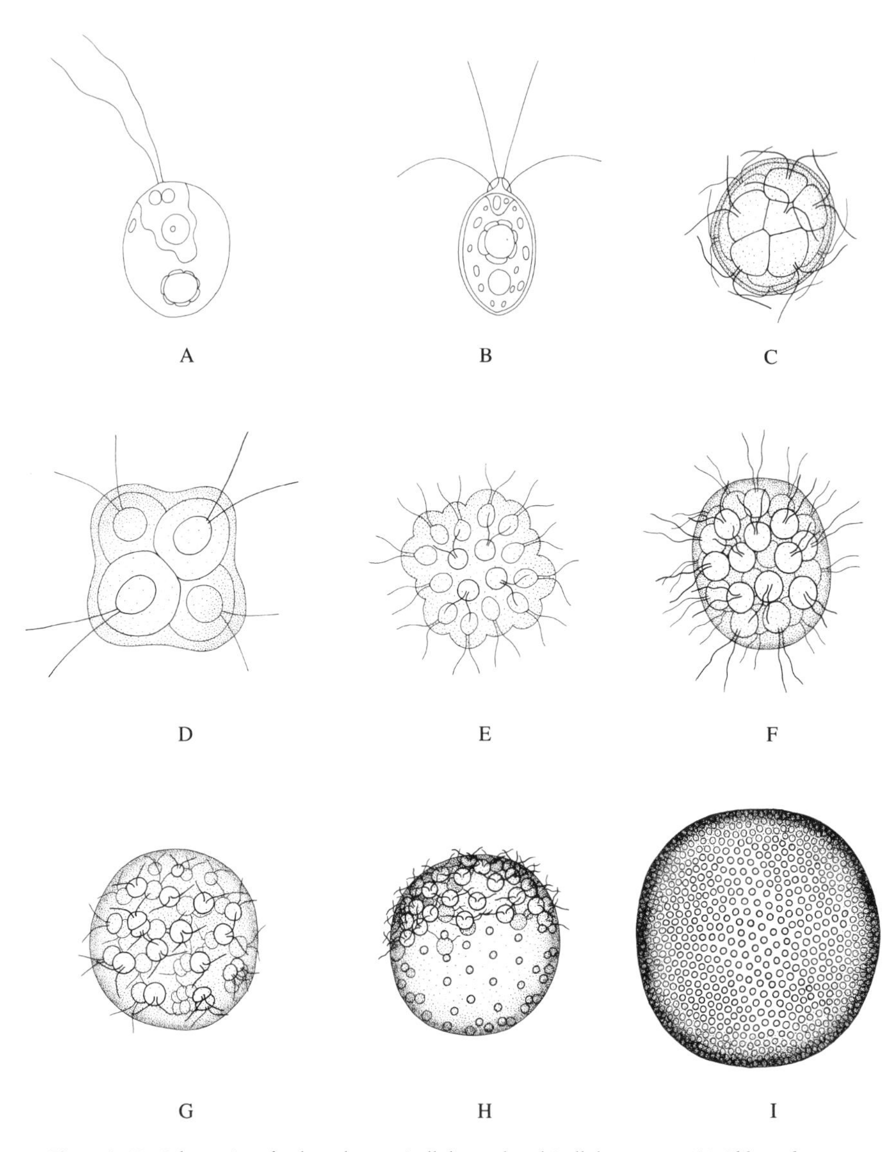

Figure 2.15. Schematics of volvocalean unicellular and multicellular genera. (A) *Chlamydomonas;* (B) *Carteria;* (C) *Pandorina;* (D–E) *Gonium;* (F) *Eudorina;* (G–H) *Pleodorina;* (I) *Volvox* (numerous cilia extending from plant surface not shown). In general terms, A → I has been adduced to be an evolutionary sequence.

Although we are in the unenviable position of treating the scaling relations for a monophyletic group of plants in which genera may be paraphyletic, the view taken here is that a spheroidal geometry likely imposes physiological and mechanical limitations and that, regardless of their precise phyletic relationships, the volvocalean algae illustrate a natural evolutionary experiment in which these limitations have been resolved in a fashion convergent among species. This view can be expressed qualitatively as follows. We have come to anticipate that the passive transport of CO_2 dissolved in water depends on the ratio of external surface area to volume $S:V$ of plants regardless of size or shape. And we know from our treatment of geometric similitude that this ratio decreases as absolute size increases among geometrically identical objects. Since the magnitude of S influences the transport rate of CO_2 and since the magnitude of V influences the rate at which CO_2 is consumed to form primary metabolites by photosynthesis, we anticipate that shape would change in relation to size such that functional similarity (reflected by the ratio of S to V) is either maintained or increased. Likewise, we know that the intensity of light declines as a function of the distance it passes through any optically translucent medium. Thus the cells at the center of a spherical mass become increasingly more shaded as cells continue to proliferate. To achieve larger plant sizes and at the same time maintain more or less equivalent rates of mass transport and reasonable levels of light intensity, shape must be modified. The volvocalean algae appear to support these predictions. Regardless of the phyletic relationships among genera, the largest multicellular plants are hollow spheroids composed of a monostromatic (one cell thick) layer of tissue surrounding a water-filled cavity (e.g., *Volvox*). Thus every cell is in immediate contact with the external fluid medium, the surface area of each plant is magnified with respect to a metabolically inert fluid-filled volume, and overall self-shading is minimized. Likewise, it is worth noting that a sphere always projects the same surface area toward the direction of incident light. For a suspended plant whose orientation can vary, therefore, a spherical morphology is a good geometry for light interception. Additionally, since plant cell densities typically exceed that of water, the largest plants, like *Volvox,* because of their water-filled cores, have the lowest bulk densities. Since a spherical geometry is one of the least streamlined and therefore generates substantial drag, particularly for the very low Reynolds numbers that define the world of small aquatic plants, the low bulk densities and high drag coefficients of these plants reduce the settling velocity and the metabolic energy required to keep individuals passively suspended in the water column. Conversely, greater energy must be invested in moving a sphere as opposed to a more streamlined object through the water toward regions of optimal light intensity. However, since the energy required to move toward the light comes from the acquisition of light, the expenditure of a fraction of the energy produced by photosynthesis seems a small price to pay for what otherwise must be considered one of the most beautifully designed groups of organisms to surface after some 500 million years of evolutionary experiment and innovation. We also must note that all the multicellular volvocalean genera are determinate in growth (i.e., each is a coenobium). Each individual plant reaches a final mature size, and then vegetative cell division ceases. The inability to grow indefinitely in size is an advantage for a spherical volvocalean plant regardless of whether it is constructed as a hollow or a solid spheroid.

Consider the relation between the external surface area S and the biomass volume V of some representative small unicellular and large multicellular volvocalean algae. Figure 2.16 plots S versus V for thirteen individual genera. Bearing in mind that we cannot be sure these genera provide independent data points to determine the scaling of S with respect to V until the precise systematic relations within the Volvocales are re-

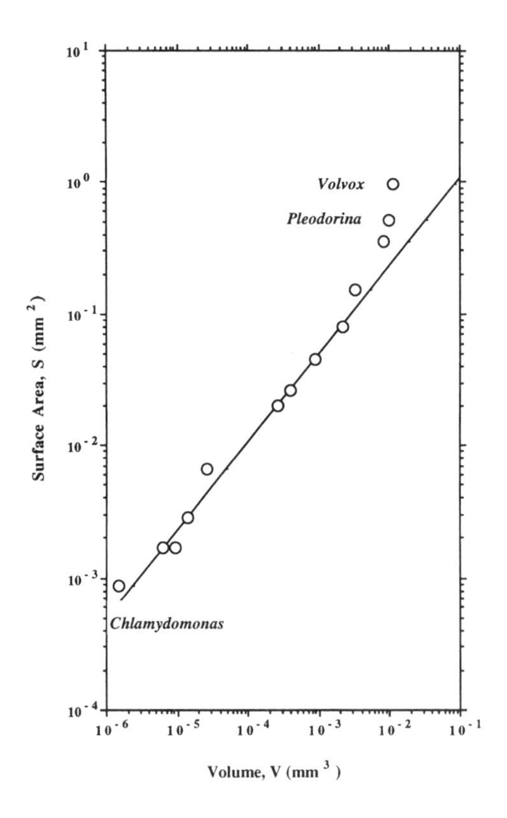

Figure 2.16. Log-log (base 10) plot of plant surface area S versus volume V for volvocalean taxa (for convenience, data points for *Chlamydomonas, Pleodorina,* and *Volvox* are identified). Diagonal line denotes regression curve obtained from LS regression analysis assuming an isometric relation between S and V.

solved, least squares regression of these data yields $S = 12\ V^{0.745}$ ($r^2 =$ 0.98). Since $\alpha_{RMA} = 0.76 \pm 0.04$, the scaling of S for these genera appears to differ from that predicted by geometric similitude (for which $\alpha \approx$ 0.667, as shown by the solid line in the figure). This is hardly surprising, since large multicellular genera such as *Volvox* are hollow and, by virtue of their *empty* volume, must have higher values of S than those predicted from geometric similitude. How far a particular genus deviates from the expectations of the null hypothesis is indicated by the divergence of its S value from the regression line plotted in figure 2.16. Note that the S value for some of the smaller as well as some of the larger genera (e.g., *Chlamydomonas* and *Volvox*) plot above the solid line predicted for a series of solid spheres differing in size for which $\alpha_{RMA} \approx 0.667$. Although they do not have a hollow morphology, many of the smaller volvocalean genera attain a prolate geometry for which S is larger than for a sphere with equivalent V. Of the largest genera among the thirteen considered in figure 2.16, *Volvox* is the only one for which S plots above the 95% confidence intervals (not shown) for the regression curve defined by the formula $S = 12\ V^{0.745}$. *Volvox* stands out statistically among its volvocalean counterparts because over 70% of its bulk volume is not occupied by cells or the extracellular matrix that helps bind cells together. Clearly, within the volvocalean algae, shape is not a conserved feature. Figure 2.17 illustrates this point by plotting shape versus size for the thirteen genera considered in figure 2.16. In this graph, shape is expressed in terms of the dimensionless aspect ratio $S^3{:}V^2$, and size is measured in terms of the radius of a sphere with the same surface area as each genus (the equivalent radius Δe). Note that these two variables are autocorrelated because $S \propto (\Delta e)^2$. Therefore we can legitimately determine neither a scaling exponent nor a regression formula for $S^3{:}V^2$ versus Δe. Nonetheless, plotting shape versus size permits us to identify two comparatively small genera ($\Delta e < 0.1$ mm) with S values significantly higher than expected. These high values result from the nonspherical geometry of these taxa. Additionally, four large genera ($\Delta e \geq 0.1$ mm) with high S values are notable. For these taxa, high values for S are a consequence of hollow multicellular construction. Once again, the genus *Volvox* segregates from its volvocalean counterparts by the extent to which its volume is unoccupied by cellular materials. Multicellularity provides a mechanism to alter the value of S with respect to V by creating an internal "noncellular volume."

The null hypothesis of geometric similitude predicts that the ratio of surface area to volume will scale as the reciprocal of the radii of spheres differing in size. Thus, $S{:}V \propto (\Delta e)^{-1}$. Such is not the case for the volvocalean algae for which data are plotted in figures 2.16 and 2.17. For these organisms, the ratio of surface area to biomass volume scales roughly as the negative 2/3 power of radius ($S:V \propto \Delta e^{-0.663}$). Thus morphological deviations from geometric similitude mitigate somewhat the size-dependent reduction of surface area with respect to body volume. Is this beneficial? Recall that the ratio $S:V$ plays an important role in determining the time required for passive diffusion (see sec. 1.10, eq. 1.12 and sec. 2.5, eq. 2.16). Accordingly, the anisometry revealed in figure 2.16 could confer a metabolic benefit. We cannot evaluate this possibility directly. With the aid of a few simple calculations, however, we can estimate the relative importance of changes in shape to the time required to absorb materials essential to growth from the external environment.

Figure 2.18 plots $S:V$ versus t, the estimated time required for the intracellular carbon dioxide concentration of volvocalean plants to reach one-half the ambient CO_2 concentration (see eq. 1.12b). The solid line in this figure represents $S:V$ versus t calculated for a series of solid spheres differing in size. For the volvocalean genera, t_{max} = 2 sec (indicated by a dashed vertical line) because the largest genera have a monostromatic tissue construction and therefore have a higher $S:V$ than do their solid spherical counterparts. In terms of genera with a solid spherical morphology,

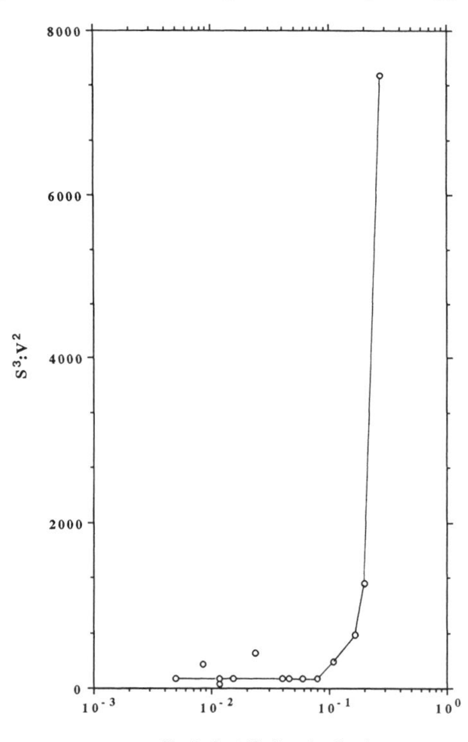

Figure 2.17. Plant aspect ratio $S^3:V^2$ plotted against the equivalent radius Δe of volvocalean taxa.

the minimum $S{:}V$ occurs somewhere near 40 mm^{-1}. Conceivably, this value may be the critical threshold below which the rate of passive diffusion cannot keep up with the demands of photosynthesis within a solid spherical mass of volvocalean cells. We can consider this hypothesis quantitatively, provided the flux density of CO_2 and the photosynthetic rate P are known. That is, the lowest value of $P{:}J_{CO_2}$ is that which can sustain steady-state photosynthesis, assuming that CO_2 is the only factor limiting growth. Unfortunately, empirical data for J_{CO_2} and P are not available. Therefore I can only illustrate the logic that would be used to explore this hypothesis. Taking the concentration of CO_2 at the surface of a plant during steady-state photosynthesis as 7 mmol m^{-3} (Raven, Beardall, and Griffiths 1982), the ambient concentration of CO_2 in fresh water at 20°C as 11.7 mmol m^{-3}, and the diffusion coefficient of CO_2 in water as 1.8×10^{-9} m^2 s^{-1}, we still need to specify the distance x in eq. (2.6) between the external and internal CO_2 concentrations. For the sake of argument, x will be taken to be one order of magnitude less than the length of the cilia (roughly 100 μm) that agitate the external layer of water coating the surface of a multicellular volvocalean. Accordingly, we find that J_{CO_2} =

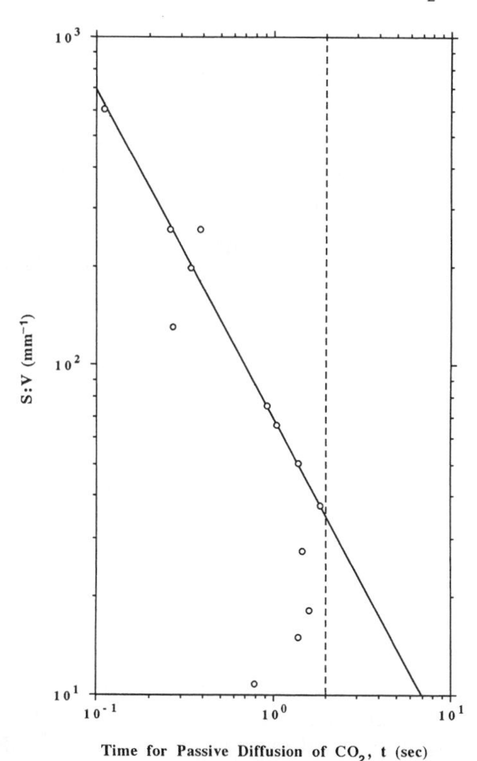

Figure 2.18. Log-log (base 10) plot of the ratio of surface area to volume $S{:}V$ versus the estimated time t required for the intracellular carbon dioxide concentration of volvocalean plants to reach one-half the ambient CO_2 concentration. The diagonal solid line denotes the LS regression curve obtained for $S{:}V$ versus t calculated for a series of solid spheres differing in size. The vertical dashed line denotes the maximum t for these data.

$D(\Delta C/x) = (1.8 \times 10^{-9}\,m_2\,s^{-1})(0.0117\,mol\,m^{-3} - 0.007\,mmol\,m^{-3})/(100 \times 10^{-6}\,m) = 8.46 \times 10^{-7}\,mol\,m^{-2}\,s^{-1}$. Considering next the magnitude of P, the photosynthetic rate at any given light intensity will vary with $[CO_2]$ approximately in accordance with the well-known Michaelis-Menten equation for enzyme kinetics. Casting this equation into the parlance of photosynthesis gives the formula

$$(2.23) \qquad P = \frac{P_m[CO_2]}{K_m + [CO_2]}, \qquad \text{(Michaelis-Menten equation)}$$

where P_m is the maximum photosynthetic rate that can be obtained at saturating $[CO_2]$ and K_m is the dissociation constant for the ribulose diphosphate carboxylase-CO_2 complex (see Kirk 1983, 262–69). In terms of an ideal enzymatic reaction, P_m has twice the magnitude of K_m. Empirically determined values of K_m range between 0.02 mol and 0.170 mol (Kirk 1983, 262), an unfortunately large range from which to select a *typical* value. Selecting an intermediate value of $K_m = 0.1$ mol (corresponding to that of the green alga *Nitella flexilis,* a charophyte), we see that $P = P_m\,[CO_2]/(K_m + [CO_2]) = 2K_m\,[CO_2]/(K_m + [CO_2]) = 0.2$ mol CO_2 mol CO_2 $m^{-2}s^{-1}$ (0.007 mol CO_2 m^{-3})/(0.1 mol CO_2 + 0.007 mol CO_2 m^{-3}) = 0.021 mol CO_2 $m^{-3}\,s^{-1}$. Taking the ratio $P{:}J_{CO_2}$ = (0.021 mol CO_2 $m^{-3}s^{-1}$):(8.46×10^{-7} mol CO_2 $m^{-2}s^{-1}$), the critical $S{:}V$ threshold is estimated to be $2.48 \times 10^4\,m^{-1}$ or roughly 25 mm^{-1}. This is roughly one-half the value of $S{:}V$ suggested by figure 2.19 ($\approx$ 40 mm^{-1}). This is not surprising considering the crude estimates of physiological variables used in our calculations, particularly since the value of K_m can range between two orders of magnitude for freshwater plants, whereas the rate of CO_2 diffusion through the external ciliated surfaces of plants like *Volvox* can only be crudely estimated. Significantly higher or lower threshold values for $S{:}V$ therefore would be anticipated if the actual volumetric consumption rates of CO_2 and diffusive flux densities of CO_2 differ among taxa.

The reduction in V with respect to S by the formation of a water-filled core surrounded by a photosynthetic rind of spherically organized cells is not without further benefit to a photosynthetic organism whose source of energy attenuates dramatically as water depth increases. The hollow construction of larger plants is easily demonstrated to reduce the bulk density ρ_o of the plant body. Likewise, the increase in unit radius R of spheroidal plants by the "ballooning" of the monostromatic layer of tissue reduces the rate at which these plants descend through the water. The reduction of ρ and increase in R are easily shown to quantitatively reduce the settling velocity U_T compared with plants with a solid core of tissue.

The settling velocity of a spherical object is given in table 2.8. For a very small sphere (i.e., at very low *Re*) with unit radius R and density ρ_o, however, Stokes's law indicates that

$$U_T = \frac{2R^2}{9\mu} g(\rho_o - \rho_f), \quad (2.24a) \quad \text{(Stokes's law)}$$

where μ is the dynamic viscosity of the fluid medium with density ρ_f. Equation (2.24a) gives the same results for very small spheres as its counterpart in table 2.8 without the trouble of dealing with the drag coefficient. Also, eq. (2.24a) reveals the *strategy* of a hollow plant, particularly when we realize that the bulk density of a plant ρ_o is the sum of the density of the biomass and the density of the water-filled core, if present. That is, $r_o = \rho_t V_t + \rho_w V_w$, where V is volume fraction and the subscripts t and w refer to tissue and water. Thus,

$$U_T = \frac{2R^2}{9\mu} g[\rho_t V_t - \rho_w (V_w + 1)]. \quad (2.24b)$$

For unicellular and some multicellular genera, $V_w = 0$. For genera like *Volvox*, however, V_w can be near 85% the entire volume of the plant. According to eq. (2.24b), as the volume fraction of the central core of water increases, the settling velocity will decrease. Also note that, since the volume of the water-filled core increases as a function of the cube of R, a small increase in R results in a disproportionately large reduction in U_T. Figure 2.19 plots empirically determined U_T versus the equivalent radius Δe of fourteen volvocalean species.[5] The straight line in this figure shows the scaling of U_T predicted from eq. (2.24a) for solid spheres ($U_T \propto R^{2.0}$). Figure 2.19 reveals that most volvocalean algae have lower settling velocities than their solid spherical counterparts. As size increases, the extent to which empirically determined values of U_T deviate from predicted U_T increases. For genera like *Volvox*, which have a hollow construction, the difference between predicted and observed U_T is a consequence of the increase in the volume fraction of the water-filled core. For other multicellular genera lacking a water-filled core (e.g., *Gonium* and *Pleodorina*), the reduction of U_T comes from the addition of an extracellular matrix binding cells to one another. The density of this matrix is very near that of water.

5. Plants were killed with a brief exposure to formaldehyde, resuspended in their culture medium, and stroboscopically photographed as they fell in the water column. U_T was calculated based on the distance between successive stroboscopic images divided by the interval of time between successive flashes of light. The data presented in figure 2.19 are the mean values of fifty measurements for each of ten conspecific plants.

In theory, the shape of a self-propelled plant plays an important role in dictating the speed of active transit through the water column, because the energy required to move is proportional to the drag force F_D resulting from active movement, which in turns depends on the surface area S_P projected against the advancing fluid. If the strategy is to conserve energy, then a reasonable biological tactic is to reduce S_P. For better or worse, however, spherical plants are geometrically isotropic in terms of S_P. That is, a sphere unavoidably projects the same cross-sectional area regardless of its orientation relative to the direction of motion. This is advantageous for reducing the settling velocity. But in terms of active motion, spherical plants are cumbersome. By contrast, prolate and oblate spheroids are anisotropic in terms of S_P. Each has two different elliptical cross sections with which to confront the advancing fluid. Of the two geometries, the prolate spheroid is the more desirable in terms of mobility. Not surprisingly, most unicellular volvocaleans are typically prolate in geometry, preferentially moving perpendicular to their minor axis and thereby projecting their minimum cross-sectional area as they advance through the water. For example, the ratio of the projected to total surface area S_P/S of

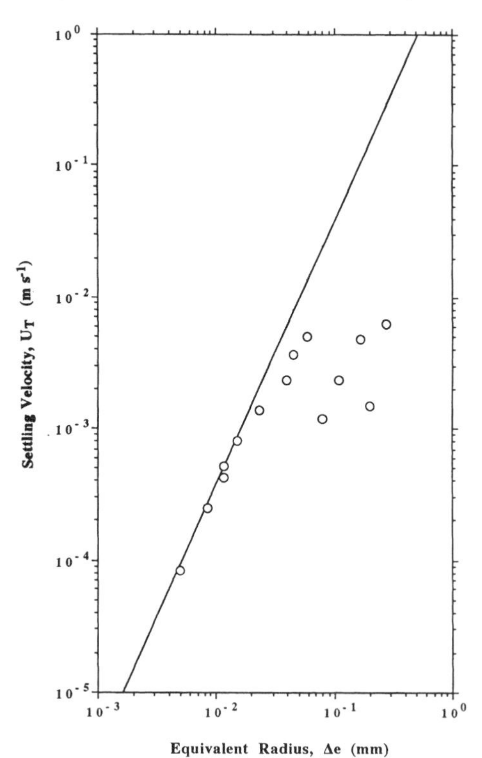

Figure 2.19. Log-log (base 10) plot of empirically determined settling velocity U_T versus the equivalent radius Δe of fourteen volvocalean algae. The solid diagonal line denotes the scaling of U_T predicted from eq. (2.24a) for solid spheres ($U_T \propto R^{2.0}$).

Chlorogonium elongatum, which has a prolate morphology, equals 0.03, whereas that of *Volvox* species equals 0.25, as is true for all spheres. The advantages and disadvantages of different geometries and shapes with respect to mobility are easily shown by plotting S_P/S versus the empirically determined maximum speed U of self-propelled plants with spheroidal morphologies. Figure 2.20 provides these data for thirty-five ciliated unicellular and multicellular genera selected to minimize as far as possible the phyletic effect. The relation between P/S and U is exponential and negative: $P/S = (1.53)\ 10^{-0.29U} (r^2 = 0.90)$. That is, as the ratio P/S decreases, U exponentially increases. The slowest-moving plants have an oblate geometry (e.g., *Gonium pectorale*). The fastest-moving plants have a prolate geometry (e.g., *Chlorogonium elongatum,* which projects its minor axis in the direction of motion). A curious footnote to the scaling relation shown in figure 2.20 is that smaller plants move relatively faster than their larger counterparts, although in absolute terms larger plants clearly move faster.

It is comparatively easy to show that the amount of energy expended to move even comparatively large volvocaleans through the water column

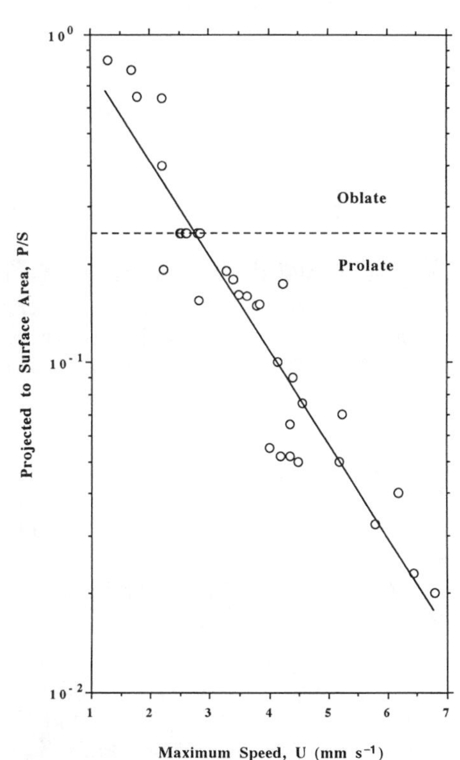

Figure 2.20. Log-log (base 10) plot of the ratio of projected to total surface area *P/S* against the maximum speed *U* reported for representative ciliated unicellular and multicellular genera. Data from a variety of sources.

is a trivial fraction of the total energy produced by photosynthesis. Indeed, a remarkably small price is paid for self-propulsion, particularly since the motion of cilia continuously mixes the thin layer of water that sticks to the external surfaces of cells, which is depleted of its oxygen and carbon dioxide by respiration and photosynthesis. Based on adenosine triphosphatase activity in a variety of ciliated organisms, the amount of energy (expressed in joules, J) consumed in motion per cell per second is roughly 2×10^{-7} J cell^{-1} s^{-1} (Gibbons 1966; Brokaw 1967). Based on an average cell diameter of 10^{-5} m, the average volvocalean cell volume is 5.46×10^{-6} m^{-3}. The energy consumed per cell volume per second, therefire, is roughly 3.66 J m^{-3} s^{-1}. From the calculations used to assess CO_2 supply and demand, the average rate of photosynthesis P for these algae was estimated to be 0.0139 mol CO_2 m^{-3} s^{-1}. Since each mole of CO_2 yields 4.5×10^5 J, the energy produced by photosynthesis is (0.0139 mol CO_2 m^{-3} s^{-1}) (4.5×10^5 J mol^{-1} CO_2) = 6.5×10^3 J m^{-3} s^{-1}. The ratio of energy consumed to that produced (3.66 J m^{-3} s^{-1}:6.5×10^3 J m^{-3} s^{-1}), therefore, indicates that 0.06% of the energy produced by photosynthesis is expended per cell per second for mobility.

2.7 The Mechanics of Ciliary Motion

I continue the discussion of self-propulsion with a treatment of the mechanics of ciliary locomotion. Only some of the salient features will be discussed, however, because ciliary propulsion is a complex affair.

I have three concerns: the mechanics of an individual cilium, the difference between the velocity profile around a self-propelled body and an inert body, and the consequences of ciliary motion on mass or energy transfer. Figure 2.21 diagrams some aspects of the movement of cilia and the difference between the fluid velocity profiles around a self-propelled body *SP* and an inert body *IB* (passively settling in the water column). The beat of a cilium involves two distinct phases—a power (or effective) stroke and a recovery stroke. During the power stroke, the cilium operates as a rigid prismatic bar that is fully extended from the surface of the cell it is attached to and swings sharply backward by pivoting at its base (steps 1–3 in the lower left insert of fig. 2.21). The recovery stroke involves the propagation of a flexure from the base to the tip of the cilium. The cilium initially *hugs* the cell surface and subsequently extends progressively outward and forward. The propagation of flexure reduces the surface area projected in the direction of the cilium's recovery motion (steps 4–6 at the lower right of fig. 2.21). Generally, the recovery stroke takes longer than the effective stroke. Both are active processes involving the expenditure of energy. The collaborative action of cilia causes the organism to swim or

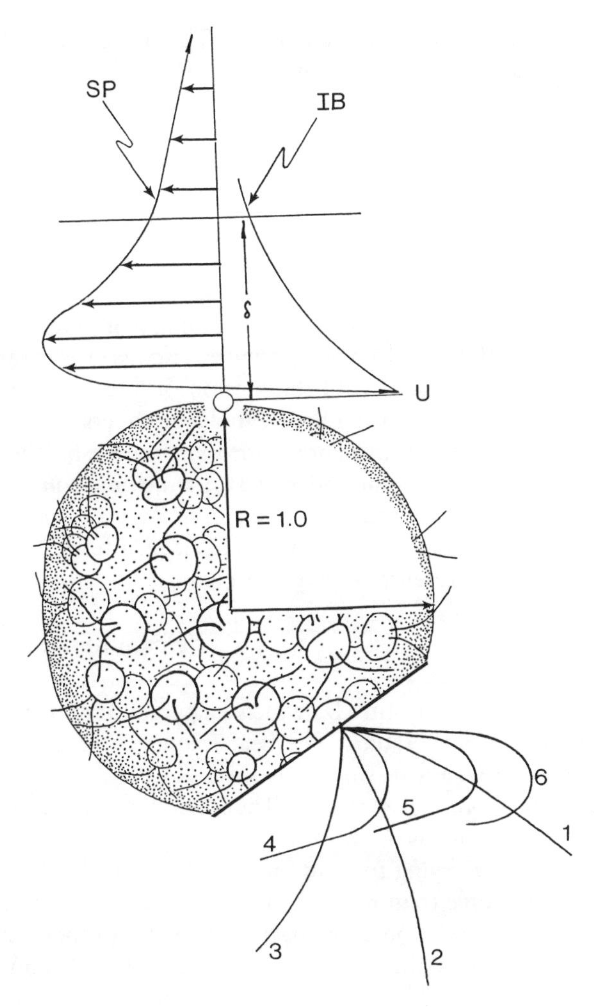

Figure 2.21. Schematics for the movement of cilia and the difference between the fluid velocity profiles around a self-propelled body *SP* and an inert body *IB* (passively moving) in the water column. Both bodies are moving (to the right) with a speed *U* relative to the ambient fluid surrounding them. The lower right portion of the figure diagrams the power or effective stroke (steps 1–3) and a recovery stroke (steps 4–6) of a cilium on the *SP* body. In the upper portion of the diagram, the velocity profile of an *SP* body is shown by a series of arrows. The direction of these arrows indicates the motion of the ambient fluid relative to the *SP* body; the lengths of the arrows reflect relative fluid speeds. The overall velocity profile is shown as a solid curved line bounding the series of arrows. In terms of the inert body, the thickness of the boundary layer is denoted by δ. The decrease in relative fluid flow within the boundary layer is indicated by the curved solid line.

maintain its position in the water column. The movements of cilia on multicellular organisms minimize mutual interference by means of metachronic waves of activity that superficially appear much like the motion of grass stalks blown in the wind. Four main patterns of metachronism have been recognized and named by Knight-Jones (1954; see also Blake and Sleigh 1974, figs. 2–3). Although flagella and cilia have the same internal structure and general function, they achieve their function differently. Flagella move fluid parallel to their axes (their bending wave is symmetric along their length); cilia move fluid perpendicular to their axes (their bending pattern is highly asymmetric along their length).

The propulsion of microorganisms involves viscous effects rather than the inertial effects associated with the motion of large or fast animals. This may be shown by comparing the Reynolds numbers associated with the movements of volvocalean plants with that for a fish. The fish may have a Reynolds number on the order of 10^6; a moving *Volvox* typically has $Re \leq 10^{-2}$. The Reynolds number of a cilium Re_c is even less than that of the organism the cilium is attached to. This is seen from the following equation and some representative data for cilia:

$$Re_c = \frac{\phi L r}{\upsilon}, \tag{2.25}$$

(cilium Reynolds number)

where ϕ is angular velocity (in Hz, which has units of s^{-1}), L is length, and r is the characteristic radius. The beat frequency of cilia can range between 1 and 40 Hz; L can vary between 5×10^{-6} and 10×10^{-6} m; and r is generally about 0.1×10^{-6} m. The Reynolds number of an *average* cilium Re_c, therefore, is on the order of 10^{-7} to 10^{-5}, which is comparable to that of a fish trying to swim in asphalt. Accordingly, cilia must exert considerable force to move in their highly viscous medium.

We might expect a relation between the maximum propulsion velocity U_P of ciliated organisms and the product of the average cilium length L and maximum angular velocity ϕ, particularly since the radii of cilia vary little among taxa and the kinematic viscosity υ of water varies only with significant changes in fluid temperature. The angular velocity of cilia can be determined by means of a microscope equipped with a stroboscopic light source, although ϕ and U_p cannot be measured simultaneously, for rather obvious reasons. Figure 2.22 plots U_P versus ϕL for six animal and eight plant species. Since all the data points for "plants" come from the Volvocales, the scaling relation between these two variables is less meaningful than desirable because of phyletic effect. With this caveat, the relation between the velocity of propulsion and ϕL is anisometric and positive. Specifically, $U_P = 0.002\ (\phi L)^{1.11}$ ($r^2 = 0.95$); $\alpha_{RMA} = 1.14 \pm 0.03$). Although an increase in either the angular velocity or the length of cilia correlates with a absolute increase in the propulsion velocity, we cannot claim that this correlation is the result of cause and effect, since the effects on U_P of the number of cilia per organism as well as shape were neglected.

Earlier I said that the velocity field around an inert body differs from that around an actively moving body (see upper portion of fig. 2.21). Both the inert plant and the self-propelled plant have a layer of fluid adjacent to the external surface of the plant body, moving at the same velocity as the plant. This phenomenon is called the no-slip condition. A macroscopic counterpart is seen for water flowing through tubes—the velocity gradient increases parabolically from the tube's inner surface (where the no-slip condition prevails) toward the center of the tube (where the maximum flow speed is achieved). A detailed consideration of the force distribution required to satisfy the no-slip condition (which need not concern us here) reveals that the velocity field near an inert plant decreases as the logarithm of the perpendicular distance from the plant (see fig. 2.21). Thus the fluid velocity in the flow field immediately adjacent to the plant is identical to the settling velocity. This has been called the *shielding* or

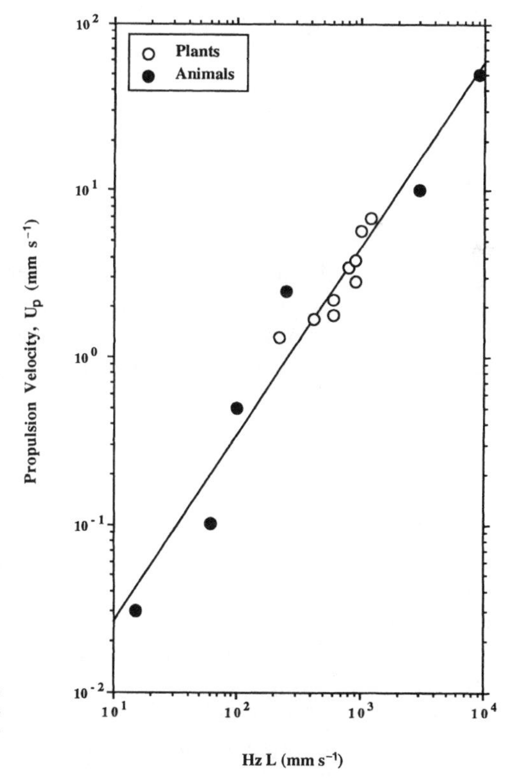

Figure 2.22. Log-log (base 10) plot of propulsion velocity U_p versus the product of the angular velocity ϕ and length L of cilia for unicellular and multicellular plants and animals. Solid diagonal line denotes the regression curve obtained for these data by LS regression analysis.

capturing effect. As we move progressively farther from the surface of the plant, the velocity field logarithmically decays to zero. By comparison, the velocity field around a self-propelled plant is influenced by the effective and recovery strokes of cilia. During the rapid effective stroke, each cilium disturbs a comparatively large volume of fluid; during the slower recovery stroke, each cilium shears off the fluid binding to its surfaces. The disturbance of the velocity field owing to the movements of cilia is restricted to a relatively thin shell of fluid enveloping the plant. Beneath this zone, however, immediately at the surface of a self-propelled plant, the no-slip condition still prevails. That is, a thin layer of fluid is pulled along with the plant. Accordingly, right at the surface of the plant body, the velocity fields around a passively settling plant and an actively moving one are identical. Farther from the surface of the actively moving plant, however, the fluid is carried in the opposite direction from the direction of movement of the plant (see upper portion of fig. 2.21). Each cilium possesses a region of influence (with height δ from the plant's surface) that counteracts the no-slip condition, and the velocity of fluid flow decays as a function of inverse distance from the surface of the plant. Beyond the region of influence, the efficacy of cilia motion diminishes and the velocity drops as a function of the inverse of the square of the distance from the surface of the plant (i.e., above the region of influence, the effects of the external surface of the plant once again dominate fluid flow). There is no sharp boundary delimiting the zone of influence of cilia from the zone of influence of the plant's external surface, although Blake and Sleigh (1974) suggest that the distance δ of the boundary is equal to approximately one-half the projected diameter of a self-propelled body.

Comparing the velocity fields around inert and mobile plants leads us to suspect that the rate of mass or energy transfer is enhanced by ciliary motion. The thickness of the film of fluid traveling with a plant is reduced by the effective strokes of cilia. Recall from our discussion of Fick's law of diffusion (see sec. 2.4, especially eq. 2.6) that the flux density of mass is inversely proportional to the distance x between the ambient and surface concentration of the molecular species in question. This distance, which is a measure of the resistance to passive diffusion, is now seen to be reduced by ciliary movements in a very sophisticated way. All things being equal, the resistance to diffusion should be proportional to cilia length L and inversely proportional to cilia angular velocity ϕ. However, the product of L and ϕ is positively correlated with the propulsion speed (see fig. 2.22). Thus, if the strategy of an organism is to move at moderate speeds and benefit physiologically from reduced resistance to passive diffusion, then a good tactic is to increase ϕ and reduce L. A complementary tactic is to increase the number of cilia to accommodate the reduction of L.

Perhaps it is not coincidental that large ciliates also have a large number of cilia.

2.8 Hydrodynamic Forces and the Mechanics of Attached Aquatic Plants

In this section we examine the hydrodynamic forces exerted on plants attached to a substrate and the way aquatic plants cope mechanically with these externally applied forces. The objective is to appreciate the different force components that constitute the net hydrodynamic force operating on an attached plant as well as the mechanical consequences of these forces on the plant body.

Figure 2.23 summarizes the hydrodynamic forces exerted on an aquatic plant attached to a substrate. The net force exerted on the plant body is the vector resultant of two orthogonally opposed forces, the lift force F_l, which operates perpendicular to the direction of water flow, and two horizontal force components, the pressure force F_D and the accelera-

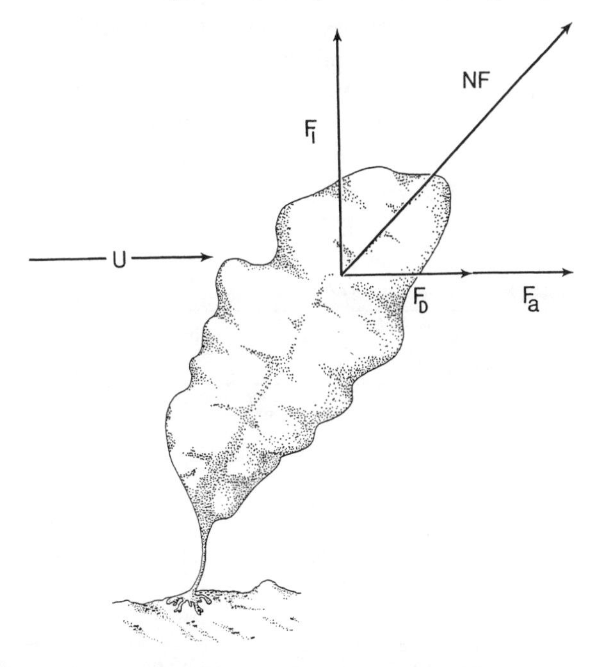

Figure 2.23. Schematic of the hydrodynamic forces exerted on an aquatic plant attached to a substrate. The net force *NF* exerted on the plant body as a consequence of fluid flow *U* is the vector resultant of two orthogonally opposed forces, the lift force F_l, which operates perpendicular to the direction of water flow, and two horizontal force components, the pressure force F_D and the acceleration force F_a.

tion force F_a. The following equations reveal the variables that account for these three forces:

(2.26a) $$F_l = \frac{1}{2}\rho U^2 S_{plan} C_l$$ (lift force)

(2.26b) $$F_D = \frac{1}{2}\rho U^2 S_P C_D$$ (pressure force)

(2.26c) $$F_a = \rho C_m Va,$$ (acceleration force)

where $\frac{1}{2}\rho U^2$ is the dynamic pressure, S_{plan} (the planform area) is the area of the plant projected perpendicular to the direction of flow, C_l is the lift coefficient, S_p is the surface area of the plant projected against the direction of ambient flow, C_D is the (pressure) drag coefficient, C_m is the inertia coefficient, V is the volume of the plant, and a is the fluid acceleration. Table 2.9 gives values for C_l and C_m for a few simple geometries. Note that when the fluid is accelerating, F_D and F_a operate in the same direction (in the direction of the flow), as diagrammed in figure 2.23. When the fluid is decelerating, F_D continues to act in the direction of flow, but F_a acts in the opposite direction (against the flow).

To appreciate the relative importance of F_D, F_a, and F_l, we shall compare the magnitudes of these forces in seawater for a sphere and a terete cylinder whose length is oriented perpendicular to the direction of ambient flow. For convenience, we will stipulate that the volumes of these two

Table 2.9. Approximate Coefficients of Drag, Inertia, and Lift for Sphere, Vertical Terete Cylinder, and Vertical Long Flat Plate ($L/w = 10$; $w/t = 10$)

Geometry	Drag C_D	Inertia C_m	Lift C_l	C_m/C_D	C_l/C_D
Sphere	0.47	1.5	1.2	3.20	2.5
Cylinder	0.75	—	1.2	—	1.6
$L/D = 1.2$	0.75	1.62	1.2	2.16	1.6
2.5	0.75	1.78	1.2	2.37	1.6
5.0	0.75	1.90	1.2	2.53	1.6
9.0	0.75	1.96	1.2	2.61	1.6
∞	0.75	2.00	1.2	2.67	1.6
Plate	1.98	8.50	—	4.30	—

Source: Values from Hoerner 1965; Sarpkaya and Isaacson 1981; Vogel 1981; Denny 1988.
Note: Values for coefficients apply for $Re \approx 10^4$ and infinite fluid (geometries placed far from any solid surface); values for C_D for sphere and cylinder for various Re are plotted in figure 2.13.
Symbols: L = length; D = diameter; w = width; t = thickness.

geometries are equivalent ($V = 3.9 \times 10^{-5}$ m^3). Thus the sphere has a diameter $D = 0.042$ m. In terms of the shape of the terete cylinder, let the diameter $D = 0.01$ m and the length $L = 0.5$ m. For both geometries, an ambient flow speed of 5 m s^{-1}, accelerating at the rate of 1 m s^{-2}, will be assumed. Since the density of seawater is $\rho = 1{,}025$ kg m^{-3}, based on eq. (2.26), we calculate $F_D = 48$ N, $F_a = 0.08$ N, and $F_l = 2.0$ N for the cylinder (where N denotes newtons). The terete cylinder has a large projected surface area S_p relative to its planform surface area S_{plan}. Consequently, the pressure force component is large compared with the lift component. For the sphere, $F_D = 10.7$ N, $F_a = 0.06$ N, and $F_l = 21.4$ N. Since the S_{plan} of a sphere equals πR^2, the lift force component for this geometry is comparatively large.

In a more general way, two dimensionless ratios show the relative importance of the three hydrodynamic force components in terms of plant geometry and shape and the ambient flow regime (see Denny 1988, 170–71):

(2.27a) $$\frac{F_a}{F_D} = \frac{2C_m}{C_D}\frac{V}{S_P}\frac{a}{U^2}$$

(2.27b) $$\frac{F_l}{F_D} = \frac{C_l}{C_D}\frac{S_{plan}}{S_p}.$$

From table 2.9, C_m/C_D and C_l/C_D are found to be greater than one for a sphere, cylinder, or long, flat plate. For a rigid cylindrical or platelike aquatic plant, V/S_p tends to be very small (≤ 0.01). Therefore F_a will likely dominate over F_D only when $a \geq 100\ U^2$. For very flexible plants, S_{plan}/S_p tends to be very large, and therefore F_l likely dominates F_D regardless of fluid speed and acceleration.

Recall that the magnitude and direction of the net force *NF* operating on the aquatic plant diagrammed in figure 2.23 are those of the vector resultant of three forces. The magnitude and direction of the net force exerted on an aquatic plant can be estimated from the Morison equation (see Denny 1988, chap. 11):

(2.28a) $$\text{Net force} = \left[\left(\frac{1}{2}\rho U^2 S_p C_D + \rho C_m Va\right)^2 + \left(\frac{1}{2}\rho U^2 S_{plan} C_l\right)^2\right]^{0.5}$$

(2.28b) $$\text{Direction} = \arctan\left(\frac{\rho U^2 S_{plan} C_l}{\rho U^2 S_p C_D + 2\rho C_m Va}\right).$$

This equation assumes that the drag and acceleration force components operate independently (Morison et al. 1950). Although this assumption may be in error, the Morison equation is widely used and provides a reasonable approximation for an otherwise very complex situation.

Aquatic plants cope with hydrodynamic forces in a variety of ways. Most species are composed of extremely flexible and extensible tissues that permit them to bend easily, thereby reducing drag. Flexure also permits plants to absorb large amounts of energy before they break. For example, the stipe of the sea palm (*Postelsia palmaeformis*), a brown intertidal alga, has a Young's modulus (measured in tension) between 5×10^6 and 10×10^6 N m^{-2}. In spite of its low breaking stress ($\approx 1 \times 10^6$ N m^{-2}), a comparatively large amount of energy is required to break the stipe ($\approx$ 100 kJ m^{-3}) because of its extreme extensibility (20%–25%) (Holbrook, Denny, and Koehl 1991). Koehl and Wainwright (1977) report that the mean work per volume required to break the stipes of the brown alga *Nereocystis luetkeana* is 0.67 MJ m^{-3} (see also Delf 1932; Charters, Neushul, and Barilotti 1969; and Wheeler and Neushul 1981 for additional data and references). This breaking energy is very near that for wood or cast iron. Indeed, representative values of the Young's modulus (measured in tension), breaking stress, breaking energy, and extensibility of wood are 12×10^9 N m^{-2}, 1.15×10^8 N m^{-2}, 0.6 MJ m^{-3}, and 1%, respectively (Niklas 1992). Note that wood extends about 1% before it breaks compared with the 20%–25% extensibility of the algal tissue. Thus, in contrast to the very flexible tissues used to construct most aquatic metaphytes, the principal mechanical tissue of the largest vascular land plants is extremely rigid.

Figure 2.24 summarizes what is meant by Young's modulus, breaking stress, breaking energy, and extensibility. I shall explore these features of materials in greater depth in chapter 3. For now, however, observe that the Young's modulus E is the ratio of stress σ to strain ε measured in the elastic range of behavior of a material. Since an external force applied to a small piece of material will produce a greater deformation than the same force applied to a bigger piece of the same material, the behavior of materials in response to forces can be understood only if the magnitude of the externally applied force F is normalized with respect to the area A through which it acts. Engineers call this normalized force stress σ ($\sigma = F/A$). By the same token, the extensions in the dimensions of a specimen subjected to tensile forces or the contractions in the dimensions of a specimen submitted to compressive forces need to be normalized with respect to the previous dimensions of the specimen. When this is done, strain ε is computed. Strain is dimensionless. Since $E = \sigma/\varepsilon$, the Young's modulus has the same units as stress (force per area). The simplest way to express strain is the extension ration λ, which is the ratio of the altered dimension x_o to the original dimension x of a specimen ($\lambda = x_o/x$). Extensibility is λ

$\times$ 100%. Above a certain stress level, elastic materials either plastically deform or break. The stress level at which either of these responses occurs defines the *proportional limit* of a material—the limit beyond which σ and ε are not proportionally constant ($E \neq$ constant). The stress level at which an elastic material breaks is called the breaking stress σ_B. Unfortunately, we cannot adduce a priori the relation between E and σ_B. That is, materials can have very high E and very low σ_B or vice versa. Thus, the Young's modulus and breaking stress of each type of material must be empirically determined.

The importance of the extensibility of aquatic plant tissues becomes obvious once we recognize that energy is stored in materials when they deform. Highly extensible materials can store large amounts of energy before they break. Within the elastic range of a material's behavior, the stored energy is used to restore the material to its original dimensions after externally applied forces are removed. Bend a branch, and it will vibrate back to its original (equilibrium) position by expending the stored (elastic) energy when you remove your hand. Beyond a certain stress level,

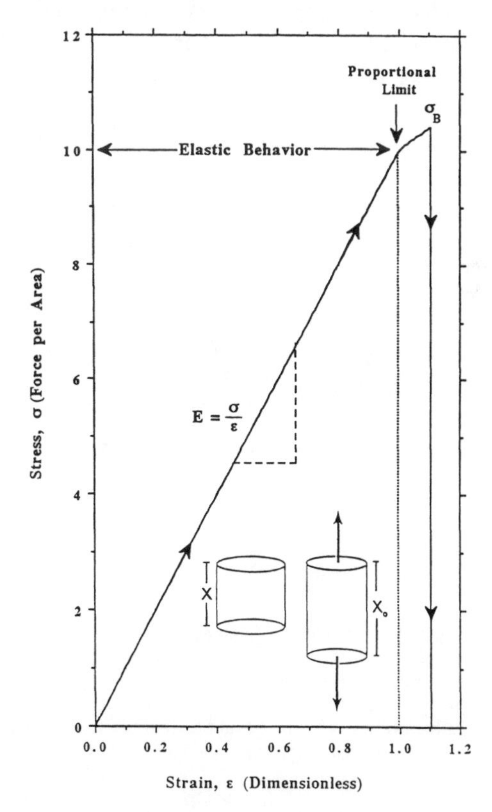

Figure 2.24. Relation between stress σ and strain ε for a hypothetical cylindrical specimen (with an original length x) of a linearly elastic material placed in uniaxial tension (see inset diagram). Stress is the applied tensile force divided by the surface area through which the force operates. Strain is computed here as the extension ratio λ (the ratio of the altered dimension x_o to the original dimension x of the specimen; $\lambda = x_0/x$). Within the material's elastic range of behavior, the quotient of σ and ε (which is the slope of σ versus ε) is constant and provides the Young's elastic modulus E of the material ($E = \sigma/\varepsilon$ = a constant). Beyond the proportional limit (denoted by the dashed vertical line), the quotient of σ and ε is shown to decline. The breaking stress σ_B is the stress at which the specimen of the material breaks.

however, a branch will break. The breaking energy B is the amount of energy stored in a material just before breakage occur (or just before a material undergoes plastic deformation). The value of B is computed from the area under a force deformation curve. Obviously, the hydrodynamic forces summarized in figure 2.23 generate mechanical forces within plant tissues. The larger the net force exerted on a plant, the larger these internal mechanical forces. Beyond a certain stress level, tissues will break and the plant may die. Since stress is force normalized with respect to area, one solution to high net hydrodynamic forces is to grow mechanical support members with compensatorially larger cross-sectional areas. We might also expect higher Young's moduli and extensibilities for plants that experience the full force of waves compared with plants that are sheltered. The sea palm, like other intertidal macrophytes, appears to observe these principles. The stipes of isolated plants have larger basal cross-sectional areas and higher E and extensibility than plants growing in clumps (Holbrook, Denny, and Koehl 1991). Isolated plants feel the full force of waves; clumped plants can mechanically insulate one another.

The sea palm illustrates the importance of the shape of the mechanical support members of aquatic plants in dealing with hydrodynamic forces. Figure 2.25 shows the morphology of the sporophyte of the sea palm, which can grow from 50 to 75 cm in height. Each plant consists of a hollow cylindrically tapered stipe, attached to the substrate at its base by

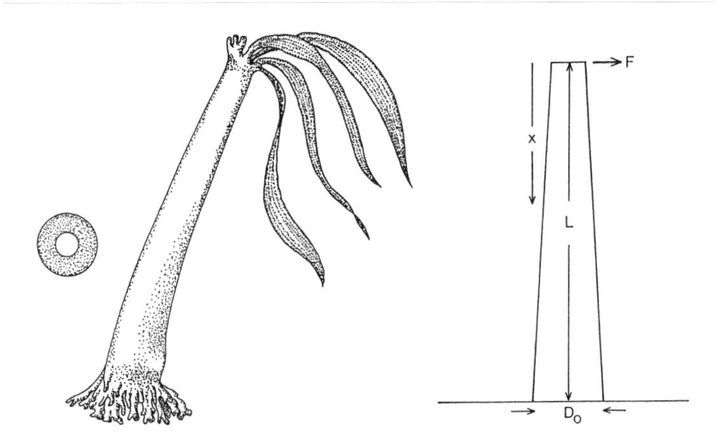

Figure 2.25. Schematic diagram of the plant body of the sea palm (*Postelsia palmaeformis*), an intertidal brown alga, subjected to waves. The sea palm consists of a tubular stipe attached to a substrate by a rootlike structure, called the haptonema, that bears a number of leaflike fronds at its distal end. The stipe of the sea palm subjected to a force F applied at its distal end mechanically operates as a tapered cantilevered beam with length L and basal diameter D_0. The bending moment applied to the stipe is the produce of F and the distance x from the point where F is applied.

a rootlike haptonema, and a crown of strap-shaped fronds at the free end of the stipe. Mechanically, the stipe operates like a cantilevered beam. It is anchored at one end and free to bend at the other. For simplicity, we shall deal with bending as if a force F is applied only to the free end of the stipe (fig. 2.25). That is, F is assumed to operate as a point load. Under these conditions, engineering theory shows that F is related to the maximum stress σ_{max} that will occur in each cross section according to the following equation:

(2.29a)
$$\sigma_{max} = \frac{FxD}{2I},$$

where x is the distance measured from the free end of the cantilevered stipe, D is the diameter of the stipe measured at distance x, and I is the second moment of area. For a hollow terete stipe, $I = (\pi D^4/64)\,(1 - k^4)$, where k is the ratio of the outer to the inner diameter of the stipe. Thus

(2.29b)
$$\sigma_{max} = \frac{32}{\pi}\,\frac{Fx}{D^3(1-k^4)} \propto \frac{Fx}{D^3},$$

which shows that, for any F, the maximum bending stress will increase as a linear function of x.[6] This is σ_{max} will increase from the tip to the base of the stipe. An ideal situation results when σ_{max} is constant along the length of the stipe and does not exceed the breaking stress. Equation (2.29b) shows that σ_{max} can be kept constant provided $D \propto x^{1/3}$. That is, the diameter of the stipe should increase as a function of the cube root of distance x. This scaling relation has been demonstrated for the fire coral, *Millepora complanata* (see Denny 1988, 206, fig. 13.8). However, the taper of the stipe of the sea palm does not comply with this scaling principle. Rather, stipe diameter increases as a linear function of the distance x (Holbrook, Denny, and Koehl 1991). Superficially, this is not a particularly good mechanical design. The mechanical behavior of the sea palm, however, capitalizes on the fact that hollow tubes crimp when subjected to large bending forces. As a hollow tube is subjected to an increasing bending force, its original terete cross section assumes a more elliptical geometry (the major axis of each elliptical cross section through the tube is aligned perpendicular to the radius of curvature). Eventually this ovalization causes the hollow tube to buckle locally, an effect called Brazier buckling. Provided the bending stresses that develop within the wall of

6. The reader familiar with solid mechanics will note that F_x is the bending moment (which has units of force times length). At the tip of the stipe, $F_x = 0$; at the base of the stipe, $F_x = F_l$, where L is the length of the stipe. Also, the maximum tensile and compressive stresses within each cross section will occur in the tissues just at the surface of the stipe.

the tube do not exceed the proportional limit of the materials used to construct the tube, the crimped tube will return to its original position when the bending force is removed. Hollow tubes will crimp at the point where the wall thickness t is the most reduced relative to the external diameter D of the tube. That is, $(t/D)_{min}$. Although t/D varies as a function of distance x, my measurements of six stipes indicate that t/D reaches its minimum value very near the base of most stipes. At this location, $t/D \approx 0.2$.

With little effort we can calculate the speed of a wave required to cause the stipe of a sea palm to crimp and deflect. This calculation is based on the fact that the critical wall stress σ_c at which crimping occurs for most hollow tubes is approximately equal to 0.5 $E(t/D)$, where t is the wall thickness and D is the outer diameter. Holbrook, Denny and Koehl (1991) reported that the average apparent Young's modulus E (measured in bending) of isolated sea palm stipes is 14.5×10^6 N m^{-2}. Thus $\sigma_c \approx (0.5)(14.5 \times 10^6 \text{ N m}^{-2})(0.2) \approx 1.45 \times 10^6 \text{ N m}^{-2}$. This value can be substituted for σ_{max} in eq. (2.29). Noting that $t/D = 0.2 = 1 - k$ and that the stipe will crimp where its wall is thinnest (typically at the base), from eq. (2.29) we find that $\sigma_c = 34.5\, FL\, D^{-3} = 1.45 \times 10^6 \text{ N m}^{-2}$. From eq. (2.26b) we see that

$$\frac{1}{2}\rho U^2 S_p C_D = \left(1.45 \times 10^6 \frac{\text{N}}{\text{m}^2}\right)\left(\frac{D^3}{3.45L}\right). \tag{2.30a}$$

Solving this equation for U, we obtain

$$U = \left[\frac{\left(1.45 \times 10^6 \frac{\text{N}}{\text{m}^2}\right)\left(\frac{D^3}{3.45L}\right)}{\frac{1}{2}\rho S_p C_D}\right]^{0.50}. \tag{2.30b}$$

For a plant whose stipe measures 0.20 m in length and 0.01 m in diameter at its base, the projected surface area of the crown of fronds S_p is near 0.02 m^2. A reasonable estimate of the drag coefficient C_D is 0.8. Thus

$$U = \left\{\frac{\left(1.45 \times 10^6 \frac{\text{N}}{\text{m}^2}\right)\left[\frac{(0.01\text{ m})^3}{3.45(0.20\text{ m})}\right]}{\frac{1}{2}\left(1{,}025 \frac{\text{kg}}{\text{m}^3}\right)(0.02\text{ m}^2)(0.8)}\right\}^{0.50} \approx 0.5\,\frac{\text{m}}{\text{s}}. \tag{2.30c}$$

Wave speeds about 10 m s^{-1} are not uncommon on many wave-swept shores. Thus, based on eq. (2.30), we would expect the stipe of a typical sea palm to deflect with almost every passing wave. Indeed, Holbrook, Denny, and Koehl comment that "watching a stand of *Postelsia* in the face of a large wave is similar to watching a field of grass subjected to a strong gust of wind: the plants bend almost prostrate as the wave hits, only to spring back to their original posture once it has passed" (1991, 40).

It is sad that we know comparatively little about the mechanical designs of other aquatic macrophytes and even less about the scaling of these organisms. More emphasis has been given to terrestrial plant biomechanics and scaling. These features are treated in the next chapter.

2.9 Summary

The biology of aquatic plants was shown to be dominated by the capacity of the water column and the materials dissolved and suspended within it to absorb light and by the high density of water, which essentially makes aquatic plants weightless.

The elaboration of photosynthesis surface areas in water therefore is desirable, because access to light limits growth. The elaboration of surface area is possible because specialized mechanically supportive tissues are typically not required for aquatic plants. The most significant attenuation of light in the water column was seen to be the result of the absorption of light by suspensions of phytoplankton. Light interception was shown to be influenced by cell size as well as cell shape, pigment concentration per cell, and the number of cells per volume of suspension. Thus, light interception by suspensions of phytoplankton is a scaling (size-dependent) phenomenon. Cell size plays a role because of the package effect. Pigments contained in discrete packages or units (chloroplasts, cells, or colonies of cells) are less effective at harvesting light than an equivalent amount of pigments in solution. Biophysical models, based on Bouguer's law, were used to show that the average absorption cross section of an algal suspension decreases with cell size. However, since pigment concentration declines as a function of plant cell size, the light absorption by an individual cell scales to the cross-sectional area of the cell. Also, algal cells absorb less light, relative to their volume, as they increase in size.

The ability to exert forces on submerged plants, supply dissolved metabolites to living cells, and thermally buffer submerged plants against rapid changes in temperature were seen to be additionally important physical properties of water. These properties were quantified by three physical laws: Fick's law of diffusion, Newton's law of viscosity, and Fourier's law of heat transport. Each describes a flux density in terms of a proportionality factor and a driving force (a concentration gradient). The transport equations were used to show that the rate of mass or energy transfer between a plant and its fluid medium depends on the rate at which ambient fluid moves with respect to the plant. Different systems of fluid flow were correlated with the aid of dimensionless groupings of numbers, among which the best known is the Reynolds number *Re*. The effect of *Re* on mass and heat transport was explored. Under zero-flow conditions ($Re \leq 0$), small oblate spheroids were shown to have the optimal geometry for mass or heat transfer. Under the conditions of creeping flow ($0 < Re \leq 1$), as when phytoplankton passively settle out of the water column, the most efficient size and geometry among those considered was shown to be the small sphere. The next most efficient was the small thin disk. The least efficient geometry was found to be the small flat plate. When actively mobile plants were considered (e.g., Volvocales), the speed of self-propulsion U_P and the product of the angular velocity ϕ and length L of cilia were seen to be anisometric and positive: $U_P = 0.002\ (\phi L)^{1.11}$ ($r^2 = 0.95$), indicating that an increase in either the angular velocity or the length of cilia results in an absolute increase in the propulsion velocity. The velocity fields around a passively settling and self-propelled plant were shown to differ significantly and to influence mass transport. The disturbance of the velocity field owing to the movements of cilia is restricted to a comparatively thin zone of fluid enveloping the plant. Beneath this zone, immediately at the surface of a self-propelled plant, the no-slip condition prevails. That is, a thin layer of fluid is being pulled along with the plant. However, farther from the surface of the actively moving plant, the fluid is carried in the opposite direction from that of movement, bringing fresh supplies of nutrients to the plant surface.

Finally, the nature of the hydrodynamic forces exerted on plants living in the intertidal zone was reviewed. These forces were related to the magnitude of the mechanical stresses that develop in plant tissues. The role of tapering of algal stipes, as well as the elasticity and extensibility of algal tissues, was treated quantitatively. The tapering of the girth of the stipe of the sea palm (*Postelsia palmaeformis*) was shown to confer a mechanical benefit through Brazier buckling when plants are struck by waves.

3 Terrestrial Plants

3.1 Introduction

This chapter deals with the scaling of terrestrial plants, much of which reflects the physiological requirements to conserve water and to deal with the influence of gravity. This stands in marked contrast to the scaling of aquatic plants, where access to light and the effects of large hydrodynamic forces tend to dominate the size-dependent variations in shape and structure.

The availability and conservation of water are important factors affecting the growth of terrestrial plants. For example, the productivity of plant communities is highly correlated with mean annual rainfall, although other factors, such as taxonomic composition, soil chemistry, length of growth season, and the like, clearly influence plant growth. Figure 3.1 plots the mean net annual primary production (*Prod,* measured in units of g m^{-2} yr^{-1}) versus the mean annual precipitation (*Prec,* measured in units of cm) for thirty-five plant communities. These data were taken from a variety of sources and likely have significant measurements of error. Nonetheless, regression of these data indicates that $Prod \propto Prec^{0.98}$ ($r^2 = 0.88$; $\alpha_{RMA} = 1.04 \pm 0.05$). Thus, on the average, the productivity of terrestrial plant communities is an isometric function of mean annual precipitation. This is hardly surprising, since plants typically lose approximately 149 mol of water for every mole of carbon dioxide they photosynthetically fix in the form of carbohydrates. The average grass, for example, loses its own weight in water every twenty-four hours in hot, dry weather.

The indirect dependency of photosynthesis and therefore growth on transpiration has played a critical role in the evolution of land plants. One of the more obvious results of this dependency is that the elaboration of external surface area is restricted. Although large ratios of external surface area to volume are desirable in terms of light interception and the passive diffusion of carbon dioxide into the plant body, large external surface area also helps water escape. The external deposit of waxy materials in the form of a cuticle reduces the extent of water loss but also significantly reduces how much carbon dioxide can be absorbed from the atmosphere. Neither nature nor the best organic chemist has developed a material that is permeable to CO_2 and at the same time impermeable to H_2O. The evolutionary solution to this biophysical dilemma was to internalize large surface areas within the land plant body in the form of aerenchyma. The exchange of gases between the atmosphere and these internal surface areas required pores or stomata flanked by guard cells whose ability to dilate and contract regulates the rate at which water and carbon dioxide are exchanged with the atmosphere.

Owing to the importance of the cuticle and stomata to plant growth and survival, I shall devote considerable attention to the scaling of these features (sects. 3.3 and 3.4). For the time, however, a few points are worth noting. Plants or stands of plants with large leaf surface areas typically transpire more than those with smaller leaf areas, although the relation of leaf area to water loss is far from simple. For example, removing part of the leaf area of an individual plant does not invariably produce a proportional reduction in transpiration. In fact, transpiration per unit of leaf area may increase when leaves are removed (Miller 1938). Apparently, removing leaves exposes the remaining leaf area to more air movement, decreasing resistance to mass transfer by thinning the boundary layer. Some authors additionally claim that the transpiration rate increases because of a rise in the ratio of the surface area of roots to shoots, which supplies more water to individual leaves. The linear and positive relation

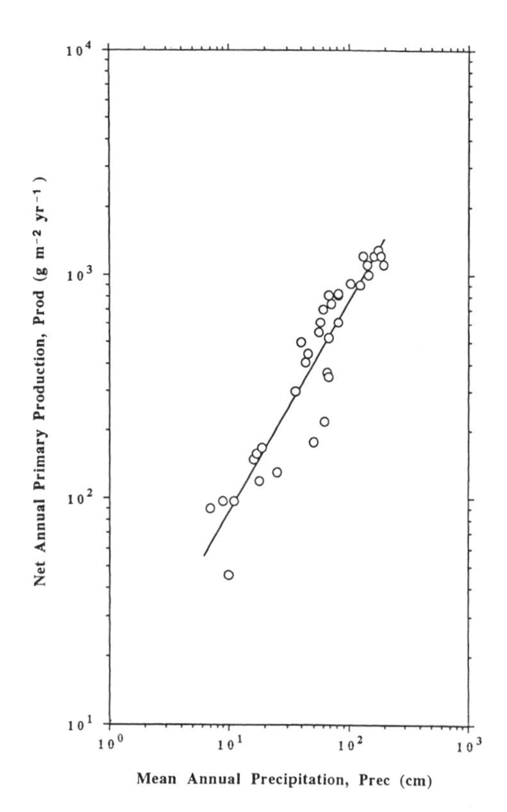

Figure 3.1. Log-log (base 10) plot of mean net annual primary production *Prod* versus the mean annual precipitation *Prec* for plant communities. Data from a variety of sources.

between transpiration rates and the ratio of the dry weight of roots to leaf area (Parker 1949; Kramer and Kozlowski 1979) is frequently used to support this claim. However, transpiration and the ratio of root dry weight to leaf area are autocorrelated, since both variables share the same denominator (leaf surface area). Be that as it may, it is extremely important to consider the ratio of subterranean to aerial surface areas when assessing the scaling of water loss from individual plants. This ratio can change over the lifetime of a plant as well as seasonally or diurnally (e.g., seasonal shedding or transient curling of water-stressed leaf laminae; see Stålfelt 1956; Kozlowski 1974; Kramer 1983).

Although the cuticle provides the principal resistance to mass transfer, water permeability is not directly correlated with cuticle thickness but is more a matter of the chemical composition and the thickness of the waxy layer within the cuticle (Raven 1984, 107; Kolattukudy 1981). Even though the cuticles of xerophytes tend to be thicker than those of mesophytic species (likewise the cuticles of sun leaves tend to be thicker than those of shade leaves borne on the same plant), the available data indicate that water permeability is reduced by approximately one-half for every additional two-carbon unit added to the paraffins within the waxy layer. The cuticles of leaves and stems have comparatively thick waxy layers composed of relatively long-chain hydrocarbons. By contrast, the cuticles covering subterranean primary roots and the cells lining intercellular air spaces within leaves typically have a low paraffin volume fraction or possess hydrocarbons with low molecular weights (hence shorter average chain lengths), thereby aiding the absorption of carbon dioxide as well as the evaporation of water while offering a modicum of resistance to water transfer. The cuticles of many plant species become more permeable to water when wet or as the temperature rises. Data on the scaling of cuticle chemistry and ultrastructure are regrettably unavailable for analysis.

In considering water conservation, we cannot neglect community structure. The mere presence of a plant reduces transpiration, since every object obstructing fluid flow produces a boundary layer whose thickness resists the transfer of mass as well as heat. Thus, plants growing in stands or that have a clumped growth habit shelter one another from water loss by creating expansive boundary layers. Isolated plants, which are directly exposed to comparatively higher wind speeds in the same habitat, are at a disadvantage in this regard, although they have greater access to light. In chapter 2 I said that the boundary layer is progressively thinned as the ambient speed of the fluid increases. Since fluid speed increases with distance from a substrate, the height of a plant plays an important role regardless of whether plants grow as isolated individuals or in stands. The dependency of local wind speeds on distance from the substrate and the density of plantings sheds light on the functional significance of the prostrate and clumped growth habits of moss and liverwort gametophytes, which typically lack stomata, contrasting with the vertical growth habit of their sporophytes, which may have stomata. By hugging their substrates as well as clumping, the gametophytes create a boundary layer that only substantial wind speeds can erode.

By contrast, vertically growing moss sporophytes extend well beyond the boundary layers generated by their subtending gametophytes. The growth habit of the sporophyte reflects the desirability of spore dispersal, which often comes at the expense of enhanced transpiration. For these plants, stomatal resistance substitutes in importance for the boundary layer resistance. Figure 3.2 plots wind speed U (measured in units of cm s^{-1}) as a function of height h above ground (measured in units of cm) measured within a clump of the moss *Polytrichum juniperinum* exposed to three ambient wind speeds (5, 10, and 15 cm s^{-1}) in a wind tunnel. A diagram of a single vertical axis of the leafy moss gametophyte subtending the cylindrical sporophyte with spore capsule is provided for convenience.

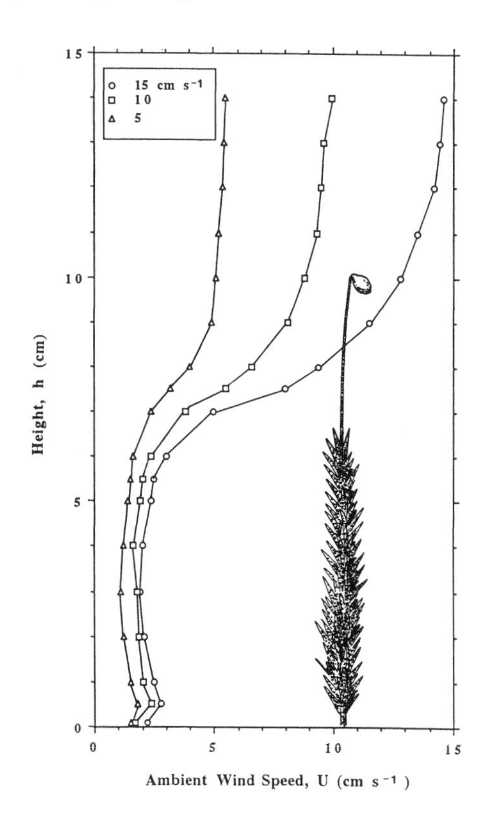

Figure 3.2. Wind speed U plotted as a function of height h above-ground measured within a clump of the moss *Polytrichum juniperinum* exposed to three ambient wind speeds (5, 10, and 15 cm s^{-1}).

For each of the three ambient wind speeds, the wind speed profile decreases nonlinearly toward the mound of leafy axes in the clump of the moss gametophyte. Regardless of which of the three ambient wind speeds we consider, the spore capsule is seen to extend well above the height at which local wind speeds diminish as a consequence of the boundary layer created by gametophytic axes. Accordingly, very short terrestrial plants (e.g., moss gametophytes, herbs, and the like) can rely to a limited extent on the boundary layer to resist water loss. Very tall plants typically cannot. As we shall see, for these plants, the resistances conferred by the cuticle and stomata are more important to transpiration than that provided by the boundary layer.

The resistance to mass transfer contributed by stomata varies as a function of aperture diameter and depth as well as number per surface area. When ambient wind speeds are comparatively high, stomatal resistance can be altered by varying aperture diameter. In still air, the resistance to mass transfer can be dominated or at least largely affected by the boundary layer, whose thickness increases as ambient wind speed decreases. We shall derive the equations that describe these reactions later. For the time, data reported by Bange (1953) for *Zebrina* leaves illustrate the effects of stomatal diameter and wind speed on the mass flux density of water vapor (i.e., transpiration). Figure 3.3 plots transpiration T (measured in units of kg m^{-2} s^{-1}) versus stomatal aperture diameter ξ (measured in units of μm) for leaves in still and moving air. Least squares regression of the data from leaves in still air yields $T = 2.1\ \xi^{0.44}$ ($r^2 = 0.81$, $N = 68$; $\alpha_{RMA} = 0.49 \pm 0.06$), while regression of the data from leaves in moving air shows that $T = 4.0\ \xi^{0.57}$ ($r^2 = 0.60$, $N = 82$; $\alpha_{RMA} = 0.74 \pm 0.05$), respectively. These intraspecific regression formulas show that leaf transpiration increases as stomata open, but that the effect is more pronounced for moving air, where the thickness of the boundary layer is significantly reduced.

Extensive vertical growth cannot rely exclusively on passive diffusion to irrigate photosynthetic cells with water. At some point, governed by a variety of factors that will be addressed later, bulk mass transport is desirable to sustain the rate at which water is lost by aerial portions of plants. The bulk flow of water requires low-resistance conduits. The hydroids of mosses and the tracheary elements within the xylem tissue fulfill this role. The protoplasts within these types of cells die, leaving behind cell wall lumens through which liquid water may pass with comparative ease. In addition to water, solutes absorbed by rhizoids or roots can pass quickly from the base of a plant to leafy photosynthetic organs, where passive diffusion serves adequately for short-distance transport. Consider that the speed of ascent of sap in xylem is on the order of millimeters per

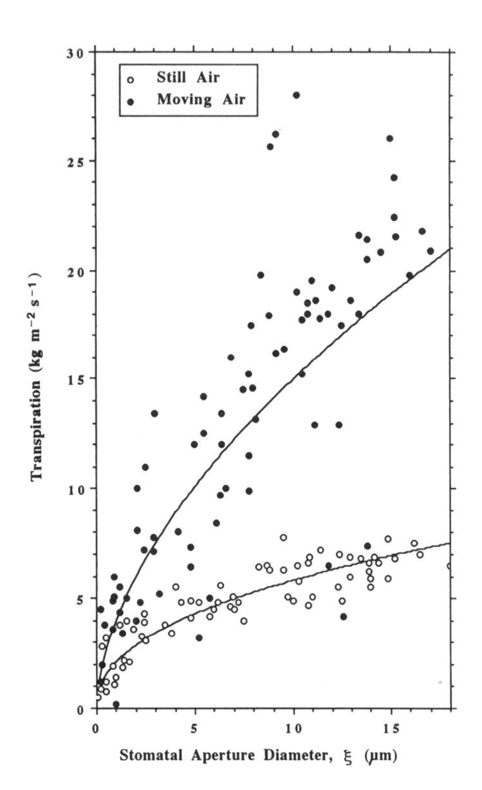

Figure 3.3. Transpiration T plotted as a function of stomatal aperture diameter ξ (for *Zebrina* leaves in still and moving air). Curved lines are the LS regression curves obtained for log-transformed data. Data from Bange 1953.

second. By contrast, the time required for 50% of a population of small solute molecules (with D_J on the order of 10^{-9} m^2 s^{-1}) to passively diffuse over a distance of 5 cm is roughly ten days.[1]

Finally, any treatment of the scaling of water acquisition, conservation, and transport must also address how plants cope with the static influence of gravity as well as dynamic wind pressures that can lead to mechanical instability. The issue of self-loading and wind loading can be dealt with in a variety of ways. In this chapter I shall emphasize the mechanical consequences of size-dependent variations in plant shape and internal structure, bearing in mind that, for organisms existing in a fluid at least one thousand times less dense then they, remaining light in weight yet strong is a primary design consideration. Not surprisingly, for its density wood is one of the strongest materials known. Unfortunately, many people fail to recognize that wooden structures are often safer than those built of

1. From Fick's law, time $= \dfrac{\text{distance}^2}{2.8\ (\text{coefficient of diffusion})} = \dfrac{(0.05\ \text{m})^2}{2.8(10^{-9}\text{m}^2\ \text{s}^{-1})} \approx 10$ days.

metal. A short historical digression makes this point. The Crystal Palace, designed in 1850 by Joseph Paxton, measured 1,848 feet long (more than a third of a mile) and 408 feet wide. It supported 293,655 panes of glass, and over the 140 days of the Great Exhibition, the Palace sheltered 6,063,986 people (roughly one-third the total population of the United Kingdom during the reign of Queen Victoria). In his diary, the historian and politician Thomas Macaulay called the Crystal Palace "a most gorgeous site; vast; graceful; beyond the dreams of the Arabian romances," yet he failed to note that the architectural glory of Victoria's reign was made of wood painted to look like steel!

3.2 Heat and Mass Transfer and Boundary Layer Resistances

I begin my treatment of terrestrial plant scalings by examining the consequences of size, shape, and geometry on the transfer of heat and mass in air. In chapter 2, the resistance to heat or mass transfer for aquatic plants was shown to be a function of the thickness of the boundary layer, a blanket of unmoving fluid enveloping an object submerged in a moving fluid (see eqs. 2.10 and 2.11). The dimensions of the boundary layer were shown to depend on the speed of ambient flow such that the thickness of the boundary layer and ambient flow speed are inversely related. This was shown indirectly by considering mass and heat transfer for passively settling phytoplankton and actively motile plants (e.g., Volvocales). The dependency of heat transfer on the rate of fluid flow is also true for terrestrial plants. The ambient speed of any fluid will affect how well plant organs absorb or dissipate heat in the fluid medium. In air, however, the effects of convection on heat transfer are more pronounced than in water, because the temperature differential between air and a land plant is typically much greater than that between water and an aquatic plant. There are three processes by which heat is dissipated: conduction, free convection, and forced convection.

Conduction involves the diffusion of heat through the random collisions of molecules. By contrast, convection involves the movement of air warmed by heated surfaces. When a plant intercepts sunlight, its surfaces warm up. The transfer of heat by conduction increases the temperature of the boundary layer. The warm air in the boundary layer has a higher temperature and therefore lower density than the ambient air, causing it to lift away from the warm surface of the plant. The lifting away and shedding of warmed layers of air is called free convection. Free convection is very noticeable on an asphalt road during hot weather. When viewed obliquely, the air hovering over the road appears to shimmer and dance upward as it is heated by the asphalt. Forced convection involves the shredding away of the air hugging heated surfaces owing to turbulence. Fluid flow speed determines whether heat is conducted freely through the boundary layer or convected by the shedding (free convection) or turbulent shredding (forced convection) of air layers in the boundary layer.

From the foregoing, it should be evident that the rate of mass transfer (e.g., the rates at which water vapor is lost and carbon dioxide is absorbed by leaves) must be correlated with wind speed. We may judge whether conduction or free or forced convection dominates heat transfer by the magnitude of the quotient of the square of the Reynolds number and the Grashof number: Re^2/Gr. Recall that the Reynolds number Re is the ratio of inertial to viscous forces. The Grashof number Gr is the ratio of buoyant forces times inertial force to the square of the viscous force. It is a measure of the tendency of a parcel of fluid to rise or fall. The dimensionless ratio Re^2/Gr, therefore, is the ratio of the inertial to the buoyant force:

$$\text{(3.1a)} \quad \frac{Re^2}{Gr} = \left[\frac{\left(\frac{\text{inertial force}}{\text{viscous force}}\right)^2}{\frac{\text{buoyant force} \times \text{inertial force}}{(\text{viscous force})^2}} \right] = \frac{\text{inertial force}}{\text{buoyant force}},$$

or mathematically,

$$\text{(3.1b)} \quad \frac{Re^2}{Gr} = \frac{\left(\frac{Ud}{\upsilon}\right)^2}{\left(\frac{agd^3\,\Delta T}{\upsilon^2}\right)} = \frac{U^2}{agd\Delta T}.$$

Essentially the ratio Re^2/Gr reflects the inertial effects of a moving fluid versus the buoyancy of the heated layer of fluid in contact with a warm surface. Experimentally, it has been shown that when $Re^2/Gr < 0.1$, free convection dominates. Conversely, when $Re^2/Gr > 10$, forced convection dominates. Consider a leaf measuring 0.05 m in length (with a temperature of 30°C) subjected to an ambient wind speed of 0.44 m s^{-1} (with an ambient temperature of 20°C). From eq. (3.1) we find that $Re^2/Gr = 0.11$, showing that the leaf's heat transfer begins to drift from free convection even at a comparatively slow ambient wind speed. By virtue of their very small size and location within or near the boundary layer hugging a plant, structures such as plant hairs (trichomes), spines, and the like may dissipate heat by conduction rather than free convection. Even for small botanical structures, however, very high wind speeds can thin the boundary layer enough so that forced convection may occur (see fig. 3.2). Additionally important for structures like plant hairs is cyclosis. Movement of the protoplasm within a long trichome can bring heat from the base to the tip of the hair, which may extend above the boundary layer.

For objects composed of materials with high conductivity, edges that have high surface areas relative to their volume may convectively cool more rapidly than surfaces that have comparatively low surface areas. For example, the center of a copper plate conducts heat laterally to its convectively cooled edges. Accordingly, the boundary layer tends to be thinner over regions of an object that have high ratios of surface area to volume and thicker over regions that are bulky. The importance of the two opposing physical processes (conduction and convection) has been emphasized by Steven Vogel, who has shown that plant organs, which are composed largely of water and organic polymers that conduct heat poorly, are in the condition of "constant heat flux" rather than in the condition of "constant temperature" that characterizes metals with high thermal conductivities (Vogel 1983, 744). Based on a series of experiments using real leaves and models of leaves made of various metals, plastics, and the like, Vogel has shown that lateral heat conduction in leaves is minimal, thereby making inferences drawn from model leaves made of metals singularly inappropriate. Vogel found a rough positive correlation between leaf thickness t and thermal conductivity k (measured in units of watts per meter per degree) and suggested that this correlation may reflect a lower volume fraction of air in thicker leaves. Figure 3.4 plots the volume fraction of spongy mesophyll V_S versus the thickness of the leaf lamina t for twenty-eight species of angiosperms. The product of leaf thickness and thermal conductivity tk (with units of watts per degree) is also plotted (data taken from Vogel 1983, table 1) for twelve of the twenty-eight species for which V_S was measured. Regression of V_S versus t yields the formula $V_S = 0.20\, t^{-0.49}$ ($r^2 = 0.87$; $\alpha_{RMA} = -0.53 \pm 0.08$), which shows that the volume fraction of spongy mesophyll, hence the volume fraction of air, anisometrically decreases roughly as the square root of leaf thickness. These data support Vogel's hypothesis about the relation between the thermal conductivity and thickness of leaves, suggesting that thicker leaves have less capacity for vertical heat conduction than thinner ones. Incidentally, although the regression of tk versus t appears to be very strong ($r^2 = 0.95$, $N = 14$), remember that t and tk are autocorrelated (although the units of tk do not contain the dimension of length). In fact, the correlation between t and k is comparatively poor ($r^2 = 0.52$, $N = 14$) and significant only at the 10% level.

For many practical applications, it is desirable to relate the resistance to heat transfer r_H to the ambient wind speed U and the thickness of the boundary layer δ. Table 3.1 (p. 133) provides the equations for r_H, the average δ, and the heat flux density H for the flat plate, sphere, and terete cylinder. These geometries were selected because they serve as reasonable geometric analogues to the leaf, fruit, and stem, respectively. Although the equations for δ given in table 3.1 apply only to air at 20°C (i.e., with a kinematic viscosity υ equal to 1.51×10^{-5} m^2 s^{-1}), it is a simple task to construct the appropriate equations for water as well as air at any specified temperature. All the equations listed in table 3.1 can be used with some measure of accuracy without understanding their derivation, but their practical application requires an understanding of the assumptions they are based on. Thus the derivation of these equations will be reviewed briefly.

The mathematical expression for the resistance to heat transfer (which has units of s m^{-1}) due to the boundary layer is obtained by assuming that all the heat flux density H measured at location x has its origin at a surface with area A_S where H is constant. Based on these assumptions, H at s is given by the relation

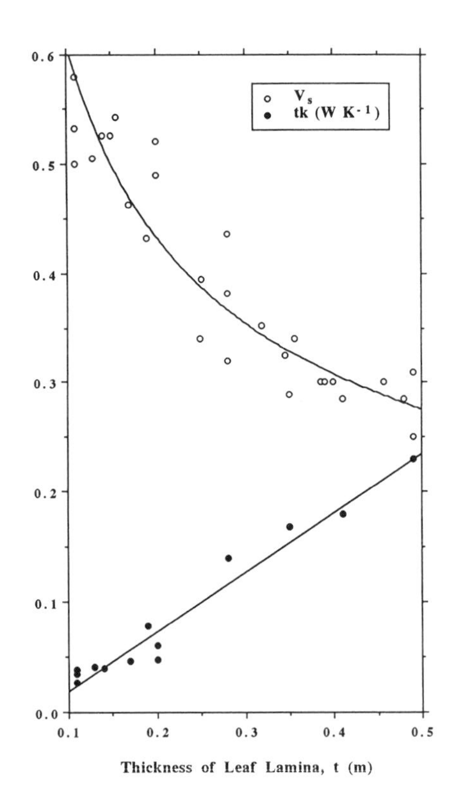

Figure 3.4. Volume fraction of spongy mesophyll V_S plotted against the thickness t of angiosperm leaf laminae. The product of leaf thickness and thermal conductivity tk is also plotted (data from Vogel 1983, table 1) for twelve of the twenty-eight species for which V_S was measured. Solid lines denote regression curves obtained from LS regression analysis.

$$H_s = \frac{A_x}{A_s} H_x = \frac{A_x}{A_s} D_H \frac{d(\rho c_p T)}{dx}, \tag{3.2}$$

where A_X is the area of the imaginary boundary. Since $H = \rho c_p(\Delta T/r_H)$ the macroscopic form of Fourier's law; see eq. 2.11), the resistance to heat transfer r_H is given by the equation

Table 3.1. Equations for Resistance to Mass and Heat Transfer, Boundary Layer Thickness, and Flux Densities for Mass and Heat That Apply to the Flat Plate, Sphere, and Terete Cylinder

	Flat Plate	Sphere	Terete Cylinder
r_H equals	$\frac{\delta}{D_H}$	$\frac{1}{D_H}\left(\frac{d\delta}{d + 2\delta}\right)$	$\frac{d}{2\,D_H} \ln\left(\frac{d + 2\delta}{d}\right)$
δ equals	$\frac{2d}{151\ (Ud)^{0.5}}$	$\frac{d}{1.23 + 152\ (Ud)^{0.5}}$	$\ln\left(\frac{d + 2\delta}{d}\right) = \frac{176^{*}}{(Ud)^{0.385}}$
H equals	$\frac{2}{\delta}\rho c_p D_H \Delta T$	$\left(\frac{d + 2\delta}{d}\right)\rho c_p D_H \Delta T$	$\frac{2\rho c_p D_H \Delta T}{d \ln\left(\frac{d + 2\delta}{d}\right)}$
r_J^{bl} equals	$\frac{\delta}{D_j}$	$\frac{1}{D_j}\left(\frac{d\delta}{d + 2\delta}\right)$	$\frac{d}{2D_j}\ln\left(\frac{d + 2\delta}{d}\right)$
J equals**	$\frac{2}{\delta}D_J \Delta\rho$	$\left(\frac{d + 2\delta}{d\delta}\right)D_J \Delta\rho$	$\frac{2D_J \Delta\rho}{d \ln\left(\frac{d + 2\delta}{d}\right)}$

Note: Equations for the boundary layer thickness apply to air at 20°C (kinematic viscosity $\upsilon = 1.51 \times 10^{-5}$ m² s⁻¹). See text for the method to derive similar equations for the boundary-layer thickness under other conditions.

Symbols: r_H = resistance to heat transfer (s m^{-1}), d = characteristic dimension (d = length of plate or diameter of sphere or cylinder, m), δ = thickness of boundary layer (m), D_H = thermal diffusivity (m^2 s^{-1}), D_j = molecular coefficient of diffusion (m^2 s^{-1}), ρ = density of fluid (kg m^{-3}), c_p = specific heat (kJ kg^{-1} °C^{-1}), ΔT = difference in temperature measured at surface T_s and some distance T_f in fluid (°C), H = heat flux density (kJ m^{-2} s^{-1}), J = mass flux density (kg m^{-2} s^{-1}), $\Delta\rho$ = difference in density (concentration) measured at surface ρ_s and some distance ρ_f in fluid (kg m^{-3}), and U = ambient fluid speed (m s^{-1}).

*Noting that $\ln(1 + x) = x - \frac{x^2}{2} + \frac{x^3}{3} - \frac{x^4}{4} \ldots$ (where $-1 < x \leq 1$), by taking only the first term of the expansion series, $\delta \approx 0.017\frac{d}{(Ud)^{0.385}}$.

**Applicable to plants lacking stomata and cuticles. For plants possessing these features, $J = \frac{\Delta\rho}{r_J^s + r_J^{bl}}$, where r_J^s is total stomatal resistance to mass transfer (see text).

(3.3) $$r_H = \frac{A_s}{D_H} \int_s^l \frac{dx}{A_x},$$

where l is the distance from the heat source. Equation (3.3) must be integrated for the flat plate, sphere, and cylinder. For a flat plat, $A_X = A_S$. Upon integration, eq. (3.3) gives $r_H = l/D_H$, where l is the sum of the thickness t of the plate and the thickness of the boundary layer d. For a very thin plate, $t \rightarrow 0$. Therefore

(3.4) $$r_H = \frac{2\delta}{D_H}.$$ (flat plate, both sides)

For a sphere, $A_X = \pi d^2$. Therefore

(3.5) $$R_H = \frac{1}{D_H}\left(\frac{d\delta}{d + 2\delta}\right).$$ (sphere)

And for the cylinder,

(3.6) $$r_H = \frac{d}{2D_H} \ln\left(\frac{d + 2\delta}{d}\right).$$ (cylinder)

Substituting these equations into Fourier's law of heat transfer, $H = \rho c_p (\Delta T/r_H)$, the following relations are found:

(3.7) $$H = \frac{2}{\delta} \rho c_p D_H \Delta T$$ (flat plate, both sides)

(3.8) $$H = \left(\frac{d + 2\delta}{d\delta}\right) \rho c_p D_H \Delta T$$ (sphere)

(3.9) $$H = \frac{2\rho c_p D_H \Delta T}{d \ln\left(\frac{d + 2\delta}{d}\right)}.$$ (cylinder)

The relation between the transfer of heat and the ambient rate of fluid flow is found by inserting eqs. (3.4) to (3.6) into the appropriate equations for the Nusselt number Nu (see eqs. 2.19 to 2.21 and table 2.4):

(3.10) $$Nu = \frac{2d}{\delta} = 0.66\, Pr^{0.33}\, Re^{0.5}$$ (flat plate, both sides)

(3.11) $$Nu = 2 + \frac{d}{\delta}(2.0 + 1.3\, Pr^{0.15} + 0.66\, Pr^{0.31}\, Re^{0.5})$$ (sphere)

$$(3.12) \qquad Nu = \frac{2}{\ln\left(\frac{d + 2\delta}{d}\right)} = 0.91\ Pr^{0.31}\ Re^{0.5}. \qquad \text{(cylinder)}$$

Although these equations apply to any fluid, the values of Pr and the kinematic viscosity υ must be considered when dealing with a particular fluid. For example, in air at 20°C, the Prandtl number Pr equals 0.70 and $\upsilon = 1.51 \times 10^{-5}$ m^2 s^{-1} (see tables 2.2 and 2.7). Thus

$$(3.13) \qquad \delta = \frac{2d}{151\ (Ud)^{0.5}} \approx 0.013\ \frac{d}{(Ud)^{0.5}} \qquad \text{(flat plate)}$$

$$(3.14) \qquad \delta = \frac{d}{1.23 + 152\ (Ud)^{0.5}} \qquad \text{(sphere)}$$

$$(3.15) \qquad \ln\left(\frac{d + 2\delta}{d}\right) = \ln\left(1 + \frac{2\delta}{d}\right) = \frac{176}{(Ud)^{0.385}} \qquad \text{(cylinder)}$$

Equations (3.13) to (3.15) show that the ambient flow speed U, the characteristic dimension d ("size"), and geometry as well as shape collectively influence heat transfer. Indeed, they reveal that the thickness of the boundary layer anisometrically decreases as a function of increasing Ud. Specifically, an increase in Ud produces a disproportionate decrease in δ. In terms of the separate effects of U and d on δ, figure 3.5 plots δ for a flat plate, sphere, and terete cylinder (d = 0.01 m) as a function of U (from 0.01 to 5 m s^{-1}). This figure shows that δ decreases as a function of U for all three geometries. For the range of wind speed considered, the flat plate consistently produces the thickest boundary layer; the sphere has the thinnest of the three geometries. Figure 3.6 plots δ versus the characteristic dimension, which is a measurement of size. These plots were obtained for an ambient wind speed of 1 m s^{-1}. They show that δ increases as the characteristic dimension of any of the three geometries increases. A curious feature of the relations shown in figure 3.6 is that the scaling of the boundary layer thickness for the terete cylinder has a higher scaling exponent than that of either the sphere or the flat plate. The crossover point for the scaling of δ for the flat plate and the terete cylinder is about 0.1 m. As we shall now see, this has a real effect on mass and heat transfer through the boundary layers around plant organs with similar geometries only when the thickness of the boundary layer is quantitatively related to the resistance to transfer.

A physical analogy between heat and mass transfer shows that the equations for the resistance to mass transfer due to the boundary layer r_J^{bl} take exactly the same form as eqs. (3.4) to (3.6):

$$r_J^{bl} = \frac{2\delta}{D_J} \quad \text{(flat plate, both sides)} \tag{3.16}$$

$$r_J^{bl} = \frac{1}{D_J}\left(\frac{d\delta}{d + 1\delta}\right) \quad \text{(sphere)} \tag{3.17}$$

$$r_J^{bl} = \frac{d}{2D_J}\ln\left(\frac{d + 2\delta}{d}\right). \quad \text{(cylinder)} \tag{3.18}$$

The rate at which water vapor diffuses from the moistened surfaces of geometric models constructed from filter paper therefore can be used to

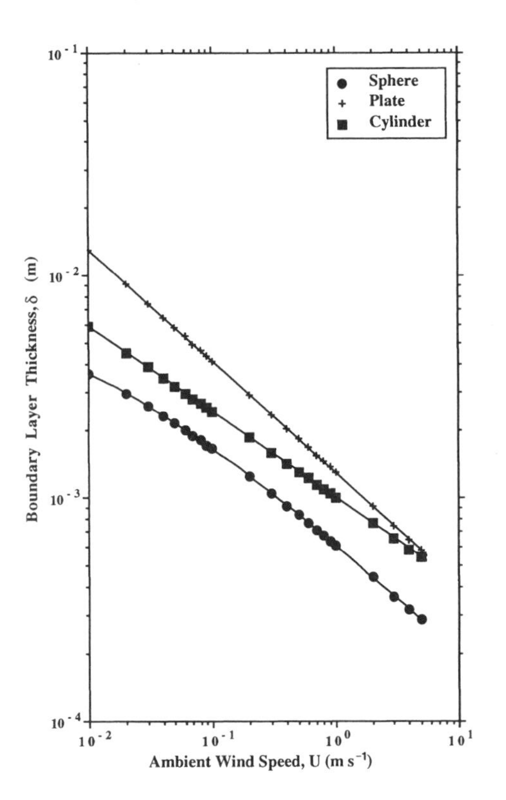

Figure 3.5. Log-log (base 10) plot of the thickness of the boundary layer δ around a sphere, a flat plate, and a terete cylinder versus ambient wind speed U. Solid diagonal lines denote regression curves obtained from LS regression analysis.

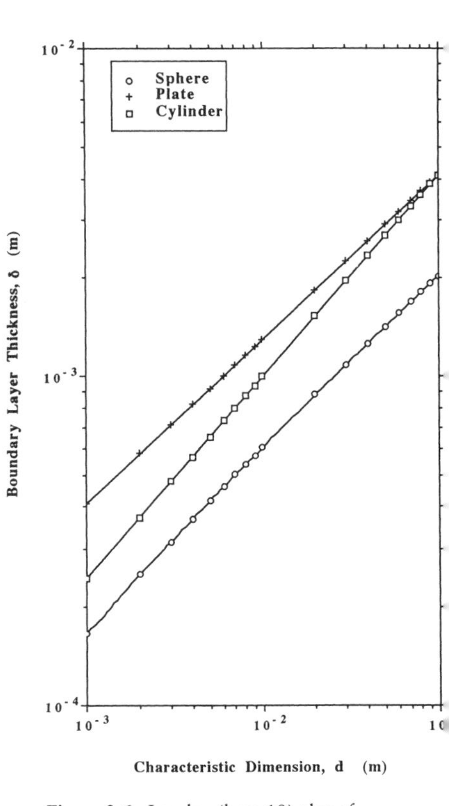

Figure 3.6. Log-log (base 10) plot of the thickness of the boundary layer δ around a sphere, a flat plate, and a terete cylinder versus the characteristic dimension d for these geometries. Plots obtained for an ambient wind speed of 1 m s^{-1}. Solid diagonal lines denote regression curves obtained from LS regression analysis.

estimate the thickness of the boundary layer, thereby testing the efficacy of eqs. (3.13) to (3.15). This experimental approach was used by Park Nobel (1974, 1975) and is based on the fact that the resistance to mass transfer r_J and the resistance to heat transfer r_H are equivalent provided the only resistance to mass transfer is r_J^{bl} (i.e., $r_J = r_J^{bl} = r_H$). Figure 3.7 plots the predicted versus empirically observed δ based on paper models of a sphere and a cylinder. These data were collected using the same format presented by Park Nobel (1974, 1975), who has led the way in this type of research. The solid line in this figure has a slope of one. How far data deviate from this line, therefore, is a measure of the failure of predicted and observed δ to correspond. The relation shown in figure 3.7 provides circumstantial evidence that eqs. (3.16) to (3.18) give reasonable approximations of δ. This is particularly true at low Reynolds numbers where the boundary layer is thick. However, as the Reynolds number increases and δ decreases, figure 3.7 shows that eqs. (3.16) to (3.18) overestimate the thickness of the boundary layer, because these equations are based on the empirically determined relations between the Nusselt and

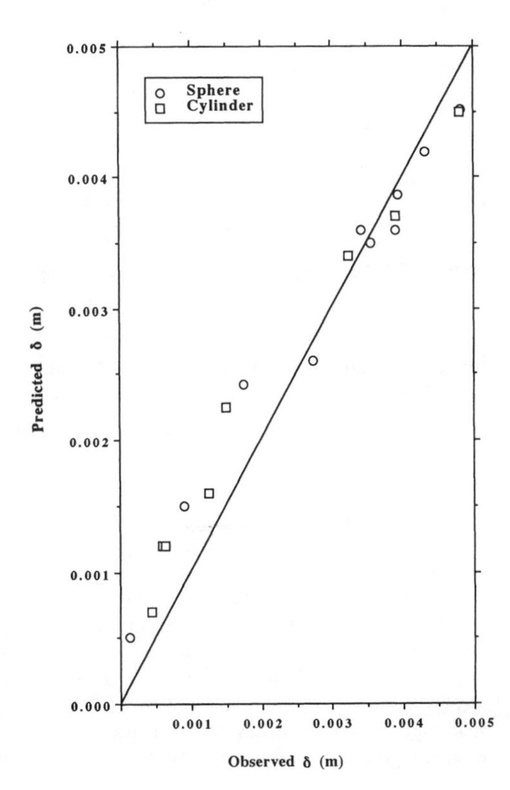

Figure 3.7. Predicted versus empirically observed boundary layer thickness δ around wetted paper models of a sphere and a cylinder. Solid diagonal line has a slope of unity and intercepts the origin of the graph.

Reynolds numbers (see eqs. 3.10 to 3.12) and the anticipated relation between *Nu* and *Re* changes as a function of *Re*. By the same token, the empirical relation between *Nu* and *Re* depends on the experimental format an investigator uses. For example, McAdams (1954) reports that for the sphere $Nu = 0.37\ Re^{0.6}$ over the range $17 \leq Re \leq 70{,}000$. For air, this leads to $\delta = d/[289\ (Ud)^{0.6} - 2]$, which differs from eq. (3.14). Likewise, some authors provide equations for the thickness of the boundary layer that reflect an ad hoc approach. That is, they begin with a derivation based on first principles and then adjust the equation to fit their experimental data. This is a valid approach, since much of fluid mechanics is an empirical affair. For example, Nobel (1974, 1975, 1983) gives the following equations for the thickness of the boundary layer around a sphere and a terete cylinder:

$$\delta_{\mathrm{mm}} = 0.33\left(\frac{d_{\mathrm{m}}}{U_{\mathrm{m\ s^{-1}}}}\right)^{0.5} + \frac{0.25}{U_{\mathrm{m\ s^{-1}}}} \qquad (3.19)\ \text{(sphere)}$$

$$\delta_{\mathrm{mm}} = 5.8\left(\frac{d_{\mathrm{m}}}{U_{\mathrm{m\ s^{-1}}}}\right)^{0.5}, \qquad (3.20)\ \text{(cylinder)}$$

while Pearman, Weaver, and Tanner (1972) give the following approximate (but useful) equation for the average boundary layer thickness next to a perfectly flat leaf[2]

$$\delta_{\mathrm{mm}} = 4.0\left(\frac{d_{\mathrm{m}}}{U_{\mathrm{m\ s^{-1}}}}\right)^{0.5}, \qquad (3.21)\ \text{(plate)}$$

where d is leaf length.

Figure 3.8, which is based on data reported by Nobel (1974, 1975), plots the resistance to water vapor r_{wv} (exclusively due to the boundary layer) for spherical fruits and cylindrical plant stems versus the characteristic dimensions of these organs (fruit or stem diameter). All the data were collected from experiments for which the ambient wind speed equals 0.20 m s^{-1}. Comparing these data with the plots provided in figure 3.6 shows

2. Unlike the equations given in table 3.1, eqs. (3.19) to (3.21) can be criticized because they give δ in units of mm, whereas the characteristic dimension is measured in units of m and the ambient wind speed is measured in units of m s^{-1}. Dimensional analysis of these equations quickly reveals that the numbers 0.33, 5.8, and 4.0 must have units of m $s^{-0.5}$, which is queer to say the very least. Although the mixing of units is generally a poor practice, typically the boundary layer around a leaf, stem, or fruit is orders of magnitude thinner than the characteristic dimension. Similarly, wind speeds are on the order of meters per second. Thus, eqs. (3.19) to (3.21) attempt to strike a reasonable balance between the conventional practice of dealing with dimensions, on the one hand, and the practical units in measuring the size of organs and the speed of wind, on the other.

that the analogy between mass or heat transfer gives very satisfactory results for real plant organs.

Finally, it must be noted that the geometries treated so far are assumed to have absolutely smooth surfaces. When surfaces are roughened or made grossly uneven, the boundary layer tends to become turbulent at lower Reynolds numbers than when the turbulent flow regime is initiated over smooth surfaces. For example, in the case of an absolutely flat plate, the transition from laminar to a turbulent boundary layer occurs on the order of $Re = 10^4$. The transition over a plate with roughened surfaces occurs when Re approaches 10^3, Interestingly, turbulent ambient airflow, which greatly increases the boundary layer turbulence over flat plates (and therefore presumably real leaves) has only a small effect on empirically measured rates of evaporation (see Grace and Wilson 1976).

3.3 Stomatal Resistances

All other things being equal, the mass flux density of a plant with stomata will be significantly less than that of a plant lacking stomata, because the

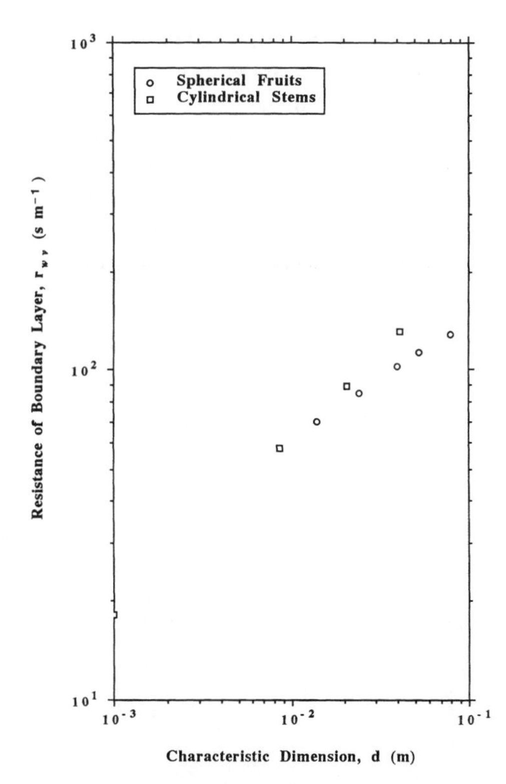

Figure 3.8. Log-log (base 10) plot of the resistance of water vapor r_{wv} through the boundary layer around spherical fruits and cylindrical plant stems versus the characteristic dimensions of these organs (fruit or stem diameter). All the data were collected from experiments for which the ambient wind speed equals 0.20 m s^{-1}. Data from Nobel 1974, 1975.

resistance to mass flow equals the sum of r_J^{bl} and the stomatal resistance r_S. Consider an aquatic plant lacking stomata and a cuticle, for which we need only substitute eqs. (3.16) to (3.18) into Fick's law of diffusion, $J = (\Delta\rho/R_J^{bl})$, to find the equations for mass flux density:

$$J = \frac{2}{\delta} D_J \Delta\rho \qquad \text{(flat plate, both sides)} \tag{3.22}$$

$$J = \left(\frac{d + 2\delta}{d\delta}\right) D_J \Delta\rho \qquad \text{(sphere)} \tag{3.23}$$

$$J = \frac{2D_J \Delta\rho}{d \ln\left(\frac{d + 2\delta}{d}\right)}. \qquad \text{(cylinder)} \tag{3.24}$$

To derive an expression for stomatal resistance, we assume that planar diffusion characterizes mass transfer within the stomatal pore. In these circumstances $r = l/D_j$ (see eqs. 3.2 to 3.3), where l (formerly taken as the distance from the source) now denotes the depth of the stomatal pore. If planar diffusion occurs outside the pore, this relation would be sufficient to describe stomatal resistance. However, the diffusion of gases just above the external opening of a stoma is nonplanar. Within the boundary layer, the contours of the concentration gradient have a hemispherical geometry and, because of the average spacing of stomata on most leaves, the contours overlap such that their maximum elevation is above the pore and their minimum elevation is at the midpoint between neighboring stomata. To account for this, an end correction factor is introduced, which, for a spherical aperture, is $\pi\xi/8D_j$, where ξ is the diameter of the opening. Since the resistance to mass transfer of a single stoma r is the sum of the resistance to planar diffusion and the end correction factor, we obtain the following formula:

$$r = \left(\frac{1}{d_j}\right)\left(l + \frac{\pi\xi}{8}\right). \tag{3.25}$$

Since the epidermis may have many stomata, the total stomatal resistance to mass transfer r_J^s is taken as the product of r and the fraction of the surface area of the epidermis occupied by stomata. Approximating the cross-sectional geometry of stomata as circles, we obtain the formula

$$r_J^s = \left(\frac{1}{D_j}\right)\left(l + \frac{\pi\xi}{8}\right)\left(\frac{4}{\pi n \xi^2}\right), \tag{3.26}$$

when n is the number of stomata per square meter of epidermis. Equation (3.26) shows that the total resistance due to stomata increases as a function of the average depth of the stomatal pore and decreases as a function of the average diameter of stomata. Also, as the number of stomata per square meter increases, the resistance to mass transfer decreases.

Much unproductive discussion concerning the role of stomatal closure and opening is found in the early literature, written before it was realized that the effect that partial closure of stomata has on transpiration depends on the quotient r_J^{bl}/r_J^s. In still air, $r_J^{bl} \geq r_J^s$. In rapidly moving air, $r_J^{bl} \leq r_J^s$. Consequently, large changes in pore diameter will have little effect on transpiration when ambient wind speeds are very slow and vice versa (see fig. 3.3). With the aid of eqs. (3.16) and (3.26), the ambient wind speed at which leaf transpiration will be influenced by stomatal diameter can be calculated. Casting the boundary layer resistance in terms of the Reynolds number, r_J^{bl}/r_J^s for one side of a platelike leaf is given by the equation

$$\text{(3.27)} \qquad \frac{r_J^{bl}}{r_J^s} = \frac{\pi n \xi^2 d}{0.587 Re^{0.5}(4l + 2\pi\xi)}.$$

If $n = 60 \times 10^6$ m^{-2}, $d = 0.05$ m, $l = 10 \times 10^{-6}$ m, and $\xi = 10 \times 10^{-6}$ m, then $r_J^{bl}/r_J^s = 0.273/U^{0.5}$ when the ambient air temperature is 20°C. Thus $r_J^{bl}/r_J^s = 1.0$ when $U = 0.075$ m s^{-1}. This conclusion is easily summarized by means of a very simple graph. Figure 3.9 plots r_J^{bl}/r_J^s versus U based on the previously assigned values for the variables in eq. (3.27). The dashed horizontal line in this figure denotes the situation where the ratio of the resistances due to the boundary layer and to stomata equals unity (i.e., $r_J^{bl}/r_J^s = 1.0$). This horizontal line intersects the dashed vertical line for $U = 0.075$, which is the ambient wind speed above which the boundary layer resistance loses its comparative importance in relation to stomatal resistance. The importance of the ratio of the resistances due to the boundary layer and to stomata $k = r_J^{bl}/r_J^s$ to mass transfer is easily shown mathematically by relating the mass flux density to the total resistance r_T, which is the sum of r_J^{bl} and r_J^s:

$$\text{(3.28)} \qquad J = \frac{\Delta\rho}{r_T} = \frac{\Delta\rho}{r_J^s + r_J^{bl}} = \frac{\Delta\rho}{r_J^{bl}\,(k + 1)}.$$

A comparison between the mass flux density of a leaf possessing stomata and a counterpart lacking stomata is informative. Consider the following circumstances:

	Air	Leaf
Length (m)	—	0.05
Speed (m s^{-1})	5	—
Concentration (kg m^{-3})		
Water vapor	1.0×10^{-2}	3.0×10^{-2}
Carbon dioxide	5.5×10^{-4}	1.2×10^{-4}
Oxygen	0.2700	0.2705

Referring to the equations given in table 3.1 for a flat plate (the geometric analogue to the leaf) the average thickness of the boundary layer equals 1.32 mm. The resistance to heat transfer, on one side of the leaf, is 31 s m^{-1}, and the heat flux density, on both sides, equals 391 J m^{-2} s^{-1} (which has units of W m^{-2}). Note that these values apply to both the hypothetical leaf with stomata and the leaf without stomata. Assuming that $\xi = 10$ μm, $d = 10$ μm, and $n = 100 \times 10^6$ m^{-2}, the following resistances are calculated:

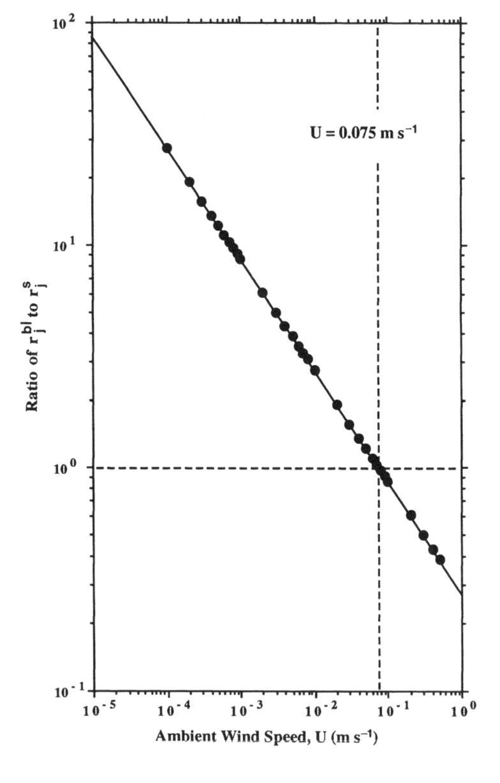

Figure 3.9. Log-log (base 10) plot of the ratio of the resistance to mass transfer offered by the boundary layer to the resistance offered by stomata r_j^{bl}/r_j^s versus ambient wind speed U (computed from eq. 3.27). The dashed horizontal line denotes $r_j^{bl}/r_j^s = 1.0$. The dashed vertical line denotes $U = 0.075$, which is the ambient wind speed above which the boundary layer resistance loses its comparative importance in relation to stomatal resistance.

	r_J^s	r_J^{bl}	$r_J^s/r^{bl} = k$
Resistance, one side (s m^{-1})			
Water vapor	73	27	2.70
Carbon dioxide	120	45	2.67
Oxygen	88	33	2.67

And based on these resistances,

	Leaf with stomata	Leaf without stomata
Mass flux density J, on one side (kg m^{-2} s^{-1})		
Water vapor	2.0×10^{-4}	7.0×10^{-4}
Carbon dioxide	2.6×10^{-6}	9.6×10^{-6}
Oxygen	4.2×10^{-6}	9.2×10^{-6}

On the average, therefore, the flux densities of the leaf with stomata are roughly one-third those of the leaf lacking these structures. Incidentally, the direction of mass transfer is such that leaves gain carbon dioxide and lose water vapor and oxygen.

In addition to the resistances provided by boundary layer and stomata, the cuticle, intercellular air spaces, and cellular components through which mass must diffuse (cell wall, plasma membranes, and the like) provide resistance to mass transport. The cuticle provides the greatest resistance. Cuticle resistances for crop plants range between 2×10^3 and 10×10^3 s m^{-1}; those of mesophytic trees and certain xerophytic species range between 5×10^3 and 20×10^3 s m^{-1} and between 10×10^3 and 100×10^3 s m^{-1}, respectively (see Nobel 1983, 393). Even a qualitative consideration of these resistances however, easily accounts for the morphological features distinguishing the leaves of hydrophytes and xerophytes. The floating leaves of aquatic plants or terrestrial hydrophytes typically have large surface areas, numerous stomata, and thin cuticles. The leaves of xerophytes typically have large surface areas relative to their volumes (but small leaves), sunken or few stomata, and thick, waxy cuticles. The boundary layers of xerophytic leaves may be thickened (to conserve water) by the dense application of plant hairs (which may reflect light and convect heat and therefore reduce leaf temperature). Crassulacean acid metabolism, CAM, where stomata tend to be closed during the day and open at night, is seen to be functionally adaptive in dry habitats as well.

3.4 The Mechanical Scaling of Foliage Leaves

The preceding sections considered the effect of size and shape on the ability of plants to exchange mass and heat with the air. Emphasis was on the resistances to mass and heat resulting from the boundary layer and stomata of photosynthetic organs. In this section I continue with the theme of scaling limitations on photosynthesis but turn to the desirability of orienting photosynthetic tissues in the direction of ambient sunlight. The focus, therefore, is on the mechanical responsibilities of foliage leaves, which are the principal photosynthetic organs of most species of terrestrial vascular plants. As we explore this topic, a number of engineering concepts relevant to subsequent topics will be developed. Among these concepts is flexural stiffness, which is a size- and shape-dependent mechanical parameter.

Figure 3.10 shows a horizontal branch of the sugar maple, *Acer saccharum,* bearing foliage leaves. Morphologically, each leaf consists of two distinct parts: a laterally expanded sheetlike structure called the leaf blade or lamina, and a prismatic structure called the petiole. From a mechanical perspective, each petiole operates as a cantilevered beam. It is rigidly fixed at its base to the subtending stem by an inflated region called the phyllopodium and free to bend at the other end at the point of attachment of the lamina. The magnitude of the bending force at the tip of the petiole equals the product of the acceleration of gravity and the mass M of the lamina. Thus large leaf laminae will impose larger bending forces than smaller laminae. Nonetheless, regardless of their position on the horizontal shoot, the petioles of the leaves shown in figure 3.10 bend and torque such that they orient laminae differing in mass more or less in the horizontal plane. This is most evident when leaves are viewed laterally (fig. 3.10A and B). It is worth noting that the planation of laminae on horizontally oriented shoots tends to maximize the capacity of leaves to intercept sunlight, particularly when neighboring trees shade one another and most of the available light comes from directly above. Most important, the planation evinced in figure 3.10 can be accomplished only by scaling petiole length and stiffness to lamina mass. This is quickly seen by means of simple engineering theory, which shows that the ability of petioles to hold leaves horizontally depends on three variables: the mass force F applied per unit length of the petiole, the length L of the petiole, and the flexural stiffness EI of the petiole. These three variables influence the magnitude of the tip deflection δ, which is the displacement from the vertical measured at the free end of the petiole where the lamina is attached:

(3.29a) $$\delta = k\,\frac{FL^4}{EI},$$ (tip-deflection formula)

where k is the constant of proportionality. Since the force per unit length at the tip of the petiole equals Mg/L, where M is the mass of the lamina

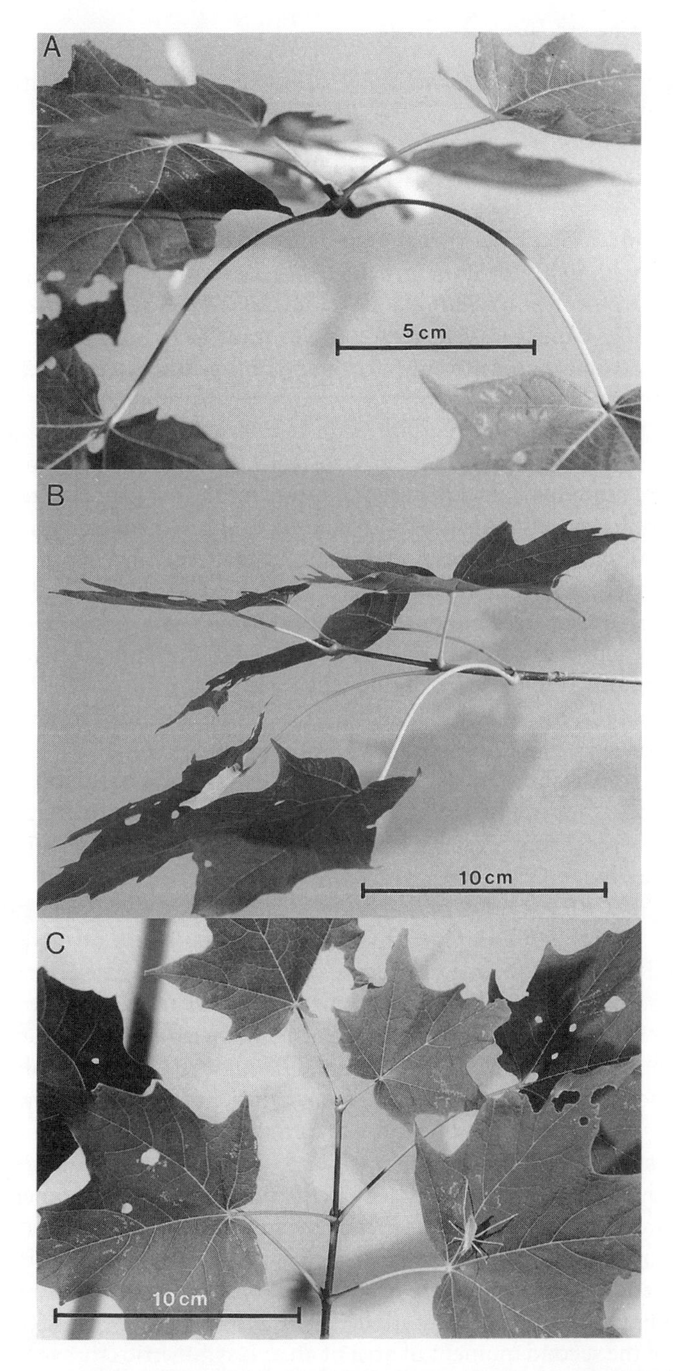

Figure 3.10. A horizontal branch of the sugar maple, *Acer saccharum,* bearing foliage leaves viewed along the length of the branch (A), from the side of the branch (B), and from the top of the branch (C).

and $g = 9.8 \text{ m s}^{-2}$ = a constant, the following proportional relation is obtained:

$$\delta \propto \frac{ML^3}{EI}. \tag{3.29b}$$

Equation (3.29) shows that the tip-deflection angles of leaves differing in size can be held constant provided the flexural stiffness of petioles is scaled in proportion to lamina mass and the cube of petiole length. Leaves with heavy lamina or very long petioles must have proportionally stiffer petioles to mechanically support their photosynthetic tissues and orient them to sunlight.

For those unfamiliar with flexural stiffness, EI describes the ability of any mechanical support member to resist bending. It is the product of Young's modulus E and the second moment of area I. In chapter 2 I said that Young's modulus is the ratio of stress to strain measured for an elastic material within its proportional limits of loading. Since strain is dimensionless, E has the units of stress—force per area (see fig. 2.24). The second moment of area mathematically quantifies the ability of a support member to resist deformation conferred by the spatial distribution of materials in a representative cross section. Table 3.2 provides the formulas for I for a variety of simple cross-sectional geometries. These formulas show that I depends on size, shape, and geometry. Although not immediately evident, regardless of absolute size or cross-sectional geometry, the second moment of area is the sum of the products of each infinitesimally small area and the square of the distance each area lies from the neutral axis of a support member (see Niklas 1992, 134). Thus I has units of length raised to the fourth power and EI has units of force times area. Flexural stiffness can be increased by using materials with large Young's moduli or by increasing the second moment of area. In terms of material properties, wood has a Young's modulus on the order of giganewtons per square meter; parenchyma has a Young's modulus on the order of meganewtons per square meter. Thus the flexural stiffness of a stem increases as its wood volume fraction increases at the expense of parenchyma. By the same token, flexural stiffness increases by adding more material to a cross section, since this increases the second moment of area. Engineered structures cannot do this, since they cannot grow. Trees do it all the time. The second moment of area of a branch or tree trunk increases by means of the amortization of secondary growth layers of wood. Another growth strategy of plant organs is to alter or adjust their geometry or shape in response to the direction in which a bending force is habitually applied. This is important because the magnitude of the second moment of area of an axiasymmetric cross section—for example, an ellipse—differs depending on the plane of bending. The petiole of the sumac, *Rhus typhina,* has an elliptical cross section with the major axis aligned with the vertical direction. This geometry resists bending owing to the influence of gravity but permits lateral bending when leaves are deflected laterally by the wind.

Returning to the scaling of leaves, eq. (3.29) assumes that the petiole is homogeneous in terms of its material properties and that it is an untapered prismatic bar (that it has uniform material properties and cross-sectional geometry and area). Also, eq. (3.29) applies only when δ is very small (the tip deflection is less than 10% of L). These assumptions severely restrict its practical application to real leaves, particularly pinnately compound leaves. Although the petioles of simple or palmately com-

Table 3.2. Second Moments of Area I and Section Moduli Z for Simple Cross-Sectional Geometries

Geometry	I	Z
Solid circle	$\frac{\pi R^4}{4}$	$\frac{\pi R^3}{4}$
Hollow circle	$\frac{\pi(R_o^4-R_i^4)}{4}$	$\frac{\pi(R_o^4-R_i^4)}{4R_o}$
Solid ellipse	$\frac{\pi ba^3}{4}$	$\frac{\pi ba^2}{4}$
Hollow ellipse	$\frac{\pi[(ba^3)_o-(ba^3)_i]}{4}$	$\frac{\pi[(ba^3)_o-(ba^3)_i]}{4a_o}$
Solid rectangle	$\frac{2hd^3}{3}$	$\frac{2hd^2}{3}$
Hollow rectangle	$\frac{2[(hd^3)_o-(hd^3)_i]}{3}$	$\frac{2[(hd^3)_o-(hd^3)_i]}{3d_o}$
Triangle	$\frac{dh^3}{36}$	$\frac{dh^2}{36}$

pound leaves tend to be untapered along their length, the petioles of pinnate leaves tend to be significantly tapered in girth. This is seen by normalizing the cross-sectional area of a petiole measured at some distance x from the tip of the petiole with respect to the basal cross-sectional area and plotting the normalized cross-sectional area An as a function of the normalized petiole length x/L. Figure 3.11, for example, plots An versus x/L for a simple leaf from the sugar maple, *Acer saccharum,* a palmate leaf from the horse chestnut, *Aesculus hippocastanum,* and a pinnate leaf from the palm *Chamaedorea erumpens.* A diagonal line with a slope of one provides the yardstick for the isometric scaling relation between An and x/L. The data from simple and palmate leaves plotted in figure 3.11 reveal that An is comparatively constant for roughly 85% the length of the petioles. The sharp increase in An toward the base of the simple and palmate leaves is due to their inflated phyllopodia. By contrast, the petiole of the pinnate leaf of the palm is linearly tapered along its entire length. That is, the data indicate that $An \propto (x/L)^{1.0}$. Indeed, the petioles of most

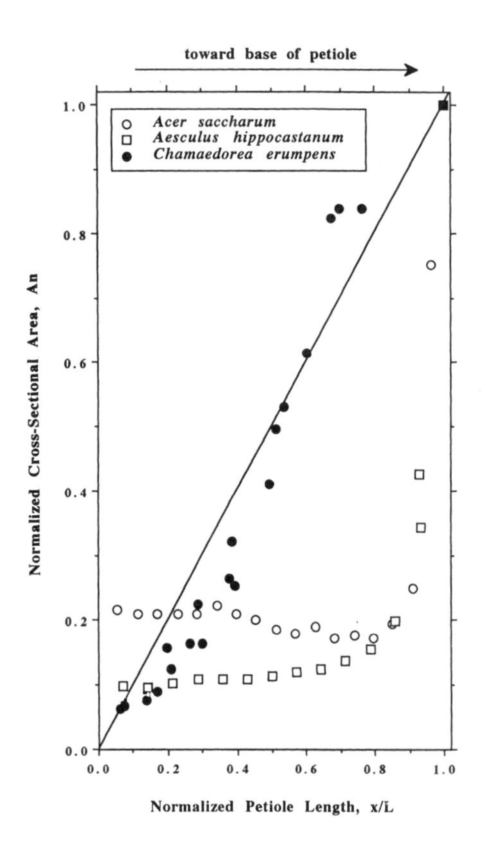

Figure 3.11. Normalized cross-sectional area An of representative petioles of three species plotted against normalized petiole length-x/L. The diagonal solid line has a slope of unit and intercepts the origin of the graph.

simple and palmate leaves can be approximated as essentially untapered prismatic cantilevered beams, whereas those of pinnate leaves are tapered cantilevered beams. Figure 3.12 shows an additional problem encountered when dealing with the entire rachis of a pinnate leaf (the rachis is the petiolelike structure to which leaflets are attached; the petiole of a pinnate leaf is the basalmost prismatic portion of the leaf lacking leaflets). The cross-sectional geometry as well as cross-sectional area of the petiole and rachis of a palm leaf vary such that second moments of area diminish acropetally. As a consequence of these morphological variations, *EI* varies in a complex way as a function of length.

Despite the limited applicability of the tip-deflection formula to real leaves, particularly pinnate leaves, the mechanical scaling of foliage leaves in general can be crudely examined in terms of the proportional relation

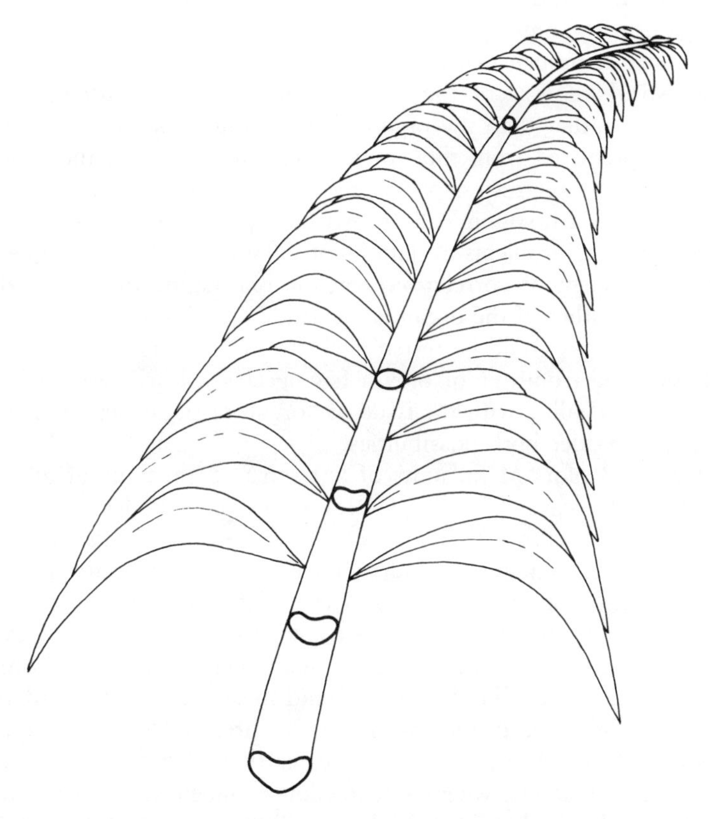

Figure 3.12. Schematic drawing of a pinnately compound leaf showing representative cross-sectional geometries of the rachis.

between M or EI and L by assuming that the simplest loading regime for each leaf is such that the mass M acting on the tip of a petiole is the mass of the lamina. From a statistical point of view, however, we are faced with an interesting question regarding how we should determine the scaling exponents for these mechanical variables. Consider that we may predict from first principles that all leaves will observe the same scaling relations. Thus the scaling exponents found for an intraspecific comparison should be statistically indistinguishable from those found for an interspecific comparison among the biomechanical variables M, EI, and L. By contrast, we might argue from first principles that scaling exponents determined for the leaves drawn from different plants will differ but that the scaling exponents determined through an interspecific comparison will comply with the proportionalities predicted by the tip-deflection formula.

Put somewhat differently, we are faced with the question: What constitutes the "individual" in our scaling hypothesis—the organ or the organism (or more precisely, the species)? The answer is far from trivial, since it influences the validity of our experimental design and the statistical analysis of the data. If the plant (i.e., species) is the "individual" rather than the organ, then we must compare the scaling exponents predicted from the tip-deflection formula with those determined from the regression of the mean values of M, EI, and L for each species in a taxonomically broad interspecific comparison; otherwise we have the problems of data point inflation (and its consequences for the coefficient of correlation), on the one hand, and the phyletic effect, on the other. If the organ is the "individual," then our analytical scaling hypothesis could be tested against any collection of leaves randomly sampled from one or only a few species. Clearly, we have no a priori way of dealing with the issue of "organ or organism" and therefore must consider both possibilities.

Figure 3.13 plots M (in units of kg) versus L (in units of m) measured for 193 leaves (86 simple leaves, 36 palmate leaves, and 71 pinnate leaves) representing a total of nineteen dicot and monocot species. On the average, ten individual leaves were collected for each species to ensure a broad range of leaf size for each species. For each species, mean M and mean L were computed. These mean values are plotted in figure 3.13 along with the individual data points from all the leaves examined. The data for M and mean M span five orders of magnitude; those for L and mean L span three orders. Least squares regression of all 193 data points indicates that $M \propto L^{1.84}$ ($r^2 = 0.87$; $\alpha_{RMA} = 1.97 \pm 0.04$), whereas regression of nineteen mean values of M versus L shows that $M \propto L^{1.88}$ ($r^2 = 0.86$; $\alpha_{RMA} = 2.02 \pm 0.03$). Analysis of covariance shows that the exponents of these two regressions do not statistically differ from one another (on the average, leaf lamina mass scales roughly as the square of petiole length) and that both differ from that predicted by the tip-deflection formula ($M \propto L^{3.0}$).

In terms of the relation between flexural stiffness and petiole length, however, the scaling of leaves is remarkably like that predicted by the tip-deflection formula. Figure 3.14 plots *EI* (in GN m^2) versus *L*. Regression of the 193 individual data points indicates the $EI \propto L^{3.09}$ ($r^2 = 0.91$; $\alpha_{RMA} = 3.24 \pm 0.05$), whereas regression of the nineteen mean values of *EI* and *L* indicates that $EI \propto L^{3.05}$ ($r^2 = 0.93$; $\alpha_{RMA} = 3.17 \pm 0.04$). Thus,

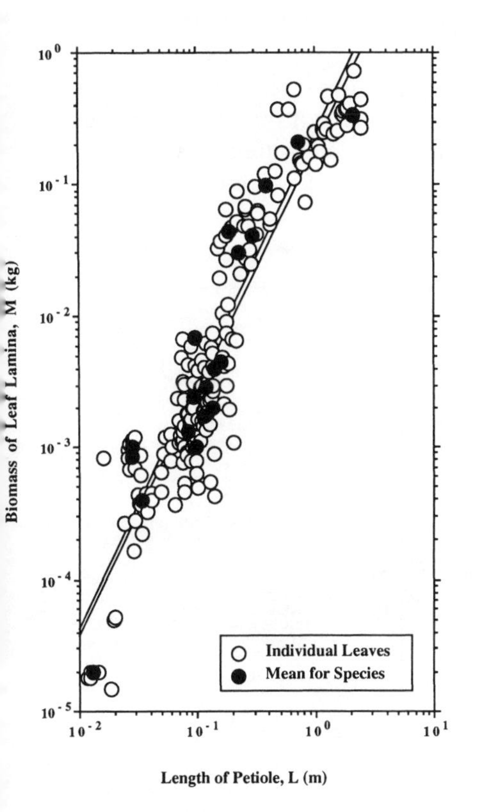

Figure 3.13. Log-log (base 10) plot of leaf lamina biomass *M* against petiole length *L* for individual leaves and for the mean values of *M* and *L* obtained for the species represented in the data set. Solid diagonal lines denote the regression curves obtained from LS regression analyses.

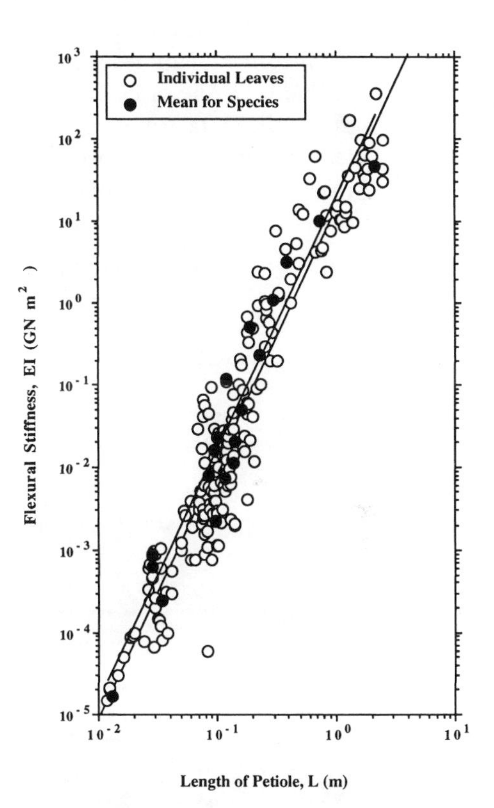

Figure 3.14. Log-log (base 10) plots of flexural stiffness *EI* versus petiole length *L* for individual leaves and the mean values of *EI* and *L* obtained for the individual species represented in the data set. Solid diagonal lines denote the regression curves obtained from LS regression analyses.

both regression formulas indicate that flexural stiffness scales very roughly as the cube of petiole length, which is compatible with the tip-deflection formula. But the scaling relations of M and EI with respect to L are theoretically incapable of maintaining a constant tip deflection among leaves from different species. Consider that, in terms of the exponents determined from reduced major axis regression, we would expect $\delta \propto ML^3/EI \propto (L^2)L^3/L^3$, $\delta \propto L^2$. In other words, among the nineteen monocot and dicot species examined, the tip deflection of leaves is expected to proportionally increase roughly with the square of petiole length.

Given the tapering of the petiole of pinnate leaves, it is reasonable to ask whether these leaf structures scale M and EI differently than do simple or palmate leaves. Figure 3.15 and 3.16 plot M versus L and EI versus L for the same data set shown in figures 3.13 and 3.14 but distinguish the data for simple and palmate leaves from those for pinnate leaves. For simplicity, we will consider only the scaling relations for the mean values. Regression of the data from simple/palmate leaves shows that $M \propto L^{1.74}$ ($r^2 = 0.75$; $\alpha_{RMA} = 2.01 \pm 0.05$) and $EI \propto L^{2.91}$ ($r^2 = 0.87$;

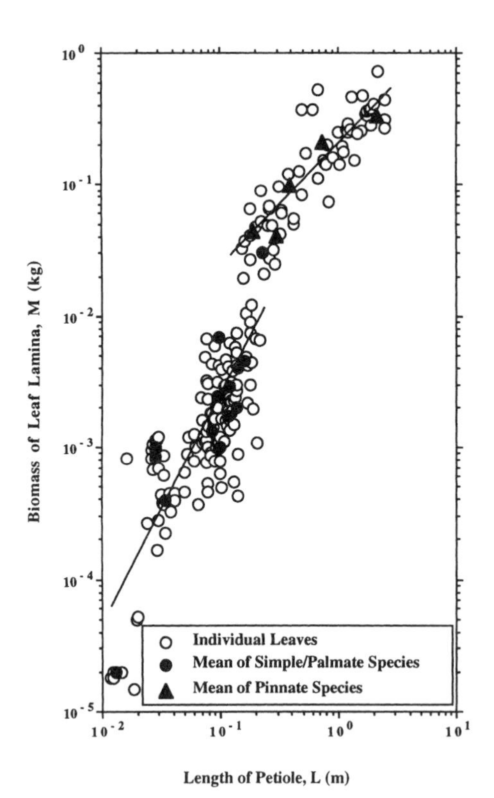

Figure 3.15. Log-log (base 10) plot of leaf lamina biomass M against petiole length L for individual simple/palmate and pinnate leaves and for the mean values of M and L obtained for the species with these two different kinds of leaf morphology represented in the data set. Solid diagonal lines denote the regression curves obtained for species mean values from LS regression analyses.

$\alpha_{RMA} = 3.13 \pm 0.05$), whereas regression of the data from pinnate leaves shows that $M \propto L^{0.94}$ ($r^2 = 0.89$; $\alpha_{RMA} = 0.99 \pm 0.03$) and $EI \propto L^{1.88}$ ($r^2 = 0.98$; $\alpha_{RMA} = 1.90 \pm 0.03$). The α_{RMA} values of these interspecific comparisons indicate that simple and palmate leaves scale M and EI differently than do pinnate leaves. Surprisingly, however, these differences may cancel out in terms of size-dependent variations in the tip-deflection angle. Consider that $\delta \propto ML^3/EI \propto L^2L^3/L^3 \propto L^2$ for simple and palmate leaves, whereas $\delta \propto ML^3/EI \propto LL^3/L^2 \propto L^2$ for pinnate leaves.

It is remarkable that essentially the same scaling exponents were obtained when we regressed the data from 193 individual leaves or the mean values of laminar mass, flexural stiffness, and petiole length from nineteen species. As a general rule, this is not the expected result. Indeed, the curious relation between the plant organ and the whole plant should not tempt us to infer that it matters little whether we consider the organ or the organism as the "individual." And it is extremely dangerous to infer interspecific scaling relations from an intraspecific comparison, even in the context of foliage leaves, as we see by considering Boston ivy, *Parthen-*

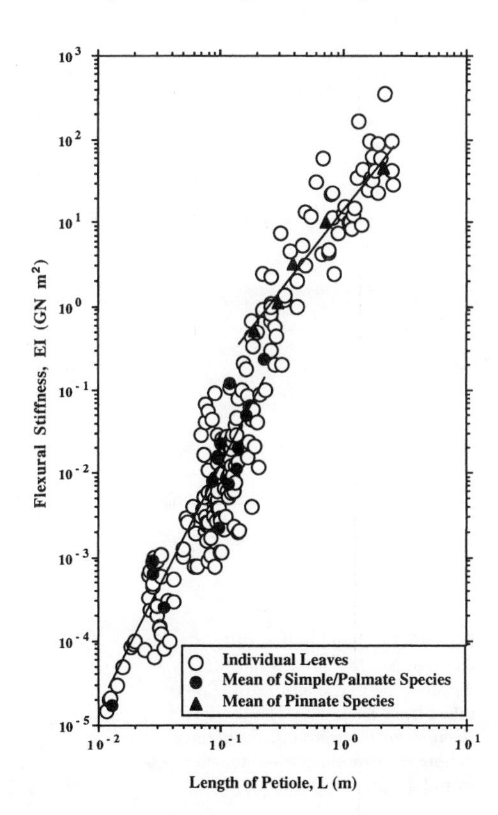

Figure 3.16. Log-log (base 10) plot of leaf lamina biomass EI against petiole length L for individual simple/palmate and pinnate leaves and for the mean values of M and L obtained for the species with these two different kinds of leaf morphology represented in the data set. Solid diagonal lines denote the regression curves obtained for species mean values from LS regression analyses.

ocissus tricuspidata. The foliage leaves of this species are simple and have deeply lobed lamina. Based on previously determined interspecific comparisons among species with simple and palmate leaves, we would expect $M \propto L^2$ and $EI \propto L^3$ for Boston ivy. This is not the case, however. Figure 3.17 plots M and EI versus L for twenty leaves taken from five Boston ivy plants. Regression of these data shows that $M \propto L^{2.59}$ ($r^2 = 0.97$; $\alpha_{RMA} = 2.63 \pm 0.01$) and $EI \propto L^{4.13}$ ($r^2 = 0.85$; $\alpha_{RMA} = 4.48 \pm 0.01$). I cannot, however, resist pointing out that in theory $\delta \propto ML^3/EI \propto L^3L^3/L^4 \propto L^2$ for these Boston ivy leaves. Thus it seems that, regardless of how M and EI scale with respect to L, the tip-deflection angle intra- and interspecifically scales roughly as the square of petiole length.

Finally, the scaling relations among leaf area LA, leaf biomass M_l, and stem diameter D and stem mass M_S measured proximal to LA permit us to examine the investment made in photosynthetic leaf tissues relative to mechanically supportive stem tissues (M_S/M_l). Figure 3.18 plots LA (in units of cm^2) versus D (in units of cm) for forty-six North American deciduous species with unlobed and lobed simple leaves, palmate leaves, or

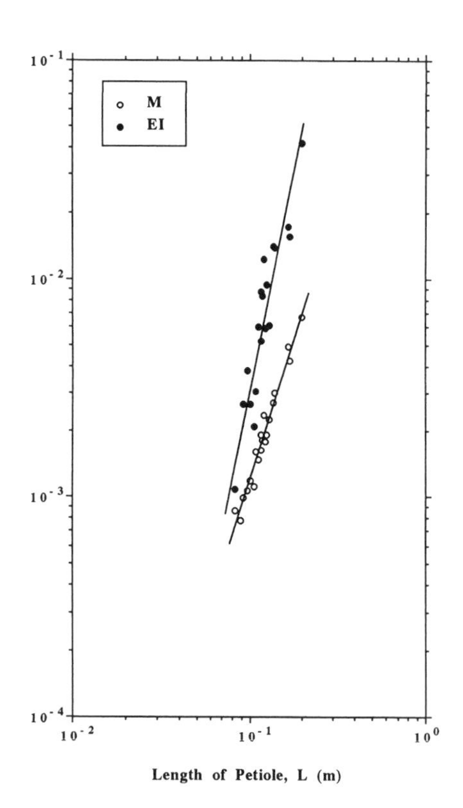

Figure 3.17. Log-log (base 10) of leaf biomass M and flexural stiffness EI versus petiole length L for leaves taken from Boston ivy plants (*Parthenocissus tricuspidata*). Solid lines denote regression curves obtained from LS regression analysis.

pinnate leaves. Regression of these data yields $LA = 917\ D^{1.84}$ ($r^2 = 0.86$; $\alpha_{RMA} = 1.98 \pm 0.05$), which indicates that leaf area increases nearly as the square of stem diameter. For these same data, the relation between D and leaf lamina biomass M_l is nearly isometric: $D \propto M_l^{1.02}$ ($r^2 = 0.98$; $\alpha_{RMA} = 1.03 \pm 0.04$). For the same forty-six species; regression of M_S versus D shows that $M_S \propto D^{2.37}$ ($r^2 = 0.95$; $\alpha_{RMA} = 2.43 \pm 0.03$; data not shown). And finally, regression of stem diameter against the quotient of stem and leaf biomass gives $D \propto (M_S/M_l)^{1.75}$ ($r^2 = 0.84$; $\alpha_{RMA} = 1.91 \pm 0.05$). Evidently the investment of biomass for mechanical support relative to the investment in photosynthetic tissues scales roughly as the square of stem diameter. Based on more comprehensive interspecific comparisons for representative forest communities, Peter White (1983) shows that branching density decreases and the percentage of tree species with lobed or compound leaves increases as the total leaf area increases. In general, these interspecific scalings comply with Corner's rules: (1) leaf morphology becomes more complex with larger (thicker) plant axes; and (2) branches and leaves become smaller with increasing frequency of branching (see Hallé, Oldeman, and Tomlinson 1978, 82).

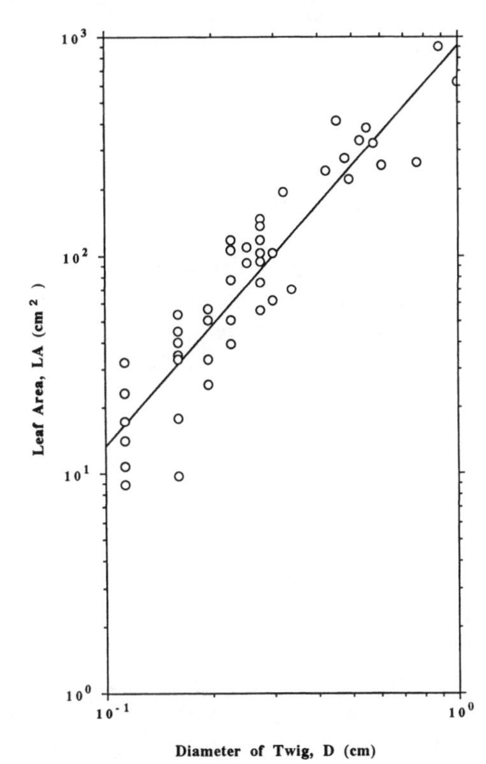

Figure 3.18. Log-log (base 10) plot of leaf area *LA* versus stem diameter *D* for North American deciduous species with unlobed and lobed simple leaves, palmate, or pinnate leaves. Solid line denotes regression curve obtained from LS regression analysis.

3.5 Mechanical Design: General Considerations

In the previous section we considered the mechanical scaling of petioles and saw that the flexural stiffness of these support members changes with respect to the weight of leaf laminae and petiole length in a way that lets leaves orient photosynthetic tissues toward sunlight. Since a petiole is composed of more than one type of tissue, its flexural stiffness can be increased by using stiffer tissues, or by increasing the volume fraction of the stiffest available tissue (either of which would increase Young's modulus), or by adding more tissues in cross section (thereby increasing the second moment of area). Clearly, the material properties, size, and shape of a plant can be modified simultaneously by growth and development. From first principles, however, limits exist on plants' ability to incorporate more materials in the construction of their organs, because adding biomass increases the load a structure must support against the influence of gravity or wind pressure. The dilemma of self-loading is overcome to a degree by the use of low-density but extremely strong and stiff materials. For this reason, engineers typically compare different materials according to two parameters: density-specific strength, which is the ratio of the critical breaking stress σ to density (σ/ρ), and density-specific stiffness (or modulus), which is the ratio of Young's modulus to density (E/ρ). The objective in designing a structure that must cope with large external forces and resist flexure is to use materials that have the highest density-specific strength and the highest density-specific Young's modulus. It is reasonable to suppose that this tactic has evolved in plants, although it is clear that the principle of minimum weight varies among plant species.

The importance of E/ρ and σ/ρ is shown by considering a prismatic untapered support member with length L and cross-sectional area A subjected to a compressive force F resulting from its own weight W. Noting that $W = \rho AL$, we see that the quotient of F and W is given by the formula

$$\frac{F}{W} = \frac{\sigma A}{\rho AL} = \left(\frac{\sigma}{\rho}\right)\frac{1}{L}, \tag{3.30}$$

where, as previously noted, σ is the critical breaking stress of the material used to fabricate the support member. Since L is constant for any particular shape and size, it is evident that the relation between strength and weight is governed solely by σ/ρ. The situation is only slightly more complex when bending is considered, because the breaking stress is given by the equation $\sigma = (FLy)/I = M/Z$, where F is the bending force applied over distance, L, y is one-half the depth of the cross section through the member, I is the second moment of area, M is the bending moment (the product of F and L), and Z is the section modulus ($Z = I/y$). Since $W = \rho AL$ and $F = \sigma I/Ly$, the quotient of F and W is given by the formula

$$\frac{F}{W} = \frac{\sigma I}{\rho AL^2 y} = \left(\frac{\sigma}{\rho}\right)\frac{Z}{AL^2}. \tag{3.31}$$

For any particular size and shape, Z/AL^2 equals a constant, and therefore the relation between the breaking force and self-weight is governed solely by σ/ρ.

The importance of E/ρ is illustrated by considering Euler's equation for the buckling of a vertical column. This equation obtains the critical load P_{crit} that will cause a column with length L and flexural stiffness EI to deflect from its original vertical orientation:

$$P_{crit} = k\frac{EI}{L^2}, \qquad \text{(the Euler formula)}[3] \tag{3.32}$$

where k is a constant that depends on boundary conditions (e.g., for a column anchored at its base and free to deflect at its tip, $k = \pi^2/4 \approx 2.47$). Taking the quotient of the critical load and weight, we find

$$\frac{P_{crit}}{W} = \frac{kEI}{\rho AL^3} = \left(\frac{E}{\rho}\right)\frac{kI}{AL^3}, \tag{3.33}$$

which shows that, for any given size and shape, kI/AL^3 equals a constant and therefore the relation between the critical load and weight is governed by E/ρ. The advantage of using a low-density material is self-evident.

Plants have evolved in response to the principle of minimum weight construction. Wood is extremely strong and stiff yet comparatively very light, and in tension cellulose is the strongest naturally occurring material for its density. This may account for the fact that plants tend to be the tallest terrestrial organisms that have ever existed (see fig. 1.6). In this regard it is useful to compare different species of wood in terms of the trade-off between σ/ρ and E/ρ. Figure 3.19 plots the mean values of σ/ρ versus E/ρ for clear grained wood samples collected from sixty species of angiosperms and twenty-six species of gymnosperms. Since the magnitude of the Young's modulus and breaking stress of any species of wood varies as a function of tissue moisture content, all the data apply to wood specimens with a 12% moisture content. Regression of the data obtains $\sigma/\rho \propto (E/\rho)^{0.45}$ ($r^2 = 0.53$; $\alpha_{RMA} = 0.62 \pm 0.14$). Note that, although it is useful from an engineering perspective to normalize stiffness and strength

3. The irrational number 2.7182818, which serves as an alternative base number for logarithms, is symbolized by e in honor of Leonhard Euler (1707–83), in much the same fashion that π is used in honor of Pythagoras.

with respect to density, the correlation between σ/ρ and E/ρ is suspect because both variables share the same denominator and therefore the coefficient of determination may be inflated as a consequence of autocorrelation. The effect of autocorrelation is easily removed, however. Figure 3.20 plots a σ versus E for the data set shown in figure 3.19. Regression yields $\sigma \propto E^{0.99}$ ($r^2 = 0.75$; $\alpha_{RMA} = 1.14 \pm 0.09$), indicating that these two properties have a nearly isometric relation ($\sigma \propto E$). It is of further interest that regression analysis gives $\rho \propto \sigma^{0.82}$ ($r^2 = 0.80$; $\alpha_{RMA} = 0.92 \pm 0.05$) and $\rho \propto R^{0.71}$, $r^2 = 0.45$; $\alpha_{RMA} = 1.06 \pm 0.09$). Thus, in general terms, the bulk density of wood increases roughly isometrically with increases in either strength or stiffness (Niklas 1993a).

Comparing the density-specific strength and the density-specific stiff-

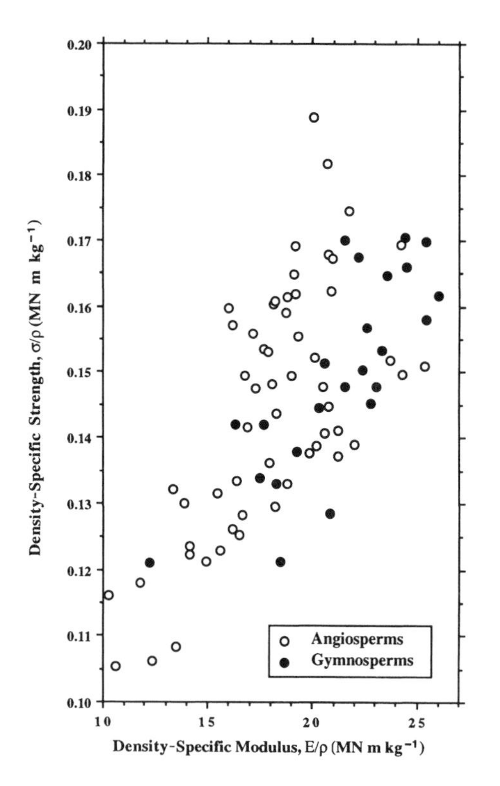

Figure 3.19. Mean density-specific strength σ/ρ plotted against mean density-specific modulus (stiffness) E/ρ of clear-grained wood samples from different species of angiosperms and gymnosperms.

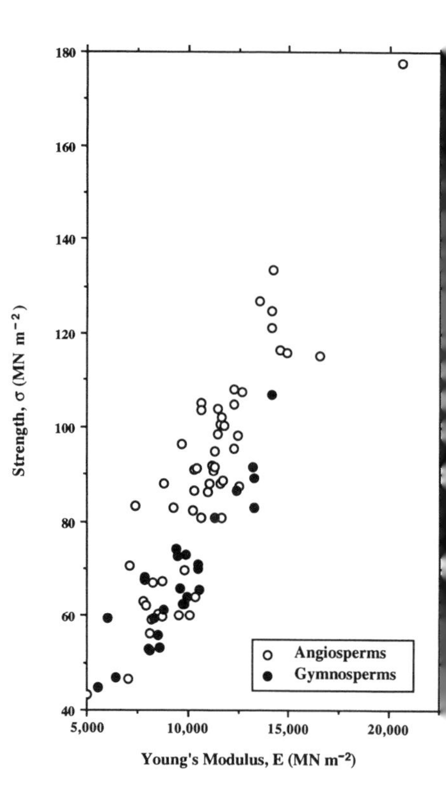

Figure 3.20. Mean value of strength σ plotted against mean stiffness E of clear-grained wood samples from different species of angiosperms and gymnosperms
(see fig. 3.19).

ness of other plant tissues is equally instructive. Parenchyma provides a good example. For potato tuber parenchyma, $\sigma/\rho \propto (E/\rho)^{0.23}$ ($r^2 = 0.95$, $N = 10$; $\alpha_{RMA} = 0.24$). In contrast to wood, $\rho \propto \sigma^{2.55}$ ($r^2 = 0.80$; $\alpha_{RMA} = 2.85$) for this tissue. Thus, for parenchyma even a modest increase in stiffness results in a dramatic increase in density, and hence weight *W*. Indeed, since $W \propto \rho$, we adduce $W \propto \sigma^{2.55}$ for parenchyma. By comparison, $W \propto \sigma^{0.82}$ for wood. Thus, for any given magnitude of strength the weight of an organ composed of parenchyma is almost twice that of an organ composed of an equivalent volume of wood. One of the consequences of this is that plant stems composed mostly of parenchyma are mechanically limited in how high they can grow before they mechanically fail under their own weight. This is shown by considering the relations among organ weight, tissue density, and tissue stiffness in light of the Euler formula for the buckling load of a column (see eq. 3.32). With the aid of a modified version of this formula (see Greenhill 1881) and empirically determined values for *E* and ρ, we can calculate the maximum slenderness ratio ($\mathfrak{R} = L/R$) that a terete cylinder composed of each type of tissue can reach before it undergoes elastic buckling. The objective is to crudely estimate the relation between the stiffness of different plant tissues and the height an organ composed of each of these tissues can attain before it begins to deflect from the vertical under its own weight. Figure 3.21 plots $\mathfrak{R}$ versus *E* for pure cellulose and isolated tissues from different organs of an elderberry plant, *Sambucus canadensis*. Data from a single plant were collected because the magnitude of *E* for a tissue type will vary among individuals of the same species. For comparison the estimated $\mathfrak{R}$ for the theoretical value of *E* for cellulose is also plotted.

Figure 3.21 illustrates a number of points. First, relatively thin-walled tissues composed primarily of water (parenchyma and collenchyma) are the poorest materials with which to maximize vertical growth, in large part because they have very low density-specific stiffness. Nonetheless, when surrounded by stiffer materials like sclerenchyma, as they are in many primary stems, parenchyma and collenchyma are excellent tissues in terms of dealing with compressive loadings (because they are essentially incompressible). Second, thick-walled tissues (vascular fibers and sclerenchyma) can achieve substantially greater slenderness ratios than thin-walled tissues like parenchyma. The density-specific stiffness of these tissues is comparatively high, and therefore stiffness disproportionately increases with respect to self-loading. Third, sclerenchyma is a better material than wood for maximizing vertical growth. Fourth, although *E* spans five orders of magnitude, the slenderness ratio spans only three orders, suggesting that dramatic increases in the magnitude of *E* yield diminishing returns for the maximum elevation a columnar plant can aspire to. In fact, regression of the data shows that $\mathfrak{R} \propto E^{0.50}$, indicating that the scaling of the slenderness ratio is very anisometric (an increase in E achieves a disproportionately smaller benefit in height). The final point is that if plants were designed solely to elevate themselves, they would be built mostly of cellulose. In fact, cellulose is the strongest known material for its density. It is also one of the stiffest materials. Incidentally, since dry cellulose is stiffer than wet cellulose, we come to recognize one of the functional roles of lignin, which is hydrophobic. Obviously, a plant built of pure cellulose is a biological impossibility. But many small terrestrial plants (e.g., mosses and liverworts) have extremely high cellulose volume fractions. These plants are composed of parenchymatous thin-walled tissues that mechanically operate as hydrostats. That is, each cell works much like an inflated balloon or tire. As the hydrostatic pressure within thin-walled cells increases, the cell walls are increasingly placed in tension and increasingly stiffen. Since these cell walls are composed of cellulose and since cellulose has a very high Young's modulus measured in tension, hydrostatic tissues are reasonable support materials for plants that have

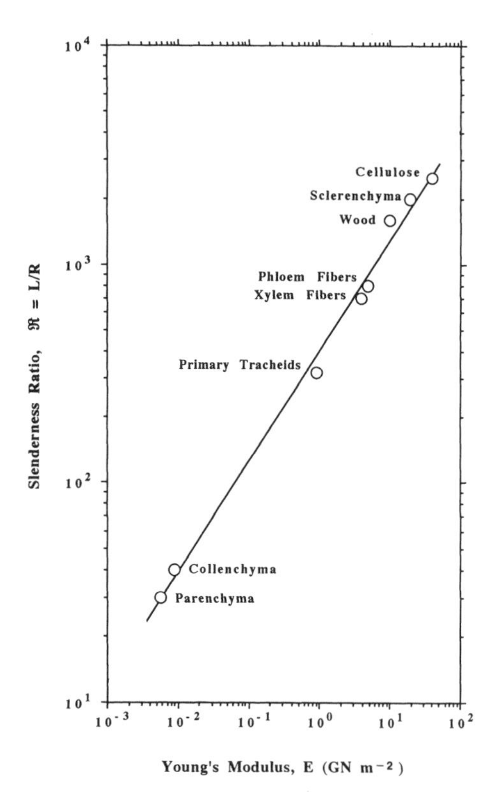

Figure 3.21. Log-log (base 10) plot of the slenderness ratio $\mathfrak{R}$ (the quotient of length L and radius R) of a terete cylinder composed exclusively of the materials indicated versus the Young's elastic modulus E of the materials. Solid line is the regression curve obtained from LS regression analysis.

ready and constant access to water. By contrast, under conditions of episodic water stress, hydrostatic tissues are unreliable for mechanical support, and thick-walled tissues are the preferred support materials. Once the mechanical behavior of different types of tissues is understood, it comes as little surprise that plants adapted to wet habitats (hydrophytes) are composed mostly of thin-walled living tissues (parenchyma and collenchyma), whereas plants adapted to very dry habitats (xerophytes) possess large volume fractions of thick-walled dead tissues (sclerenchyma and wood).

Thus far we have made the dangerous assumption that ρ, E, and σ are constant among different samples of the same tissue type. This assumption was pedagogically useful, and in some circumstances it is valid. For example, ρ and E can be measured for samples of wood isolated from sequentially older growth layers GL in the trunk of an individual tree. Regression of data (not shown) of wood from twenty-five growth layers from the trunk of a silver maple, *Acer saccharinum,* shows that $\rho \propto GL^{0.016}$ ($r^2 = 0.73$, $N = 25$; $\alpha_{RMA} = 0.019$) and $E \propto GL^{0.093}$ ($r^2 = 0.78$, $N = 25$; $\alpha_{RMA} = 0.11$). These regression formulas show that ρ and E increase only modestly (or not at all) as a function of the age of a wood sample, presumably as a consequence of the progressive accumulation of secondary metabolites within cell lumens, which increases the bulk density of the wood sample. Sapwood growth layers (which have tracheary elements largely unoccluded with secondary metabolites) have lower ρ and E than heartwood growth layers. For whatever reason, however, from the previous regression formulas it appears that E increases with respect to GL at a greater rate than ρ. Consequently, the density-specific stiffness of wood varies to a small degree among growth layers. Figure 3.22, for example, plots E/ρ versus GL for the twenty-five specimens of wood collected from a trunk of *Acer saccharinum.* Regression of these data yields $E/\rho \propto GL^{0.08}$ ($r^2 = 0.75$; $\alpha_{RMA} = 0.10$). It is difficult to evaluate the significance of this trend without other information. It is worth noting, however, that the volume fraction of heartwood in a trunk increases with the age of the tree. The trunks of comparatively young trees, which are composed of sapwood, therefore are less stiff for their density than those of older trees, which have proportionately more heartwood. The increase in E/ρ of growth layers from the pith toward the bark of the silver maple is the result of a disproportionate increase in E with respect to increasing ρ, even though the magnitudes of both variables increase from pith to bark. In this regard Wiemann and Williamson (1989), who measured pith to bark specific gravity SG trends in eighteen tropical dry forest and six montane rain forest species, report statistically significant increases in SG in most cases. These authors show that the number of species evincing a radial increase in SG declines along a transect from tropical wet forests to mesic temperate forests (Wiemann and Williamson 1989, 927 fig. 1). Unfortunately, radial changes in E were not reported by Wiemann and Williamson, although the complex scaling relation between ρ and E would have shed light on many important ecological effects, such as the demographic differences in wind-induced mechanical failure seen among tree species (Putz et al.1983).

We need to consider minimum weight design further in terms of variations in plant size and shape. As shown in eqs. (3.30) to (3.33), an increase in the girth of a mechanically supportive stem can be as important as a change in cross-sectional shape. The size and shape dependency of mechanical design can be examined in a variety of complementary ways, but the scaling of the section modulus Z and the second moment of area I serves as one of the more useful tools, since the numerical values of these parameters are both size and shape dependent. Table 3.2 provides formulas for Z and I. Since $Z = I/y$, the numerical values of the section modulus and the second moment of area are interdependent. The effect

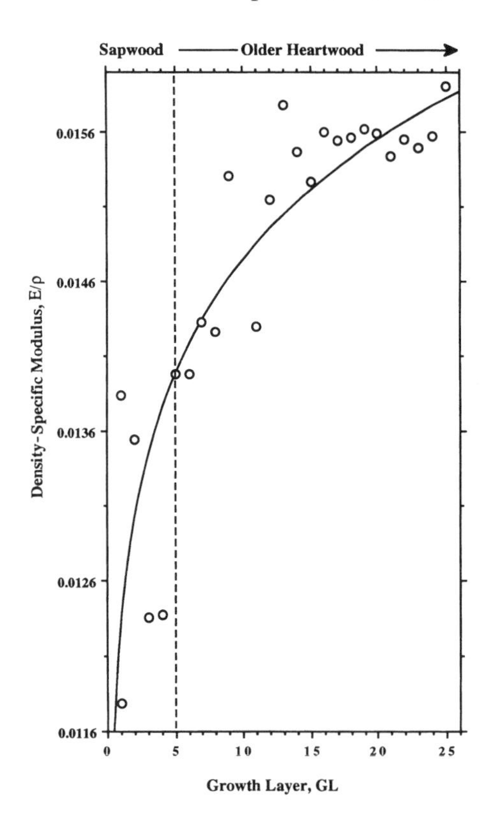

Figure 3.22. Density-specific modulus (stiffness) E/ρ plotted as a function of the relative position of growth layers GL of wood within the trunk of a tree. Dashed vertical line indicates the boundary between sapwood and heartwood GL. Curved solid line is the LS regression curve for the log-transformed data.

of geometry on mechanical behavior is seen by comparing Z for different prismatic elements with equivalent cross-section area A. In terms of geometry, when $A = 10$ (arbitrary units), the section modulus of a terete cylinder equals 4.46; Z equals 2.63 for a square prismatic bar; and Z equals 2.48 for a three-sided prism having sides of equal length. An elliptical prismatic bar is even more interesting, since the magnitude of its section modulus depends on the orientation of the elliptical cross section with regard to the plane of bending. That is, an elliptical cross section has two section moduli. We shall denote the section modulus when the plane of bending is perpendicular to the minor axis of the elliptical cross section as Z_{minor} (as shown in table 3.2); the section modulus when the plane of bending is perpendicular to the major axis of the elliptical cross section is denoted as Z_{major} (the reverse of that shown in table 3.2). When $A = 10$ and the ratio of the major to minor axes is 2, we calculate that Z_{minor} equals 3.14 while Z_{major} equals 6.28. Thus $Z_{major}/Z_{minor} = 2.0$. A change in the aspect ratio of the elliptical cross section has a proportional effect on the difference between Z_{major} and Z_{minor}. For example, when the ratio of the major to the minor axis equals 4, $Z_{major} = 8.9$, $Z_{minor} = 2.2$, and $Z_{major}/Z_{minor} \approx 4.0$.

Plant growth and development can adjust Z and I in response to the prevailing direction of externally applied forces. For example, when a sapling is caused to sway in one direction, the cross section of its trunk assumes an elliptical geometry whose wider dimension is aligned parallel to the plane of sway (Knight 1811). As previously noted, this growth response to mechanical perturbation decreases the bending stress (owing to a preferential increase in Z) and increases stiffness (owing to a preferential increase in I). Likewise, when a tree is guy wired, portions of the trunk below the wires, which are less mechanically perturbed than those above the point of restraint, grow less rapidly in girth than the younger, more acropetal portions of the trunk that are free to sway and bend (Jacobs 1954). In general, the stems and leaves of mechanically perturbed plants tend to grow more in girth and less in length than those of plants free from dynamic loadings. These growth responses illustrate that plant growth and development can alter the size, shape, and orientation of organs to minimize mechanical stresses and increase stiffness.

To appreciate the importance of Z and I, consider a stem bent under its own weight. Under these conditions, the stress σ that develops within each of its cross sections is given by the quotient of the bending moment M and Z. In other words, $\sigma = M/Z$. Since M is the product of the force F of body weight (i.e., self-loading) and the distance l over with F is applied, and since F equals the product of the mass m of the stem and the acceleration due to gravity g(9.8 m s^{-2}), the bending moment equals $Fl =$

$mgl = \rho ALgl$, where A is cross-sectional area and L is stem length. Thus the bending stress is given by the following formula:

$$\sigma = \frac{M}{Z} = \rho g \frac{ALl}{Z}. \tag{3.34a}$$

The maximum stress σ_{max} will occur when l equals the length L of the stem:

$$\sigma_{max} = \rho g \frac{AL^2}{Z}. \tag{3.34b}$$

Equation (3.34b) shows that σ_{max} will increase as the square of L or A/Z. Assuming that the stem is indeterminate in growth in length and that the bulk density of the stem is constant, a reduction of the maximum bending stress requires minimizing A/Z.

The influence of size and shape on mechanical behavior is readily seen for any simple untapered prismatic support member, for example, a terete cylindrical column. Table 3.2 shows that $Z = \pi R^3/4 = AR/4$ for this geometry, where A is cross-sectional area. Inserting this formula into eq. (3.34b) yields

$$\sigma_{max} = \rho g \left(\frac{V}{A}\right) \Re. \tag{3.34c}$$

This equation contains the slenderness ratio $\Re = L/R$ and a size- and shape-dependent term (V/A). It reveals that the maximum bending stress decreases as cylinders become more slender, or as their cross-sectional area increases, or as the whole size (volume) of the cylinder decreases. Provided the volume of the cylinder is conserved, an increase in the slenderness ratio requires a decrease in cross-sectional area. Thus, from eq. (3.34c), bending stresses can be greatly reduced if the bulk of the material within each cross section is orthogonally aligned to the direction of an applied bending force. Table 3.2 also shows that the magnitude of the second moment of area I is influenced by the orientation of support members with axiasymmetric cross sections. This is important because flexural stiffness can be increased by increasing I regardless of the Young's modulus of the material used to construct the member.

3.6 The Mechanical Scaling of Tree Height

The scaling of the height of gymnosperm and dicot trees with respect to trunk diameter D has been examined principally in terms of two scaling models (McMahon and Kronauer 1976): stress similarity and elastic similarity. The stress similarity model assumes that a constant maximum stress level is maintained throughout the length L of a branch or tree trunk. This assumption requires L to scale to the square root of D (i.e., $L \propto D^{1/2}$). The elastic similarity model assumes that the deflection at the free end of a stem must remain constant relative to L. Based on Euler's formula, this assumption requires L to scale to the 2/3 power of D (i.e., $L \propto D^{2/3}$). Both the stress similarity and elastic similarity models assume that the density of stems differing in size is constant ($\rho \approx$ a constant). Therefore both models predict an isometric relation between mass and volume ($M \propto V^{1.0}$). This assumption leads to additional predictions about the relation between stem mass M and L or D. In terms of stress similarity, since $M \propto V \propto D^2L$ and $D \propto L^2$, we see that $M \propto (L^2)^2L$, and therefore $L \propto M^{1/5}$ and $D \propto M^{2/5}$. In terms of elastic similarity, since $M \propto V \propto D^2L$ and $D \propto L^{3/2}$, we see that $M \propto (L^{3/2})^2L$ and therefore $L \propto M^{1/4}$ and $D \propto M^{3/8}$. The null hypothesis to the elastic similarity and the stress similarity models is geometric similitude, which predicts that L scales in direct proportion to D (i.e., $L \propto D^{1.0}$). Since $M \propto V \propto D^2L$, geometric similitude predicts that $L \propto M^{1/3}$ and $D \propto M^{1/3}$. Table 3.3 summarizes the predicted relations among L, D, and M for the three scaling models. Note that all three models assume $M \propto V^{1.0}$.

Because the elastic similarity and stress similarity models predict very similar scaling exponents for plant height with respect to stem diameter, statistical evaluations of how well empirical data comply with these two models can be very difficult. For example, the elastic similarity model predicts that plant height should scale as the 2/3 power of stem diameter ($\alpha \approx 0.67$), while the stress similarity model predicts that height will scale as the 1/2 power of stem diameter ($\alpha = 0.50$). The slopes of regression curves for data from real plants often obtain very broad 95% confidence intervals that encompass both of these predicted scaling relations. Another difficulty with all three models is that each assumes that material properties, specifically density and Young's modulus, are independent of size (in trees, size equates with age). Previous scaling analyses reveal that this assumption is not always justified (see fig. 3.22). Provided tissue den-

Table 3.3. Scaling Exponents α for Three Allometric Models Predicting the Relations among the Diameter D, Length L, and Mass M of a Mechanical Support Member

	Stress Similarity	Elastic Similarity	Geometric Similitude
$L \propto D^\alpha$	0.500	0.666	1.00
$L \propto M^\alpha$	0.200	0.250	0.333
$D \propto M^\alpha$	0.400	0.375	0.333

sity and Young's modulus are size-dependent physical parameters, the scaling exponent for plant height with respect to stem diameter can deviate from those expected by the three models summarized in table 3.3. Consider the elastic similarity model. Based on Euler's buckling formula, the critical height H_{crit} to which a vertical tree trunk could be elevated before it undergoes elastic buckling is given by the formula

$$H_{crit} = C\left(\frac{E}{\rho}\right)^{1/3} D^{2/3}, \tag{3.35}$$

where C is the constant of proportionality. When the force inducing elastic buckling acts on the support member's center of mass, $C = 0.851$; when the force is distributed over the full extent of the member, $C = 0.792$ (Greenhill 1881). For trees, King and Loucks (1978) show that C depends on the ratio k of the weight of the crown of a tree to the weight of the trunk such that $C = [(0.007601 + 0.08655\,k + 0.334\,k^2)/(0.001427 + 0.02907\,k + 0.1695\,k^2 + 0.6125\,k^3)]^{0.333}$. Although eq. (3.35) obtains the proportionality $H \propto D^{2/3}$ when $E/\rho \approx$ a constant, consider the possibility that the density-specific stiffness of wood is size dependent such that $D^{\alpha 1} \propto (E/\rho)^{1/3}$. If so, then $H = CD^{\alpha 1+2/3=\alpha}$, where α can assume a value greater or less than 2/3 depending on the value of α_1. Thus, the evaluation of the three contending models requires data on the relation between ρ and E. Typically these data are not reported in the literature.

McMahon (1973, 1975; see also McMahon and Kronauer 1976; Rich et al. 1986; Norberg 1988; Bertram 1989) asserts that the elastic similarity model describes the mechanical design of dicot and gymnosperm trees. McMahon (1973, 1975) plotted height H versus trunk diameter D for the largest known living specimen of nearly every American dicot and gymnosperm tree species ($N = 576$) and reports that the scaling relation between H and D conforms to that predicted by the elastic similarity model. Unfortunately, the data were not examined statistically to determine the scaling exponent for H versus D. Rather, a line with $\alpha = 2/3$ was "drawn by eye" (McMahon and Kronauer 1976, 456, fig. 1) through the center of mass of a log-log plot of the data points. This approach did not test whether the actual scaling exponent agrees with that predicted by either the stress similarity or elastic similarity model (whether α equals 1/2 or 2/3, respectively). Therefore McMahon's conclusions are subject to well-reasoned criticism (LaBarbera 1986, 1989).

McMahon (1973) calculated theoretical critical buckling height (see eq. 3.35), based on H and D for each of the 576 species, by assuming that $C = 0.792$ and $E/\rho \approx$ a constant. Unfortunately, the actual value of E/ρ used in these calculations cannot be reliably determined because the value reported by McMahon does not yield a regression curve with the same Y_1-intercept reported in this study (1973, 1202).[4] Nonetheless, since the numerical value of E McMahon used was lower than for the wood of most species, how far the record heights of tree species fall below estimated critical buckling heights is underestimated. Therefore McMahon's conclusion regarding the safety factor of tree height is very conservative, although the assumption that $E/\rho \approx$ a constant may be in error and therefore the slope of the regression curve for estimated critical buckling heights may differ from that predicted from the proportionality $H \propto D^{2/3}$.

Given the circumstances just outlined, it seems desirable to reevaluate the relation between tree height H and trunk diameter D as well as the relation between H_{crit} and D. For this purpose, data for H and D(N = 480 species) were taken from the references used originally by McMahon (1973) (i.e., Chittenden 1931; Royal Horticultural Society's *Conifers in Cultivation,* 1932; Pomeroy and Dixon 1966; and Social register of big trees 1966, 1971). Additionally, the theoretical critical buckling height was calculated using eq (3.35) assuming $C = 0.792$ based on data for ρ and E, available for fifty-six species (see table 3.4).

Figure 3.23 shows that the scaling of tree height obtains equivocal results regarding which among the three contending scaling models is the most appropriate (Niklas 1993b). Regression of the pooled data from gymnosperm and angiosperm trees shows that $H = 20.6\ D^{0.535}$ ($r^2 = 0.54$, N = 480; $\alpha_{RMA} = 0.73 \pm 0.02$). The 95% confidence intervals for the RMA scaling exponent are 0.69–0.77, indicating that stress similarity is very unlikely, whereas elastic similarity is "just within grasp." The problematic nature of the scaling exponent for tree height is seen further when we consider the relation between H and D separately for dicots and for gymnosperms (most of which are conifer species). Regression of H versus D from dicot trees shows that $H = 19.1\ D^{0.474}$ ($r^2 = 0.47$, $N = 375$; $\alpha_{RMA} = 0.69 \pm 0.03$). The 95% confidence intervals for the RMA scaling exponent are 0.63–0.75. Thus, the elastic similarity model seems reasonable for scaling the height of very old dicot trees. By contrast, the situation for conifers indicates that $H = 27.8\ D^{0.430}$ ($r^2 = 0.25$, $N = 105$; $\alpha_{RMA} = 0.87 \pm 0.08$). The 95% confidence intervals for the RMA scaling exponent are 0.70–1.03. Thus the null hypothesis (geometric similitude) cannot be rejected for gymnosperm trees, whereas the stress similarity and elastic similarity models appear to be inappropriate for the pooled data from all tree species.

4. The value given, $E = 1.05 \times 10^5$ kg m^{-2} is said to be the average value for green wood. Nonetheless this value is evidently too low, since E for most species of wood is on the order of 10^8 kg per square meter. In a subsequent paper (McMahon and Kronauer 1976), we find that the Young's modulus used to compute the critical buckling height is that of pine wood.

Table 3.4. Data for Fifty-six Tree Species Used to Calculate the Theoretical Buckling Height H_{crit}

Species	ρ	*E*	*D*	*H*
Abies amabilis	415	1079	2.08	56.7 g
Abies balsamea	414	879	0.679	35.4 g
Acer nigrum	620	1141	1.35	35.4
Acer rubrum	546	1155	1.49	41.5
Acer saccharinum	506	805	2.22	27.4
Acer saccharum	676	1290	1.92	35.4
Aesculus octandra	383	829	1.54	25.9
Betula lenta	714	1520	1.47	21.3
Betula papyrifera	600	1119	1.06	29.3
Betula populifolia	552	797	0.703	18.3
Castanea dentata	454	870	1.52	27.4
Chamaecyparis thyoides	352	655	1.50	26.5 g
Cornus florida	796	1085	0.517	9.14
Diospyros virginiana	776	1443	1.26	24.4
Fagus grandifolia	655	1180	1.79	27.7
Fraxinus americana	638	1246	2.16	24.4
Fraxinus nigra	526	1126	1.48	26.5
Gleditsia triacanthos	666	1165	1.82	28.0
Juglans cinerea	404	830	1.14	25.9
Juglans nigra	562	1185	1.96	32.9
Juniperus virginiana	492	612	1.18	23.2 g
Larix occidentalis	587	1188	2.33	36.6 g
Liriodendron tulipifera	427	1058	2.57	25.3
Magnolia acuminata	516	1276	1.78	38.1
Picea glauca	431	1001	1.02	35.4 g
Picea mariana	428	1069	1.14	22.9 g
Picea rubens	413	1001	1.34	32.4 g
Pinus banksiana	461	868	0.631	27.7 g
Pinus echinata	584	1345	1.03	44.5 g
Pinus palustris	638	1445	1.04	34.4 g
Pinus resinosa	507	1264	0.857	38.1 g
Pinus rigida	542	965	0.800	29.6 g
Pinus strobus	373	898	1.74	46.0 g
Pinus taeda	593	1354	1.60	39.0 g
Populus balsamifera	331	716	1.16	27.1
Populus deltoides	433	972	2.49	39.9
Populus tremuloides	401	838	1.12	21.3
Prunus pennsylvanica	425	8920	0.517	13.7
Prunus serotina	534	1046	2.26	31.1
Pyrus malus	745	894	0.218	7.93
Quercus alba	710	1251	2.68	28.9
Quercus bicolor	792	1446	1.76	18.3
Quercus chrysolepis	838	1149	3.15	20.4
Quercus laurifolia	703	1182	1.94	25.6
Quercus macrocarpa	671	723	2.01	37.2

(Continued)

Table 3.4. (*Continued*)

Species	ρ	*E*	*D*	*H*
Quercus prinus	756	1247	2.16	28.9
Quercus stellata	738	1063	1.33	24.4
Quercus velutina	669	1153	2.16	38.1
Robinia pseudoacacia	708	1448	1.54	25.9
Salix nigra	408	513	2.53	25.9
Sequoia sempervirens	436	958	4.27	112 g
Thuja plicata	344	819	6.41	39.6 g
Tilia americana	398	1029	1.75	32.0
Tsuga canadensis	431	846	1.92	35.9 g
Tsuga heterophylla	432	1012	2.64	38.1
Ulmus americana	554	948	2.39	48.8

Note: See figure 3.23. Data for record diameters and heights from Pomeroy and Dixon 1966 and Social register of big trees 1971.

Symbols: ρ = density (kg m^{-3}), *E* = Young's modulus (10^6 kg m^{-2}), *D* = diameter (m), *H* = height (m), and *g* = gymnosperm species.

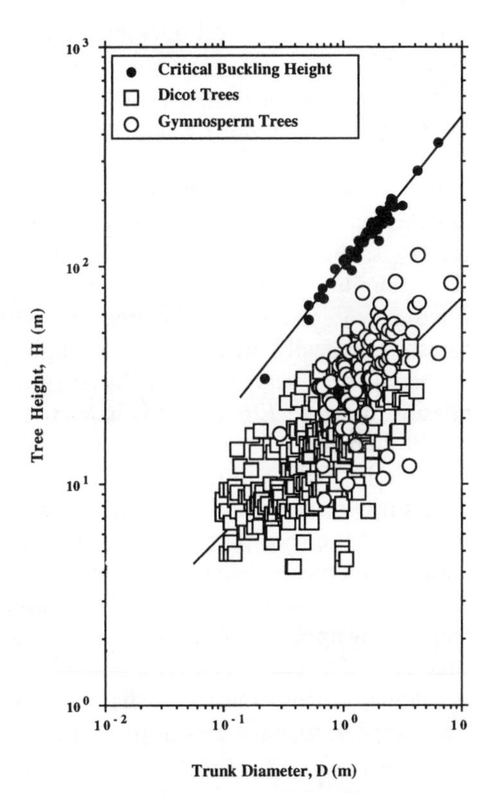

Figure 3.23. Log-log (base 10) plot of the estimated critical buckling height and the height *H* of dicot and gymnosperm tree species versus trunk diameter *D*. The data for *H* and *D* are for the largest known specimens of each species. Data points for the critical buckling height are for those species for which *H*, *D*, and the density-specific stiffness *E*/ρ of wood are known (see table 3.4).

As a brief aside, the scaling of tree height is subject to the phyletic effect, since the slope of the regression curve for tree height versus trunk diameter differs for "dicot" and "gymnosperm" data sets. As I noted in chapter 1, the phyletic effect is most evident in data sets that evince taxonomic bias. This is clearly the case for the data just reviewed. It also is painfully evident in the literature treating "trees." For example, Whittaker and Woodwell (1968) report mean values for the height, diameter, and dry mass of seven woody species, such that $H \propto D^{0.838}$ ($r^2 = 0.99$; $\alpha_{RMA} = 0.84 \pm 0.03$) and $H \propto M^{0.341}$ ($r^2 = 0.99$; $\alpha_{RMA} = 0.34 \pm 0.03$). From these scaling relations we might adduce geometric similitude as the most appropriate model (see table 3.3). However, among the seven species treated by Whittaker and Woodwell, three are oaks (*Quercus coccinea, Q. alba,* and *Q. ilicifolia*) and two are blueberries (*Vaccinium angustifolium* and *V. vacillans*). The species composition of this data set biases the scaling exponents in favor of geometric similitude, and therefore claims regarding broad interspecific scaling relations based on these data are erroneous (e.g., LaBarbera 1986, 76).

Turning to the critical buckling height of trees, figure 3.23 shows that $H_{crit} = 97.7\ D^{0.689}$ ($r^2 = 0.97$; $\alpha_{RMA} = 0.70 \pm 0.02$). The confidence intervals of the RMA scaling exponent are 0.66–0.74, indicating that critical buckling height roughly scales as the 2/3 power of trunk diameter. But this is hardly surprising, since H_{crit} is computed from Greenhill's formula (which contains the parameter $D^{2/3}$ and therefore invariably obtains $\alpha = 2/3$ provided E/ρ is size independent). Curiously, the empirical value of the H_{crit} scaling exponent is just slightly greater than that predicted by the elastic similarity model. There are two plausible explanations for this: (1) E/ρ may scale in a slightly positive way with respect to D, and (2) the way foresters typically measure D may slightly overestimate the diameter of the trunk of a very old tree. Based on the data given in table 3.4, the relation between E/ρ and D is isometric. Consequently, the way breast-height diameter is measured may be more relevant. As the trunks of trees increase in girth, and therefore as trees get older, the location of breast height progressively shifts in relation to the transition region between the trunk and the root butt (Norberg 1988). In this transition region trunk diameter increases disproportionately in relation to the taper in girth that characterizes most of the trunk. Consequently, larger and older tree trunks may appear to have disproprotionately larger diameters than their smaller and younger counterparts. Since the theoretical buckling height is calculated from the trunk diameter measured at breast height, older trees appear to have larger theoretical buckling heights than younger trees, largely as an artifact of how and where diameter is measured. Despite this artifact, inspection of figure 3.23 shows that all the data points for tree height fall below the regression curve for H_{crit}. Since the two regression curves shown in this figure were determined from independent data sets, it is statistically legitimate to take the quotient of critical buckling height and tree height to determine the mechanical "safety factor" for very old trees. Based on reduced major axis regressions, $H_{crit} = 97.7 D^{0.70}$ and $H = 20.6 D^{0.73}$. Therefore $H_{crit}/H = 97.7 D^{0.70}/20.6 D^{0.73} = 4.74 D^{-0.03}$. At issue is whether $D^{-0.03} \approx D^{0}$. If we take this leap of faith, the ratio of the critical buckling height to actual tree height, on the average, roughly equals five. Although McMahon asserts that the safety factor is closer to four, his conclusion that the vertical growth of trees attains a comparatively high safety factor against elastic buckling is justified, although the ancillary claim that the elastic similarity model categorically describes the interspecific scaling of H versus D is statistically questionable.

Each of the three scaling models summarized in table 3.3 is derived analytically from "first principles." Thus each hypothesis expects every individual taxon, regardless of its phyletic affiliation, to scale height according to the scaling exponents it predicts based on the first principles used to derive the model. Thus far we have examined these models only in the context of broad interspecific comparisons based on data from extremely old specimens of dicot and gymnosperm tree species. As we have seen, these interspecific comparisons give ambiguous results. The scaling exponent for the pooled data from dicot and gymnosperm trees was higher than predicted by either the stress similarity or elastic similarity model and significantly lower than that predicted by geometric similitude. Conversely, we could not reject the hypothesis of geometric similitude for the gymnosperm data set, whereas the slope of the regression for dicot trees did not differ from that predicted by elastic similarity. No model, therefore, was found to be more robust than its two counterparts in terms of these interspecific comparisons. Much the same conclusion is reached when the three models are examined working from intraspecific comparisons. For example, Whittaker and Woodwell (1968) found $\alpha_{LS} = 0.68$ ($r^2 = 0.95$; $\alpha_{RMA} = 0.70$) for *Quercus coccinea*. Based on the statistical data these authors provide, this intraspecific scaling exponent is consistent with that predicted by the elastic similarity model. Yet for the congeneric *Q. alba*, the same authors report $\alpha_{LS} = 0.56$ ($r^2 = 0.95$; $\alpha_{RMA} = 0.58$), which is compatible with the stress similarity model (see table 3.3). By the same token, Dean and Long (1986) report that the stress similarity model best describes the intraspecific scaling of *Pinus contorta* var. *latifolia* as well as the petioles of *Trifolium pratense*. Yet these authors report that the elastic similarity model may be statistically valid when height is regressed against the trunk diameters of only very large, fully mature pine trees. Finally, Bertram (1989) reports that the length of branches on a silver maple tree (*Acer saccharinum,* measuring 13.2 m in height) scales to diameter as predicted by the elastic similarity model ($\alpha_{LS} = 0.652$, $r^2 = 0.87$, $N = 183$), whereas the scaling exponent of the lengths of peripheral twigs versus diameter conforms to the geometric similitude model ($\alpha_{LS} = 1.08$, $r^2 = 0.60$, $N = 218$) (see Bertram 1989, table 2). In terms of RMA regression, the scaling exponent for the branches Bertram examined is 0.70, whereas that of twigs is 1.39 (much higher than that obtained by geometric similitude).

In light of these intra- and interspecific comparisons, none of the three scaling models summarized in table 3.3 appears to be a valid analytical scaling hypothesis. This view considers further that none of these models takes into account the effects of stem ontogeny on the functional relation between stem length and girth. The trunk of a large (old) tree functions principally as a compressive support member that must support an extremely large biomass. By contrast, the young tree trunk, as well as the younger branches and twigs of even very old trees, principally operates as a cantilevered beam that mechanically supports disproportionately less biomass than the trunk of an old tree. The ontogenetic status as well as the mechanical design considerations of the trunk, on the one hand, and young branches and twigs, on the other, clearly differ. For these reasons, very young trees can grow according to the geometric similitude model and still resist elastic buckling despite the proportionality $H \propto D$. As a tree ages and accumulates secondary growth layers of wood, however, its mechanical configuration may gradually acquire the scaling relation predicted by the elastic similarity model until, with extreme age and size, the scaling of height may shift to assume the appearance of the stress similarity model. Essentially, the three scaling models reflect the influence of the slenderness ratio of support members on the ability of stems to sustain self-loadings. It is important to remember that this ratio is the external manifestation of the scaling of primary growth, which accounts for stem elongation, and secondary growth, which accounts for the increase in stem girth. Further, since every empirically determined scaling exponent relating stem length to girth (hence, diameter) reflects the differential growth in length compared with diameter, it is reasonable to suppose that the scaling of length to girth for support members depends on the position of the member within the branching hierarchy of an individual tree. The very few empirical studies available on the mechanical scaling of individual trees support this view. Among the best of these is that of Bertram, who makes the cogent observation that "the peripheral branches of trees, because of their small size (which matches that of shrubs), are constrained in a manner similar to shrubs and can be considered, in a structural sense, as shrubs that are simply held aloft by the tree's larger stems and branches" (1989, 252). Nonetheless, until the physical principles that mechanically "constrain" the scaling of peripheral branches of large trees or the stems of shrubs are rigorously identified, it seems more appropriate to use the geometric, stress, and elastic similarity models to describe the relation between length and girth for different phases of stem ontogeny rather than as analytical (functional) scaling hypotheses.

Finally, there appear to be correlations between the lateral spread *Sp* of the crowns of trees and dicot tree height *H* as well as trunk diameter *D*, suggesting that dicots invest a comparatively large portion of their biomass in the formation of long, cantilevered horizontal branches relative to their trunks as opposed to gymnosperms, which appear to invest more in the formation of their central trunks compared with their branches. Figures 3.24 and 3.25 plot *Sp* versus *H* and *Sp* versus *D*, respectively. Regressions of the data from angiosperm species show that $Sp = 1.19\ H^{0.877}$ ($r^2 = 0.59$; $\alpha_{RMA} = 1.14 \pm 0.02$) and $Sp = 16.6\ D^{0.599}$ ($r^2 =$

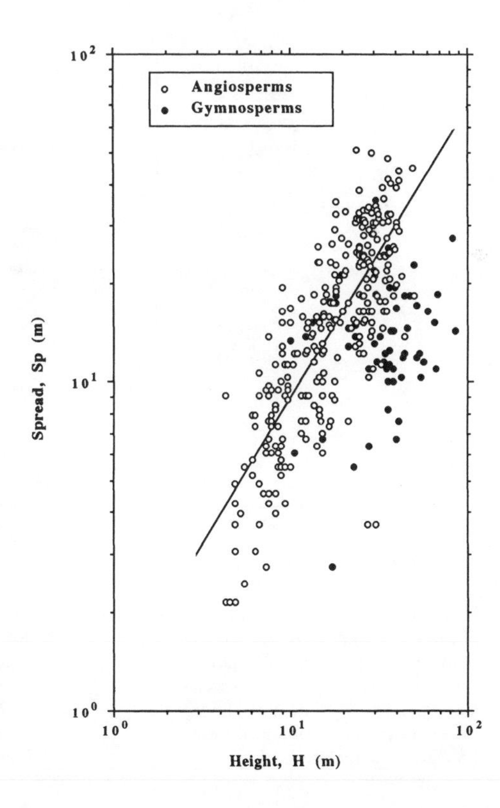

Figure 3.24. Log-log (base 10) plot of canopy spread *Sp* versus height *H* for record specimens of angiosperm and gymnosperm tree species. Solid diagonal line is the regression curve obtained from LS regression analysis.

0.58; $\alpha_{RMA} = 0.79 \pm 0.02$). Figure 3.26 plots *Sp* versus the ratio of tree height to diameter *H/D*. Data for gymnosperms are coplotted for comparison. No statistically significant relation between *Sp* and *H* or *D* was found for gymnosperms, possibly because of the pyramidal growth habit of the conifer species that numerically dominate the data set. Although no regression curve adequately predicts the relation between the lateral spread of tree crowns and tree aspect ratio, there is a slight tendency for *Sp* to decrease with increasing *H/D*. Finally, most of the data points fall below *H/D* = 50, while angiosperms hold the record for both *Sp* and *H/D*.

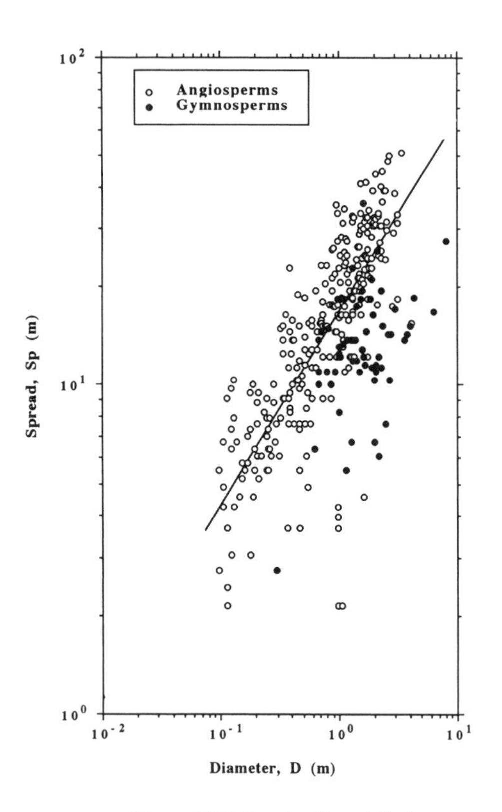

Figure 3.25. Log-log (base 10) plot of canopy spread *SP* versus diameter *D* for record specimens of angiosperm and gymnosperm tree species (see fig. 3.24). Solid diagonal line is the regression curve obtained from LS regression analysis.

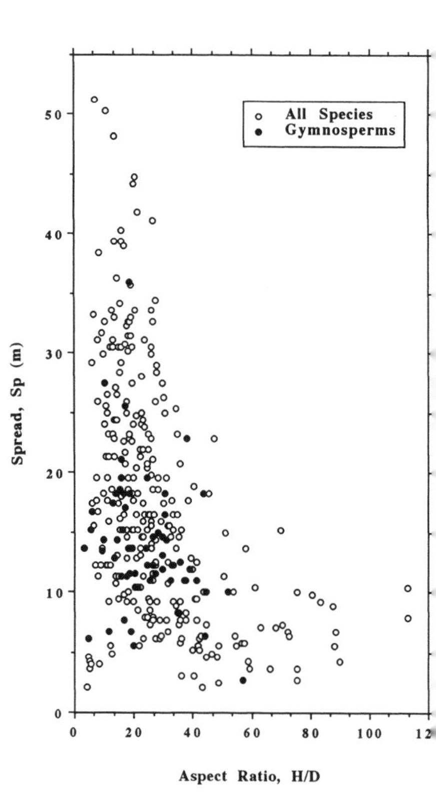

Figure 3.26. Log-log (base 10) plot of canopy spread *Sp* versus the aspect ratio (the quotient of tree height *H* and trunk diameter *D;* see figs. 3.24 and 3.25) for record specimens of angiosperm and gymnosperm tree species.

3.7 The Mechanical Scaling of Other Terrestrial Plants

In an early and important study of plant scaling, Whittaker and Woodwell (1968) report that the scaling of the height and girth of shrub species (e.g., *Quercus ilicifolia, Gaylussacia baccata,* and *Vaccinium vacillans*) complies with the geometric similitude model (on the average, $H \propto D^{1.023}$). Although the data these authors report evince a strong phyletic effect, a curious feature is that regression of the intraspecific scaling exponent versus stem size yields $\alpha_{RMA} \propto D^{-0.21}$ ($r^2 = 0.97$, $N = 6$), showing that the intraspecific scaling exponent for plant height decreases as plant size increases. Indeed, comparatively small nonwoody plants tend to scale height in much the way predicted by geometric similitude, whereas very large woody plants scale height very much as predicted by stress similarity. For example, Norberg (1988, citing Gorham 1979), concludes that geometric similitude describes the interspecific relation between the height and girth of mosses, ferns, grasses, and herbs as well as small shrubs and saplings. In this section I examine how the scaling exponent for plant height varies interspecifically by comparing very different plant clades and anatomical grades. I begin by examining the sporophytes of moss species, the smallest and anatomically perhaps the simplest land plants. I then turn to herbaceous dicot species, whose aerial stems essentially lack significant quantities of wood. This is followed by a consideration of arborescent palm species, the largest specimens of which rival dicot tree species in height because of their extremely stiff stems. I conclude the analysis by drawing an interspecific comparison among mosses, herbaceous dicots, arborescent palms, and dicot and gymnosperm trees in terms of the critical buckling heights of stems composed exclusively of different tissue types differing in their density-specific stiffness.

The sporophytes of mosses grow in length by virtue of an intercalary meristem and stop growing longer when fully mature. Each mature sporophyte consists of a basal swollen region called the foot, a prismatic columnar stemlike axis called the seta, and a terminal spheroidal sporangium called the capsule, in which numerous spores are produced. The seta is functionally analogous to the vertical stem of a vascular plant. It mechanically supports reproductive structures (the sporangium); it conducts water and other nutrients (from the foot to the capsule); and until fully mature, it provides photoassimilates. The scaling of seta length with respect to diameter therefore influences many biological functions. Figure 3.27 plots mean seta length L versus mean diameter D for forty moss species. Each datum is the mean for measurements taken from twenty mature plants of a particular species (as gauged by the presence of mature spores). Regression of the data shows that $L = 26.8\ D^{1.10}$ ($r^2 = 0.97$; $\alpha_{RMA} = 1.12 \pm 0.04$). The 95%

confidence intervals for the scaling exponent obtained from RMA regressions are 1.03–1.20, indicating that the interspecific relation between seta length and diameter more or less complies with geometric similitude. Curiously, however, this model fails to predict the scaling of seta biomass M with respect to L or D. For example, regression shows that $M \propto L^{0.61}$ ($r^2 = 0.78$; $\alpha_{RMA} = 0.69 \pm 0.08$), whereas geometric similitude predicts $M \propto L^{3.0}$ (see table 3.3). Evidently the biomass of seta, on the average, decreases relative to increasing seta length or diameter. Although not examined directly, it is likely that the bulk density of the seta accounts for this (ρ decreases relative to increasing sporophyte size).

Apparently the ability to incorporate vascular tissues into the construction of stems alters the scaling of plant height. Figure 3.28 plots mean stem height versus mean diameter for sixteen species of pteridophytes (*Lycopodium, Equisetum,* and *Psilotum*). Pteridophytes are vascular plants that reproduce without seeds by shedding their spores much like mosses. Once again, each datum represents the mean for measurements taken from twenty plants of each species. Note that these data are from

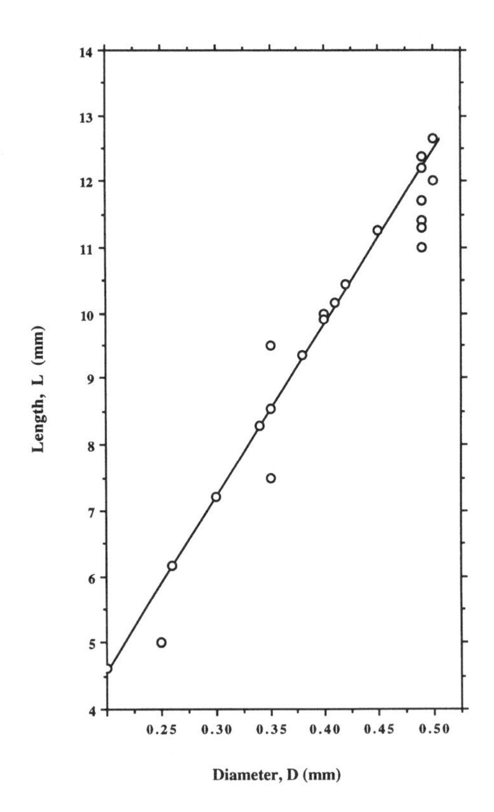

Figure 3.27. Log-log (base 10) plot of mean seta length L versus mean seta diameter D for moss species. Solid diagonal line is the regression curve obtained from LS regression analysis.

vertical stems that evince indeterminate growth in length. Regression of these data shows that $H \propto D^{1.69}$ ($r^2 = 0.85$; $\alpha_{RMA} = 1.83 \pm 0.06$). For comparison, comparable data for moss setae are co-plotted. The scaling exponent for pteridophyte stem height is statistically significantly higher than predicted by geometric similitude; the 95% confidence intervals for α_{RMA} are 1.70–1.97. On the average, therefore, the length of these vascularized stems scales roughly as the square of stem diameter. Presumably this scaling reflects the mechanical benefits conferred by the capacity to produce primary vascular tissues as well as sclerenchyma (see fig. 3.21). The mechanical consequences of vascular tissues and sclerenchyma on the scaling of stem height are further illustrated for the vertical stems of dicot herbs. Figure 3.29 plots mean H versus mean D for 117 species of dicot herbs and compares their scaling of height with those of mosses and pteridophytes (whose size ranger overlaps with that of herbaceous dicots). Regression of the dicot herb data yields $H \propto D^{1.26}$ ($r^2 = 0.74$; $\alpha_{RMA} = 1.46 \pm 0.09$, L_1–L_2 = 1.28–1.64). Although this proportionality reflects an interspecific relation from which intraspecific scalings of height may dif-

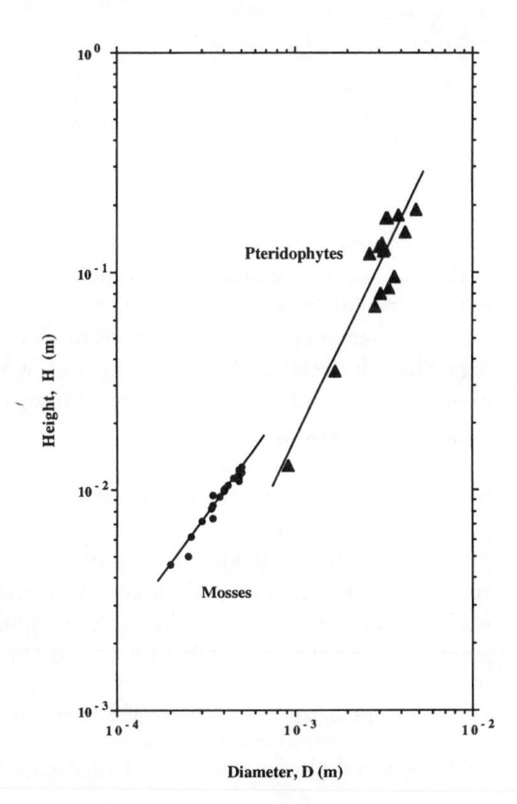

Figure 3.28. Log-log (base 10) plot of mean stem height H versus mean stem diameter D for species of pteridophytes (*Lycopodium, Equisetum,* and *Psilotum*). Data from figure 3.27 for mosses are coplotted. Solid diagonal lines are the regression curves obtained from LS regression analysis.

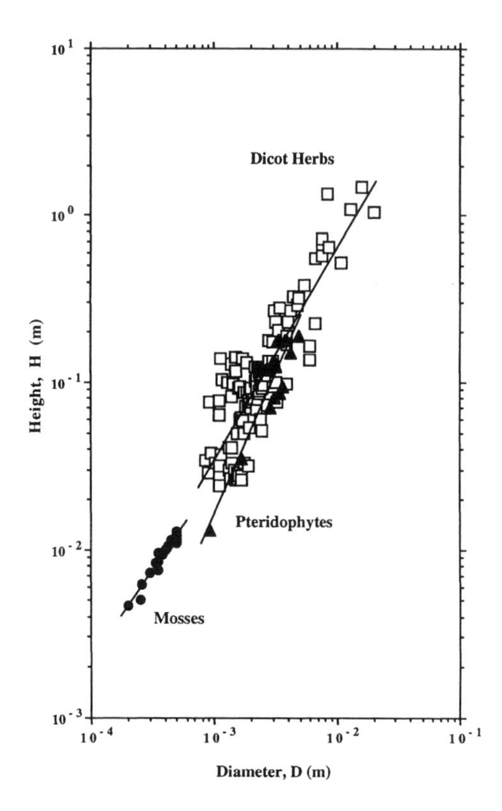

Figure 3.29. Log-log (base 10) plot of mean stem height H versus mean stem diameter D for species of dicot herbs. Data from figures 3.27 and 3.28 for mosses and pteridophytes (*Lycopodium*, *Equisetum*, and *Psilotum*) are coplotted. Solid diagonal lines are the regression curves obtained from LS regression analysis.

fer or not depending on the plant species examined, it appears that dicot herbs can elevate stems higher than can pteridophytes with equivalent cross-sectional area. Since the size range of moss setae does not overlap the size range of either pteridophytes or dicot herbs, no claim can be made regarding the relative height of mosses with respect to these vascular plants (we cannot extrapolate the scaling of moss height into the size range of tracheophytes).

How far the scaling of a particular species deviates from that expected based on comparisons among species is illustrated by considering a comparatively robust data set for the pteridophyte *Psilotum nudum*. This comparison also shows that the scaling of length depends on the position of a stem within the branching hierarchy of a plant. The sporophyte of *Psilotum nudum* produces aerial stemlike trusses that have determinate growth in length.[5] Each truss consist of a series of dichotomously branched axial ele-

5. The organographic status of the aerial portions of *Psilotum* sporophytes is a matter of debate. The interpretation that these structures are homologous with leaves (Bierhorst 1977) has been challenged (Kaplan 1977). Until this debate is resolved, a conservative view

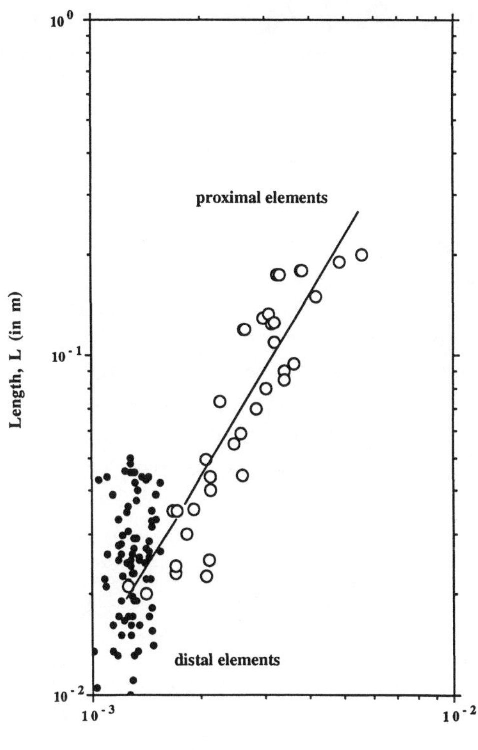

Figure 3.30. Log-log (base 10) plot of length L versus diameter D for proximal and distal axial (stemlike) elements of the pteridophyte *Psilotum nudum*. Solid diagonal line is the regression curve obtained from LS regression analysis of the data for proximal elements.

ments, of which the basalmost is the longest and has the greatest girth. This element is the principal mechanical support member for each truss. The distalmost elements serve as photosynthetic structures that also bear sporangia. Figure 3.30 plots L versus D for forty-five promixal and eighty-two distal axial elements taken from forty-five *Psilotum nudum* plants. The intraspecific scaling of L versus D for the proximal stems elements is given by the proportionality $L \propto D^{1.78}$ ($r^2 = 0.91$; $\alpha_{RMA} = 1.87$), whereas for distal elements $L \propto D^{0.90}$ ($r^2 = 0.05$). Both of these differ from the interspecific scaling shown in figure 3.28. Most important, the scaling exponents for distal and proximal elements also differ. Distal elements, which have comparatively slender vascular strands and little sclerenchyma, evince a strong positive anisometry. By contrast, no correlation exists between the length and diameter of proximal elements.

The comparatively high scaling exponent for the relation between stem

seems prudent. Consequently, each of these structures is referred to as a "truss," while each branchlike portion of a truss is called an "element" or "axis."

length and diameter found for *Psilotum* is statistically indistinguishable from that of some palm species. Figure 3.31 plots H versus D for seventeen species of palms. Regression of these data shows that $H \propto D^{1.76}$ ($r^2 = 0.94$; $\alpha_{RMA} = 1.82 \pm 0.09$, compared with $\alpha_{RMA} = 1.87$ found for the basalmost axial elements of *Psilotum*). Apparently tissues within the basalmost elements of *Psilotum* are capable of sustained lignification. These elements also increase slightly in girth by means of diffuse primary growth (Niklas 1990). Anecdotal observations suggest that the stem tissues of palms likewise become stiffer and stronger with age (Schoute 1912; Sudo 1980), while more recently Paul Rich (1986, 1987) has shown that the stiffness and strength of some palm stems increase basipetally toward the root crown, presumably because of the sustained lignification of primary vascular fibers. The palms are a remarkable group of organisms in that they achieve heights equal to those attained by many of the largest dicots and gymnosperms without the benefit of secondary xylem. Nonetheless, Rich et al. (1986) have shown that some palm species (e.g., *Welfia georgii* and *Socratea durissima*) are mechanically overbuilt when

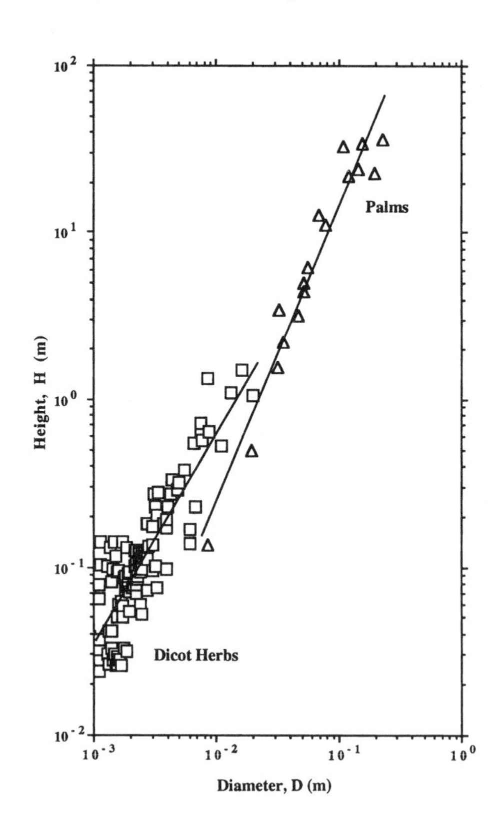

Figure 3.31. Log-log (base 10) plot of mean stem height H versus mean stem diameter D for palm species. Data from figure 3.29 for dicot herbs are coplotted. Solid diagonal lines are the regression curves obtained from LS regression analysis.

short but progressively lose their safety factor as they grow taller. That is, the slenderness ratio increases such that plants gradually approach or even exceed their theoretical buckling heights. That the stems of very tall palms appear bent in a manner that agrees well with the first mode of Euler buckling is undeniable. Whether these plants have actually exceeded their critical buckling heights, however, is conjectural because H_{crit} is typically calculated by assuming $E/\rho \approx$ a constant. As in the case of dicot and gymnosperm trees, this assumption is suspect for palms until proved.

Finally, we come full circle by juxtaposing the scaling of all the types of the plants previously considered and comparing these scaling relations with those of plant stems composed exclusively of different types of tissues with significantly different density-specific stiffnesses. Figure 3.32 contrasts the interspecific scaling of mosses, pteridophytes, herbaceous dicots, and arborescent palms ("nonwoody" species) with that of the tallest gymnosperm and dicot trees dealt with in the previous section ("woody" species). This figure also provides the regression curves for the

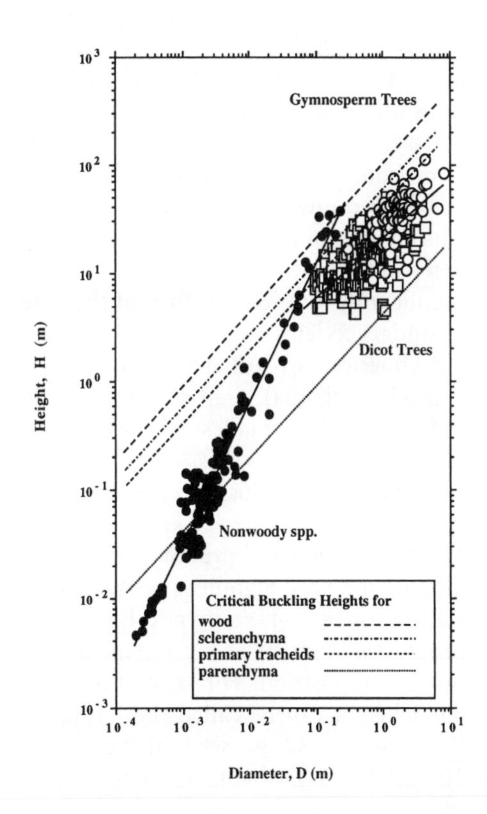

Figure 3.32. Log-log (base 10) plots of plant height H versus stem diameter D for nonwoody species (mosses, pteridophytes, dicot herbs, and palms) and dicot and gymnosperm trees (see fig. 3.23) and log-log (base 10) plots of the critical buckling heights versus D of hypothetical stems composed exclusively of wood, sclerenchyma, primary tracheids, and parenchyma (see inserts). Solid diagonal lines through the data points for nonwoody and tree species are the regression curves obtained from LS regression analysis.

critical buckling heights of hypothetical stems composed exclusively of different tissue types. These curves were computed based on Euler's formula and the density-specific stiffness of the different tissue types. Consequently each of these curves has a scaling exponent equal to 2/3.

Regression of the pooled data from all 670 species indicates that $H = 21.9\ D^{0.896}$ ($r^2 = 0.95$; $\alpha_{RMA} = 0.92 \pm 0.008$, L_1–L_2 = 0.90–0.94). Although the RMA scaling exponent for this broad interspecific comparison is compatible with geometric similitude, statistical analysis (or simple visual inspection) of the data plotted in figure 3.32 reveals that a single regression curve does not adequately describe the relation between H and D across the full range of terrestrial plant size. A pronounced "break" in slope is evident at $D \approx 10^{-1}$m. This change in the scaling exponent occurs in the size range with the appearance of stems capable of producing secondary xylem ("woody" or tree species versus "nonwoody" plant species; $D \geq 10^{-1}$m and $D \leq 10^{-1}$ m, respectively). In this regard, regression of the data for nonwoody species shows that $H = 240\ D^{1.29}$ ($r^2 = 0.97$, $N = 190$; $\alpha_{RMA} = 1.32 \pm 0.02$, L_1–L_2 = 1.29–1.37), while $H = 20.6\ D^{0.535}$ ($r^2 = 0.54$, $N = 480$; $\alpha_{RMA} = 0.73 \pm 0.02$, L_1–L_2 = 0.67–0.77) for woody species. Analysis of covariance is not required to show that the RMA scaling exponents as well as the Y_1-intercepts of these two regressions differ significantly.

Turning to the regression curves for the critical buckling heights of stems composed of different tissue types, figure 3.32 highlights the extent to which nonwoody and woody plants have mechanically capitalize on altering the material properties or the shape of their vertical stems. Note that most species for which $D \leq 10^{-1}$m are developmentally incapable of significantly altering the girth of their stems because they cannot produce secondary xylem. Nonetheless, by virtue of phyletic differences in the volume fractions of primary tissues (e.g., parenchyma, primary tracheids, and sclerenchyma) plant stature ascends five decades of H over three decades of D. The smallest of these nonwoody plants (mosses) have stems composed almost exclusively of parenchyma (which has a comparatively low Young's modulus and therefore a low density-specific stiffness); the largest plants lacking secondary xylem (palms) construct their stems out of sclerenchyma and primary xylem fibers (which have very high Young's moduli and very high density-specific stiffness). Thus the ascent of plant stature for these taxa by and large reflects modifications in the material properties of stems as a consequence of developmental modifications in tissue composition. For lack of a better phrase, this solution to scaling plant height can be called the "Young's modulus strategy" (E-strategy). From figure 3.32, we see that the E-strategy appears to have a limit ($D \approx 10^{-1}$m) beyond which the material properties of even the stiffest plant tissues cannot extend plant stature. This is because, at some point, indeterminate vertical growth requires a modification in the shape of stem cross sections. We may call this the "second moment of area strategy" (I-strategy). Clearly the E-strategy and the I-strategy are not mutually exclusive. We have seen that the flexural stiffness of a stem is the product of E and I (sec. 3.5). But for any value of E, stems that continue to grow vertically can avoid elastic buckling only by increasing I. Once the maximum E for the stiffest plant tissue is reached, a phyletic increase in plant stature requires the I-strategy as reflected by the capacity to produce and amortize secondary growth layers of xylem. Because an increase in I causes an increase in weight, the I-strategy is less efficient in many respects. Consider that with the advent of secondary growth, plants gained barely one additional decade of height at the expense of two decades in stem diameter (fig. 3.32).

If these conjectures have substance, we would predict that a woody plant undergoes a mechanical transformation as it grows and develops (and gradually amortizes secondary xylem in its stems) such that the initial juvenile scaling relation between height and stem diameter complies with the interspecific relation that obtains for nonwoody plants, whereas the adolescent scaling relation converges toward the interspecific scaling relation observed for mature dicot and gymnosperm trees. Figure 3.33 shows that this transformation occurs. This figure plots height versus stem diameter for fourteen specimens of the black locust, ranging in height from 5 cm to 2.6 m. The data for nonwoody and woody species are coplotted (see fig. 3.32). As predicted, the data for very small (young) black locust plants fall along the regression curve for nonwoody species, whereas the data from the largest black locust plants appear to converge on the regression curve for large (mature) trees. The intraspecific pattern for black locust plants is not idiosyncratic; similar measurements taken from seedlings, saplings, and adolescent specimens of twenty-three other tree species reflect essentially the same phenomenology. Thus there is good circumstantial evidence that tree species comply with the E-strategy in early development and later adopt the I-strategy.

3.8 The Scaling of Self-Thinning

I conclude the treatment of the scaling of terrestrial plants by considering the relation between plant size and number of plants in a population. This topic is appropriately dealt with here because analytical attempts to derive what has been called the "$-3/2$ self-thinning rule" often rely on the scaling relations predicted from stress similarity, elastic similarity, or geometric similitude (see McMahon 1973; Norberg 1988).

The $-3/2$ self-thinning rule was proposed Yoda et al. (1963) to describe the relation between average plant biomass $\overline{M}$ and population density N (number of plants per unit area A) for competition in even-aged stands of plants. Yoda et al. (1963) offered the following formula for this relation:

$$\overline{M} = kN^{-3/2}, \tag{3.36}$$

where k is a constant of proportionality whose numerical value is permitted to vary among species. Equation (3.36) states that the average plant biomass for a population (in dynamic equilibrium at some crowding density N) is proportional to the $-3/2$ power of N. Thus as N decreases, presumably as a consequence of plant mortality resulting from growth-related competition, $\overline{M}$ will increase anisometrically. Equation (3.36) sets an upper absolute limit or boundary on the distribution of data points for the relation between average plant biomass and the population density. Data are predicted to converge on this boundary but not exceed it as N decreases because, when the average plant is small, individuals are

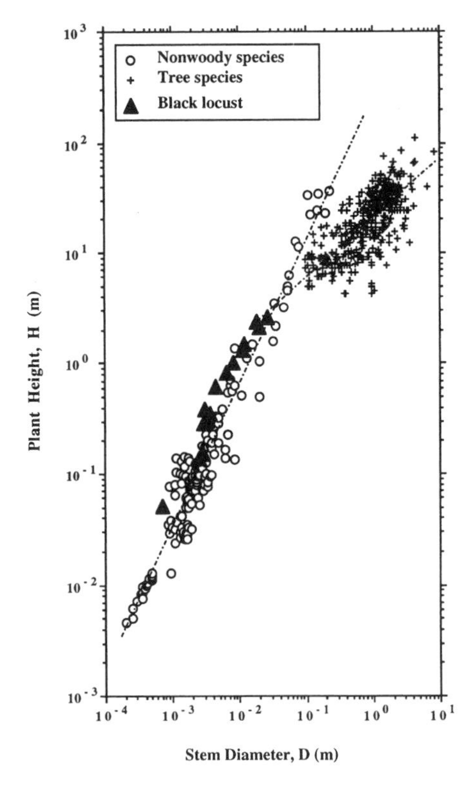

Figure 3.33. Log-log (base 10) plots of plant height H versus stem diameter D for nonwoody species (mosses, pteridophytes, dicot herbs, and palms) and dicot and gymnosperm trees (see fig. 3.32) and log-log (base 10) plot of height H versus stem diameter D for black locust trees. Dashed diagonal lines are the regression curves obtained from LS regression analysis.

predicted to experience little or no growth-related competition. As the population density decreases, however, individual plants are expected to die, and the average size of the surviving individuals is predicted to increase. The upper limit of eq. (3.36) therefore is attained only when the effects of crowding are felt. In this regard, the scaling exponent for the relation between $\overline{M}$ and N is generally found to be close to -1.0 for uncrowded monospecific stands of morphologically and taxonomically diverse species (see Norberg 1988), indicating that the standing biomass per unit area is independent of population density. Apparent confirmation for the $-3/2$ self-thinning rule quickly came from empirical studies treating diverse plant morphologies and taxa (e.g., mosses, ferns, gymnosperms, and angiosperms; see White and Harper 1970; Gorham 1979; White 1980; Norberg 1988). The ability of the $-3/2$ thinning rule to generally describe the relation between $\overline{M}$ and N was considered remarkable because it surfaced from comparisons made among successive growth stages in monotypic populations, different growth stages of plants in monotypic populations, and populations with heterogeneous species composition. This led to the claim that eq. (3.36) is a generalization worthy of the stature of an ecological "law," and analytical hypotheses were developed to explain why this was so (e.g., Norberg 1988).

We shall soon see that the $-3/2$ "rule" is in large part a mathematical artifact. For the time, however, it is instructive to explore attempts to derive this rule from "first principles." The simplest of these attempts is based on geometric similitude (see table 3.3), for which the relation between plant height and diameter and between plant biomass and volume is predicted to be isometric ($L \propto D^{1.0}$ and $M \propto V^{1.0}$). Under these conditions, average plant biomass is predicted to be proportional to the product of length and the square of diameter or, in turn, the cube of diameter (i.e., $\overline{M} \propto D^2L$ or $\overline{M} \propto D^3$).[6] Since the number of individuals per unit area A must be proportional to the reciprocal of the square of diameter ($N \propto D^{-2}$), the relation $\overline{M} \propto N^{-3/2}$ follows mathematically, which is precisely the relation sought (see eq. 3.36). In terms of growth-related competition, the diameter of a plant is argued to give a more or less adequate measurement of the ground area it covers. As the average plant diameter increases, neighboring plants are expected to compete for space and resources, increasing crowding and mortality. Clearly, this conveniently simple scenario is not directly applicable to plants whose growth scaling is more complex, for example, plants that observe stress similarity or elastic simi-

6. It should not escape our attention that $\overline{M} \propto D^2L$ also yields $\overline{M} \propto L^3$. The relation between plant biomass and body length remains an important ingredient in recent attempts to model the effects of canopy photosynthesis on the dynamics of the size structure of plant populations (see Yokozawa and Hara 1992).

larity. For these scalings the relation between L and D is not isometric, and the *effective* plant diameter, which is the spread of the crown *Sp,* must be considered. These more complex models were considered by Norberg (1988), who derived the $-3/2$ self-thinning rule based on the proportional relation between plant height and *Sp* for the elastic similarity model. In this derivation, *Sp* is accounted for by assuming that branch diameter is proportional to trunk diameter and by assuming a ratio between the average trunk diameter and the average horizontal projection of the longest branches on a tree. Norberg's derivations predict that the scaling factor in eq. (3.36) should range between -2.0 and $-3/2$. Once again we see that the upper theoretical limit is $-3/2$.

Previously I noted the superficial similarity between the predictions of geometric similitude and the scaling of plant height for nonwoody species (mosses, horsetails, lycopods, angiosperms, and even saplings). Likewise, how well the scaling of woody plant height complies with the elastic similarity and stress similarity models has been considered. Although it is fairly easy to see why empirically determined scaling exponents for plant height appear to provide circumstantial evidence for the $-3/2$ self-thinning rule, it should be evident that analytical attempts to derive the rule based on any of these three models are intrinsically flawed because the scaling of plant height does not comply with these models. Most important, however, is that empirical studies adducing the $-3/2$ self-thinning "rule" were methodologically flawed (Weller 1987; Zeide 1987; LaBarbera 1989, 106–7). As pointed out by Weller (1987), any regression of $\overline{M}$ versus N is equivalent to plotting (total plant biomass/total number of plants) versus (total number of plants/total area). Note that "total number of plants" is shared by the y and x variables. Consequently, the coefficient of correlation for the regression of $\overline{M}$ versus N is inflated, as is the slope of the regression. For regression analysis, Weller (1987) notes that the correct formulation of eq. (3.36) is

$$M = kN^{-1/2}, \tag{3.37}$$

where M is the total plant biomass *per unit area* (i.e., $M = \bar{M}N$).

Based on 448 data sets, Weller (1987) concludes that no single scaling exponent describes the relation of plant biomass and population density for every species. Rather, the scaling exponent for M depends on the ability of a particular species to tolerate shading, water deprivation, and other stress-inducing factors. Roughly 50% of the variation in the scaling exponents among species was found to be attributable to the scaling of shape and the pattern of biomass allocation to different parts of the plant body. From Weller's analyses, it appears that the intraspecific scaling exponent for M can be used to compare species in terms of how complex interactions among a variety of morphological and physiological factors influence the ability to cope with crowding.

As a final note on the self-thinning rule, let me briefly mention plants that graft as they grow. Martínez and Santelices (1992) examined the size structure of populations of the red alga *Iridaea laminaroides.* The sporelings of this species are capable of fusing as they germinate, thereby forming "individuals" that in reality are mosaics composed of two or more individuals. In their examination of the size-density relation for this alga, Martínez and Santelices found transgression of the self-thinning slope that they conclude is the consequence of the physiological integration of sporelings. A similar pheonomenology has been described by Cousens and Hutchings (1983) for the brown alga *Ascophyllum nodosum.* Like *Iridaea laminaroides,* this marine macrophyte evinces significant size inequality in the absence of self-thinning as well as physiological integration among initially separate individuals. A curious parallel exists for some terrestrial plant species that can form physiologically integrated units by means of root grafting. The way the scaling exponent for M varies for these organisms deserves investigation.

3.9 Summary

The scaling of terrestrial plants was considered in terms of heat and mass transfer, on the one hand, and mechanical design, on the other.

Unlike aquatic plants, forced convection was shown to typically play an important role in heat or mass transfer even at comparatively low ambient wind speeds. The influence of ambient wind speed on mass and heat transfer was examined by means of the equations that predict the resistance to mass transfer conferred by the boundary layer. These equations were derived for a flat plate, sphere, and cylinder and refer to the thickness of the boundary layer, for which equations were likewise derived. Mass and heat transfer were shown to be size dependent for the geometries considered as well as dependent on ambient airflow speed. The effect of stomatal resistance on heat and mass transfer was shown to be more significant than that of the boundary layer resistance for even modest ambient wind speeds. However, large changes in stomatal aperture had little effect on transpiration when wind speeds were very slow. Based on the treatment of stomatal and boundary layer resistances, the flux densities of leaflike structures with stomata were shown to be roughly one-third those of structures lacking stomata.

The scaling of mechanical design was considered in terms of the contributions made by materials (i.e., tissues) and by the geometry and size of support members (petioles, stems, branches, etc.). The scaling of the lamina mass and petiole flexural stiffness with respect to the length of the petioles of simple, palmate, and pinnate leaves was evaluted in terms of an analytical scaling model (the tip-deflection formula, derived from engineering theory). Simple and palmate leaves were shown to have different lamina mass and petiole flexural stiffness scalings than pinnate leaves. Yet in the scaling of tip deflections, these two categories of leaves converge. The general principles of mechanical design were reviewed in the context of structures that maximize their stiffness and strength relative to their weight. The density-specific stiffness and density-specific strength of different plant tissues were examined. Wood was found to have the highest density-specific stiffness and density-specific strength among the plant tissues examined. In terms of minimum-weight construction, wood was seen to be an extremely effective material with which to build stems. The scaling of tree height with respect to trunk diameter was shown to differ from gymnosperms and angiosperms ($\alpha_{RMA} = 0.87 \pm 0.08$ and $\alpha_{RMA} = 0.69 \pm 0.03$, respectively). The scaling exponent for gymnosperm tree height did not comply with that predicted by stress similarity or elastic similarity; the null hypothesis (geometric similitude model) for the scaling of gymnosperm tree height could not be rejected. For dicot trees, the stress similarity and geometric similitude models were found inappropriate. The elastic similarity model could not be rejected. Intraspecific comparisons yielded equally ambiguous results regarding the predictive capabilities of these three models. Although regression of plant height against stem diameter for a total of 690 woody and nonwoody plant species indicated that $\alpha_{RMA} = 0.90 \pm 0.008$, which was statistically compatible with the slope predicted by geometric similitude, the scaling of height versus stem diameter for plants lacking the capacity to produce secondary xylem differed significantly from the scaling for those that produce woody stems ($\alpha_{RMA} = 1.29 \pm 0.02$ and $\alpha_{RMA} = 0.73 \pm 0.02$, respectively). This was discussed in terms of the consequences of using different types of primary plant tissues to mechanically support stems versus the ability to modify the cross-sectional area of stems (the *E*- and *I*-strategies).

Finally, the "$-3/2$ self-thinning rule" was examined in terms of analytical scaling hypotheses based on geometric similitude and elastic similarity. Although these models can predict the $-3/2$ rule, they were rejected because empirical scaling exponents for plant height are only superficially in agreement with them. More important, the $-3/2$ rule has been shown to be the consequence of methodologically flawed regression analyses. The available data indicate that the scaling exponent for plant biomass at different population densities varies among species owing to interspecific differences in the capacity to deal with environmental stress as well as interspecific differences in biomass allocation in the various parts of the plant body.

4 Reproduction

4.1 Introduction

The mechanisms responsible for size-correlated variations in reproductive as well as vegetative structures are largely unknown and likely diverse. Pleiotropic effects can link variation in the size of reproductive and vegetative characters (e.g., the *S* gene in tobacco, *Nicotiana tabacum,* that affects the size and shape of foliage leaves, stamens, carpels, etc.). Consequently, if plant size is measured in terms of the number and size of vegetative organs, then large individuals are likely to bear large reproductive structures. The effect of polyploidy, which was traditionally viewed as typically producing gigas types (individuals significantly larger than their diploid ancestors), is now considered highly variable in its effect on the relative proportions of floral parts and mature reproductive and vegetative structures. Typically, polyploidy increases cell size, resulting in gigas types when the original diploid ancestor is strongly heterozygous (e.g., *Oenothera lamarckiana*). The eventual size of an organ, or whole plant for that matter, however, equally depends on subsequent cell division and enlargement, which in many instances are not affected by chromosome number.

Regardless of their underlying mechanisms, size-correlated variations in reproductive organs are often important because they bear on diaspore dispersal, on gender expression, and indirectly on the habitat preferences of species. For example, intraspecific comparisons show that seed size and offspring fitness are strongly positively correlated (Mazer, Snow, and Stanton 1986; Haig and Westoby 1988; Molau, Eriksen, and Teilmann Knudsen 1989; see, however, Marshall 1986, who shows that the intraspecific relation between seed size and offspring fitness is far less clear), particularly under competitive conditions. In theory, larger seeds can produce larger seedlings because they have proportionally larger nutritional reserves than smaller seeds. Large seed size can confer an advantage in establishment, since larger seedlings may survive longer and more frequently under stressful conditions than smaller seedlings. The significance of seed size to the success of establishment is circumstantially supported by the low phenotypic and genetic variability of this character compared with other plant characters less obviously related to fitness. To be sure, a trade-off may exist between seed size and dispersal. This is particularly true for species whose seeds are abiotically dispersed, since small seeds can be transported farther than larger seeds by wind or water. Factors that influence mean seed size are numerous but include genotype, plant vigor, and frequency of pollination. A correlation between seed size and number of seeds per fruit is typically found because of rather obvious volumetric relations—large fruits contain seeds that are either large but few or small but numerous—although a review of the available data supports the notion that the average mass of the individual seed increases nearly isometrically with increasing average fruit mass. Thus, in terms of establishment, on the average plants with larger fruits are at advantage over species with small fruits. In terms of the absolute number of potential progeny and the probability of colonization, however, plants with small fruits and small but numerous seeds may be at an advantage. Indeed, when the relation between seed and seedling size is juxtaposed with the trade-off between seed size and number of seeds per fruit, many ecological hypotheses spring to mind. For example, if it is reasonable to suppose that small seeds and fruits take less time to develop than large seeds and fruits, then species that produce small but numerous seeds may be expected to colonize and occupy open habitats more frequently than those that produce large seeds, if only because the latter lack the time for their fruits and seeds to mature. Whether this is so is a matter of empirical resolution that will inevitably rely in part on scaling analyses.

Over the course of this chapter, some of the topics dealt with relate to the scaling relations among the constituent parts of reproductive structures (e.g., parts of an individual flower and the various flowers on an inflorescence). Others concern the relation between total plant size and the size, number, or phenology of reproductive structures (e.g., fruits and seeds per branch or entire plant). The distinction between these two categories of size-dependent variations is important if we are to avoid confusion and is legitimated by the modular nature of plants.

Confusion may result from failure to distinguish between the scaling of physiologically and anatomically integrated structures and that of homologous but more or less autonomous structures borne by the same plant. Among the former, size-dependent variation tends to be the rule because constituent parts of a meristic structure share a common meristematic legacy that results in physiological and anatomical integration. Likewise, the influence of phenotypic variability among the parts of a structure is significantly reduced because constituent parts grow and develop under similar or nearly identical local environmental conditions. The physiological and anatomical integration among the parts of a plant structure permits reallocation of resources, one of whose symptoms is size-dependent variation. By contrast, size-dependent variations may not be observed among homologous structures borne on the same plant, because these structures have resulted from different meristems and therefore often possess a high degree of physiological autonomy. Likewise, homologous structures on a plant are separated in time as well as space and may develop and mature under significantly different local environmental conditions. The consequences of phenotypic variation among homologous structures that are genetically identical can obscure size-dependent variations among structures.

The legitimacy of drawing a sharp distinction between homologous but physiologically more or less autonomous structures and those that evince some metabolic interdependency was discussed in chapter 1. At that time I argued that the individual plant may be viewed as an assemblage of modules. Plants grow by adding more modules. And in species with the capacity for secondary growth, plant size also increases as individual modules increase in size. Among vascular plants, each module consists of a stem internode, its leaf or leaves, and the axillary meristems associated with these leaves. As the individual stem grows it adds more internodes, and each internode grows in size. In plants that lack secondary growth, the size of each module is determinate. In plants with secondary growth, internodes can continue to grow in volume by the accumulation of secondary tissues. Modules can be assembled into larger structures, such as a woody branch. A branch ramifies by the growth of axillary meristems; it increases in girth by the formation of wood in each internode; and it extends in length by adding more internodes along its principal axis. A size-dependent variation is rarely observed in terms of the size of reproductive organs and the size of the indeterminately growing branch that supports these organs. Because the frequency of ramification of a branch increases with increasing overall branch size, however, the girth of a branch may be highly correlated with the total biomass of reproductive organs mechanically sustained distal to where girth is measured.

Modules can be assembled into more complex levels of morphological organization. The reproductive organs of gymnosperms and angiosperms illustrate this well. The pinecone is a determinate shoot. The appendages of its primary axis are modified in the form of bracts; its axillary meristems are modified in the form of ovuliferous scales. The individual flower serves as another example of a complex morphology assembled from reiterated modules. Like the pinecone, the flower is a determinate shoot consisting of numerous short internodes bearing highly modified appendages. Some of these appendages serve in pollen donation and receipt; others are involved in the subsequent dispersal of seeds. In turn, modules can be assembled into yet more complex morphological structures such as the inflorescence. This structure consists of numerous flowers, each subtended by a stem (the pedicle), that are borne on a main stem (the peduncle). Studies show that individual flowers and inflorescences on the same plant differ in both their demographic performance (e.g., rate of photosynthesis and fecundity) and their physiological properties (e.g., extent to which resources are partitioned among constituent parts). Assimilates can be partitioned among developing ovules and seeds within an individual flower. Resource reallocation among individual flowers borne at different times of the year or on widely separate branches on the same plant is either impossible or extremely rare. By contrast, studies show that resources can be reallocated among the flowers on an individual inflorescence. For example, as the number of developing fruits per inflorescence increases, seed size and number of seeds per fruit tend to decrease.

In the absence of the detailed anatomical and physiological study required to conclusively demonstrate how far parts of a plant are interdependent, determinate versus indeterminate growth may serve as a surrogate means to estimate how much size-dependent variation is likely to exist. The logic behind this assertion rests on the notion that structures characterized by determinate growth share a common meristematic legacy that in turn leads to physiological and anatomical continuity, permitting the reallocation of assimilates. By contrast, new modules produced by plants with indeterminate growth are increasingly separated in space and time. Larger plants therefore have modules that are less capable of physiological interdependency. Circumstantial evidence for this viewpoint is seen when the scalings of annual or biennial plants are compared with those of perennial plants. Size-dependent variations among all the parts of an annual or biennial plant are commonly found. That is, larger individual plants produce larger as well as more numerous fruits and seeds, perhaps as a consequence of having proportionately larger or more numerous leaves and therefore greater metabolic resources to draw on. For perennial species characterized by indeterminate growth, the situation is more complex because plant size increases through the production of more modules (in species for which secondary growth is absent or poorly expressed), some of which can grow larger from year to year (in species with vigorous secondary growth). Perennial species tend to produce more but not necessarily larger reproductive structures as plant size increases. That larger perennial herbaceous plants produce more numerous reproductive structures is hardly surprising, since total plant size increases by adding more modules. Much more interesting is that the larger reproductive modules of perennial species produce larger flowers, fruits, and seeds, as do larger individuals of annual or biennial species.

From the foregoing, it should be apparent that the question, Do larger plants produce larger parts? as well as its analogue, Do larger plants produce larger reproductive structures? is fatuous unless framed in the context of the growth pattern of the particular species examined. I emphasize the point in this chapter by distinguishing among plant species whose individuals grow to a final mature size and those whose individuals continue to grow indefinitely. Among the latter, we shall explore the scaling of reproductive modules, which in many respects mimics that of the individual plant characterized by determinate growth.

Other size-dependent phenomena have a bearing on plant reproduction. Some of these have nothing to do with the relative size of constituent plant parts. For example, the sporophytes of a number of plant species are known to change gender as they age and increase in size. Several studies show that plants of species that contain individuals producing both pollen or ovules (cosexual species) vary in their relative male and female reproductive effort. The concept of standardized phenotypic gender has been developed to quantitatively measure this variation in male and female effort for cosexual species. The astute reader well recognizes that the sporophytes of these species do not have gender in the sense used by the zoologist. That is, the sporophytes of cosexual species do not produce male or female gametes. Rather, phenotypic gender refers to the influence of the sporophyte on the sexual expression of the gametophytes that develop from meiospores. It is the gametophyte of these species that evinces sexuality, not the sporophyte. Be that as it may, individuals of cosexual species may be dimorphic. Variation of phenotypic gender in response to environmental variation or changes in plant size or vigor is known. For situations like these, the scaling of interest is size-dependent variation in phenology rather than size-dependent variation in the proportionality of constituent plant parts.

4.2 The Scaling of Reproductive Biomass

The scaling of reproduction is most easily addressed whenever total plant size can be measured without ambiguity. One ideal circumstance obtains for plants that are determinate in growth, lack secondary tissues, consist of only one module, and reach sexual maturity in one season. Among many nonvascular and vascular plant species that fulfill these criteria are moss sporophytes, each of which grows to a certain size, matures, and dies (each is monocarpic). For what follows it is important to note that each moss sporophyte consists of a basal absorptive structure called the foot, a single prismatic supportive structure called the seta (plural setae), and a single reproductive organ called the capsule, in which spores develop. For the moss sporophyte, therefore, the question, Do bigger plants produce bigger reproductive organs? converts to, Do bigger setae bear bigger capsules? and Do bigger capsules produce more or bigger spores?

Taking the length L of a seta as an adequate measure of sporophyte size and capsule volume V as the measure of the size of the reproductive organ permits an interspecific comparison based on the maximum L reported for each moss species. Unfortunately, maximum V is not typically reported in the bryological literature, but the maximum dimensions of the capsule, together with the average geometry of the capsule reported in the literature, permit estimates of the maximum V for each species. As in previous chapters, geometric similitude provides a null hypothesis for the relation between L and V (i.e., $V \propto L^3$). As we shall see, this hypothesis must be rejected for moss sporophytes. Figure 4.1 plots V versus L for seventy-five moss species (data from the 1972 facsimile edition of Grout's 1936–39 work on the moss flora of North America). Regression of V (in units of mm^3) versus L (in units of cm) yields $V = 0.66\ L^{1.61}$ ($r^2 = 0.78$; $\alpha_{RMA} = 1.82 \pm 0.05$, 95% confidence = 1.9–1.7), indicating that the hypothesis of geometric similitude, which predicts $\alpha = 3.0$, must be rejected. The interspecific anisometry for V versus L is a consequence of the scaling of seta and capsule length. RMA regression shows that capsule length roughly scales as $L^{0.61}$ ($r^2 = 0.76$, $N = 75$).

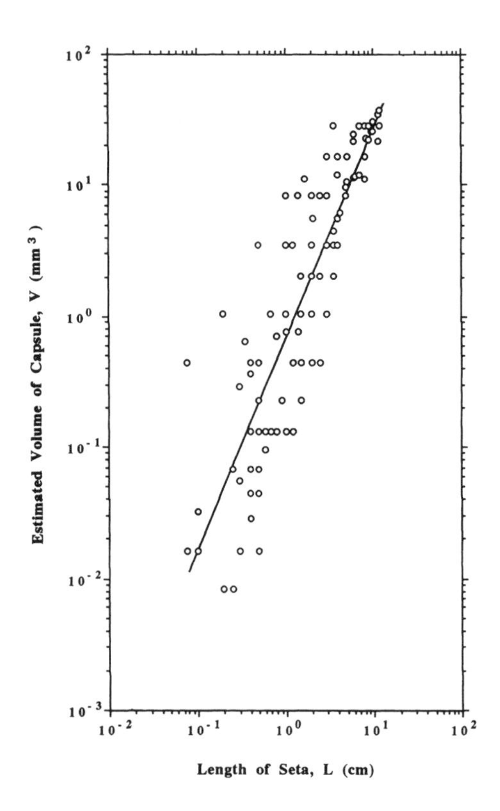

Figure 4.1. Log-log (base 10) plot of the volume V of the capsule of moss sporophytes against the length L of seta. Solid diagonal line is the regression curve obtained from LS regression analysis. Data from Grout (1972 facsimile of his 1936–39 work on the moss flora of North America).

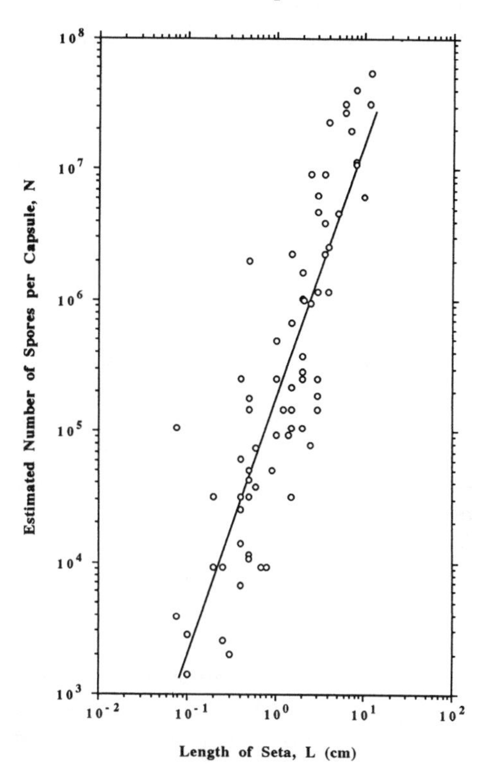

Figure 4.2. Log-log (base 10) plot of the estimated number of spores N in a moss capsule against the length L of seta (see fig. 4.1). Solid diagonal line is the regression curve obtained from LS regression analysis.

The relation between the size of the sporophyte and the size of its spores is not clear-cut; the data suggest that larger sporophytes tend to produce a greater number N of spores but that the size of individual spores declines with increasing seta length, so that total spore biomass declines with increasing size. Figure 4.2 plots N versus L. N was estimated by taking the quotient of V and the average spore volume v_s.[1] The maximum spore diameter for each of the seventy-five species reported in Grout's *Moss Flora of North America* was used to calculate v_s by assuming that all spores have a spheroidal geometry. Regression of the data shown in figure 4.2 yields $N = (1.5 \times 10^5)\ L^{2.11}$ ($r^2 = 0.76$, $\alpha_{RMA} = 2.42$

1. Clearly, the entire volume of the moss urn is not occupied by spores. The volume occupied by the wall thickness of the urn must be subtracted from the volume occupied by spores. Likewise, the spore-containing volume of the urn is further reduced by the presence of a columella, a sterile plug of tissue running the length of the capsule. Thus the relation shown in figure 4.2 is problematic and based on the presumption that the number of spores contained within an urn is proportional to the volume of the urn (i.e., the morphological features of the urn are presumed to evince isometry).

$\pm$ 0.05), indicating that the number of spores increases roughly as L^2. Since $V \propto L^{1.61}$ (α_{RMA} = 1.82 $\pm$ 0.05), we hypothesize that $N/V \propto L^{0.60}$. If true, then larger plants tend to produce more spores per unit capsule volume. This is possible because of the inverse relation between average spore volume and the size of the sporophyte. That is, $v_s \propto L^{-0.26}$ (r^2 = 0.68; α_{RMA} = 0.32 $\pm$ 0.02), suggesting that larger plants tend to produce smaller spores.

The relations established among seta length, capsule volume, and average spore size bear upon dispersal. As L increases, the distance spores may be transported from their point of origin theoretically increases because the capsule is elevated higher into regions of greater wind speed. Also, average spore size decreases with increasing plant size. This also favors dispersal, because smaller spores have slower settling velocities and therefore remain adrift longer than their larger counterparts. The increased potential for long-distance dispersal may come at a cost, however, because small spores may have smaller metabolic reserves to draw on when they begin to form gametophytes. If so, then the inverse relation between spore number and size would reflect a trade-off between fecundity and short-term survival. Data required to test this hypothesis were not available.

Capsule volume and seta length are but two among many possible measurements of reproductive and vegetative effort, and they shed no light on the allocation of biomass M between these two biological functions. Yet the partitioning of metabolic resources between the reproductive and vegetative parts of a plant is an important issue. Unfortunately, the bryological literature provides little data in this regard, despite the insights that may be gained from interspecific comparisons. The intraspecific scaling of reproductive versus vegetative effort, however, can be illustrated by data from the moss *Polytrichum commune,* which curiously complies with the interspecific scaling shown in figure 4.1. Figure 4.3 plots V versus L for forty-seven sporophytes. Regression of the data yields $V \propto L^{1.61}$ (r^2 = 0.84; α_{RMA} = 1.76 $\pm$ 0.02), which does not statistically differ from the interspecific scaling for seventy-five moss species (α_{RMA} = 1.82 $\pm$ 0.05). This may be taken as circumstantial evidence that the allocation of biomass to the capsule relative to the seta of *Polytrichum commune* is not unrepresentative of mosses in general. In this regard, regression of the mass of the capsule M_C versus the mass of the seta M_S yields the linear regression formula $M_C = 1.079\ M_S + 0.0009$ (r^2 = 0.80, N = 47). The Y_1-intercept of this regression formula very nearly equals zero. Thus the allocation of biomass to the capsule relative to the seta may be size independent. Whether this is true for other moss species is not known and bears looking into.

Similar linear relations between the partitioning of reproductive biomass M_R (measured as the dry matter of flowers, fruits, and seeds) and vegetative biomass M_V (measured as the dry matter of leaves, stems, and roots) are found among angiosperm species (Gaines et al. 1974; Kawano and Masuda 1980; for a review, see Samson and Werk 1986). That is, $M_R = mM_V + b$, where m is the slope and b is the Y_1-intercept. Based on previous discussions of the linear scaling formula (see sec. 1.5, eq. 1.2), we see that the Y_1-intercept affects how much M_R/M_V will increase or decrease with respect to absolute size. That is, when b is positive, M_R/M_V will monotonically decrease with size, and when b is negative, M_R/M_V will monotonically increase with size. And as illustrated with the data set from the moss *Polytrichum commune,* when b is zero, M_R/M_V will be size independent. All three conditions for the Y_1-intercept are observed for annual species. For some, $b > 0$ and therefore M_R/M_V decreases with size (e.g., *Polygonum cascadense* and *P. minimum;* see Hickman 1975, 1977), while for closely related species $b < 0$ and therefore M_R/M_V increases with plant size (e.g., *Polygonum douglasii* and *P. kelloggi;* see Hickman 1977).

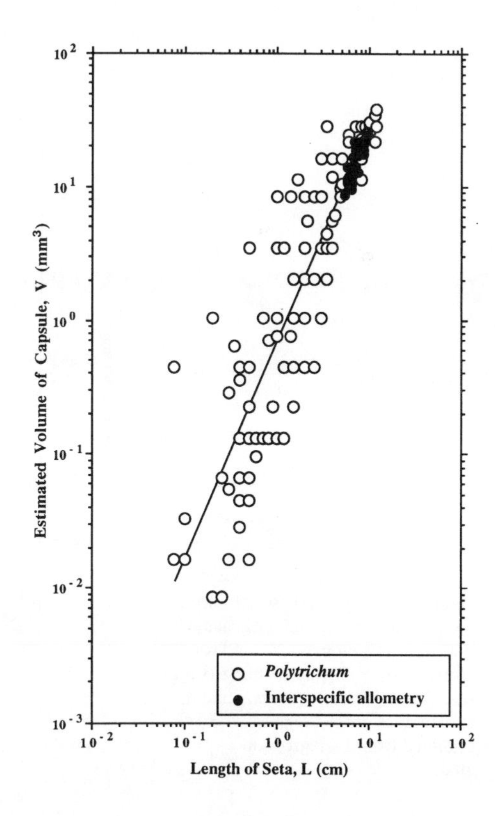

Figure 4.3. Log-log (base 10) plot of the estimated volume V of moss capsule against the length L of seta (see figs. 4.1 and 4.2). Solid diagonal line is the regression curve obtained from LS regression analysis.

Thus no clear interspecific trend is evident even among closely related species.

Many workers have related M_R to total plant biomass M_T rather than to M_V in examining the partitioning of resources between reproductive and vegetative effort. As noted by Samson and Werk (1986, 669) and by Klinkhamer, De Jong, and Meelis (1990), since $M_T = M_R + M_V$, the variables M_T and M_R are autocorrelated. This presents little problem when the ratio of M_R to M_T is very small, as for many perennial nonwoody and woody species. When M_R equals or exceeds M_V, however, as it does for many annual species, the autocorrelation between M_R and M_T poses significant problems, since the slope m decreases and the Y_1-intercept b may change when M_T is used instead of M_V. This can be illustrated by a data set for *Capsella bursa-pastoris*. Field-grown plants were removed from the soil and oven-dried for two days, then the combined mass of shoots and roots (M_V) and the combined mass of flowers, fruits, and seeds (M_R) were measured. Additionally, the biomass of the root and stem for each plant was recorded separately. Figure 4.4 plots M_R versus M_V and M_T

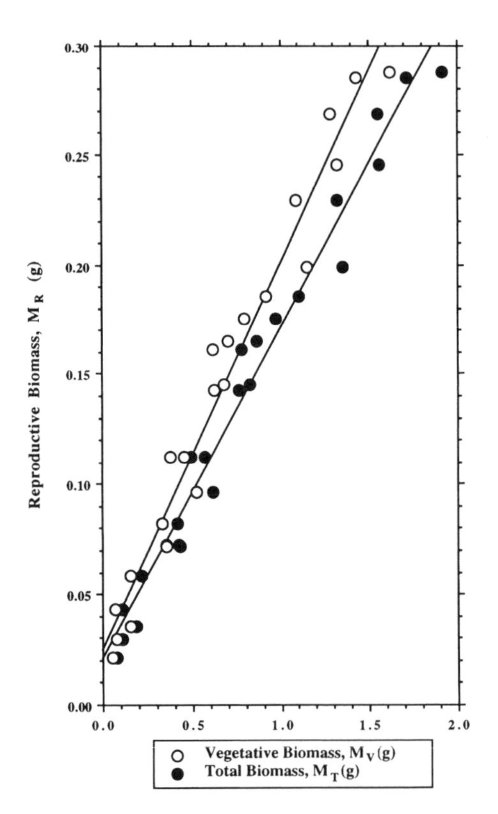

Figure 4.4. Log-log (base 10) plot of the reproductive biomass M_R against the vegetative biomass M_V and total biomass M_T of shepherd's purse (*Capsella bursa-pastoris*) plants. Solid diagonal lines are the regression curves obtained from LS regression analysis.

for twenty-three *Capsella bursa-pastoris* plants. Regression of these data yields $M_R = 0.025 + 0.176\ M_V\ (r^2 = 0.97)$ and $M_R = 0.020 + 0.150\ M_T$ $(r^2 = 0.98)$. For both regressions, $0 < b > 1$, and therefore we conclude that M_R/M_V decreases monotonically with increasing plant size. Note, however, that M_R versus M_T has a smaller Y_1-intercept and slope as well as a higher coefficient of correlation than obtained for M_R versus M_V. The decrease in the Y_1-intercept and slope is attributable to the shift in the x variable that results from adding M_R to M_V; the increase in the coefficient of determination clearly is spurious. For a small data set, these differences could lead to the conclusion that $b \approx 0$ based on M_R versus M_T alone and therefore could lead one to predict that the ratio of reproductive to vegetative effort is size independent, which the regression of M_R versus M_V shows is not true. The higher r^2 value would tend to reinforce this erroneous perception. In general, model II regression analyses should be applied to data dealing with reproductive effort, because neither this variable nor "plant size" is predetermined by the investigator. Also, as noted by Klinkhamer, De Jong, and Meelis (1990), the F-test is the most superior among the various statistical tests that may be applied to determine whether the Y_1-intercept is negative, positive, or zero.

The use of M_T as the independent variable has merit, however, particularly when dimensionless biomass ratios $\mathfrak{R}$ are used as "dependent" variables. In this regard, three dimensionless ratios are evident for the *Capsella bursa-pastoris* data set: the ratio of M_R to shoot biomass M_S, the ratio of M_R to root biomass M_{root}, and the ratio of M_S to M_{root}. Figure 4.5 plots each of these ratios versus M_T. Inspection of the three functions reveals that the smallest plants (measured in terms of M_T) have the largest ratios of reproductive to shoot biomass. We also see that the ratio of shoot to root biomass increases and then decreases as a function of increasing plant size, presumably because shoots gradually undergo a transition from their vegetative role to their reproductive role while the root system continues to mature and enlarge. The phenology of biomass allocation to reproductive organs and to roots and shoots is particularly revealing. Note that the ratio of M_R to M_{root} appears relatively uniform until $M_T \approx$ 0.8 g and then declines with increasing plant size, while the ratio of M_R to M_S tends to steadily decline in an exponential manner. These two scalings may reflect a steadily enlarging root system as well as the basipetal pattern of flower and fruit maturation such that new flowers are produced at the tips of inflorescences while mature fruits dehisce and discharge seeds toward the base of reproductive shoots. The stability of the ratio of M_R to M_{root} until $M_T \approx 0.8$ g corresponds with the observation that fruits have yet to reach maturity on plants below this size, while the ratio of M_R to M_S steadily declines because maturing plants produce and shed reproductive organs while retaining their older internodes and peduncles at the base of elongating inflorescences. In this regard, $b < 0$ for most of the few perennial species examined. For perennial nonwoody species, this may be a consequence of how biomass is allocated between roots and shoots. Nonwoody perennial species often have large storage roots. The biomass allocated to these structures may disproportionately increase with plant size and therefore will shift the ratio of M_R to M_V accordingly. The tendency for M_R/M_V to decrease with the size of woody species is more easily explained in terms of the amortization of secondary xylem in both root and stem (see sec. 4.6).

From previous discussion (sec. 3.8), we know that the intraspecific relation between M_R or M_V and plant size is expected to depend in part on local environmental conditions and population density. Species growing in shaded or dry microhabitats often obtain lower slopes for M_R versus M_V than do conspecifics grown in sunny or moist locations (see Samson and Werk 1986 and references therein). Although this superficially suggests that reproductive allocation is reduced relative to vegetative allocation for

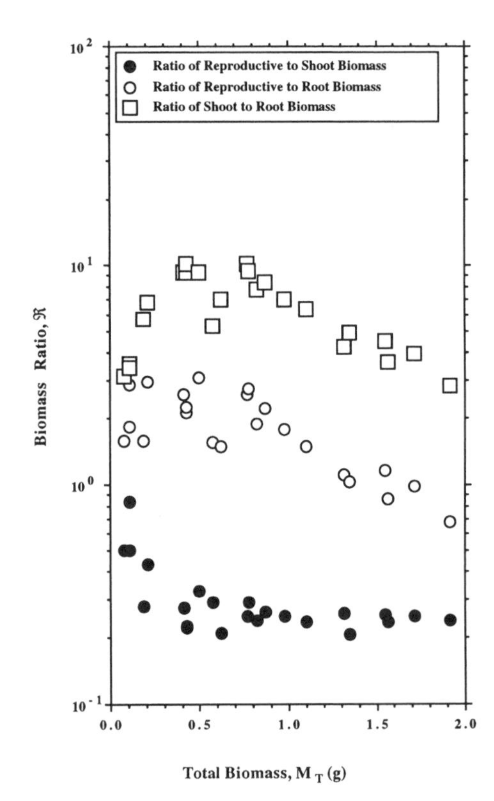

Figure 4.5. Log-log (base 10) plot of various biomass ratios $\mathfrak{R}$ (see inset) against the total biomass M_T of shepherd's purse (*Capsella bursa-pastoris*) plants. See figure 4.4 for the data used to calculate $\mathfrak{R}$.

stress-inducing local conditions, this supposition is confounded by the possibility that physiological stress delays the onset of reproductive activity. By the same token, the response of total plant yield ($M_T N$, where N is the total number of plants per population) to increasing population density (N/A, where A is unit area) typically is asymptotic—as population density increases, $M_T N$ initially increases and then, above a certain density level, levels off. By contrast, the response of total reproductive yield ($M_R N$) typically is observed to be "parabolic"—as population density increases, $M_R N$ initially increases but then may decrease with increasing population density (see Weiner 1988 and references herein). Since the data for reproductive response patterns typically are taken from populations harvested at approximately the same time, stresses induced by population density may simply delay the advent of reproduction, thereby accounting for the decline in reproductive allocation at high densities. In any circumstances, regression analysis of $M_T N$ versus N/A or $M_R N$ versus N/A yield statistically suspect results because in each case the Y and X variables share N as a parameter (see sec. 3.8). As Weiner noted (1988, 237), many plants must reach a certain minimum size before they initiate reproduction. Above this size threshold, the intraspecific relation between M_R and M_V is predicted to be linear. This, together with density-dependent mortality and plant size reduction, may account for the "parabolic" response pattern of reproductive allocation to increasing population density. Weiner presents evidence for this scaling hypothesis and concludes that "parabolic" changes in reproductive allocation with size are obtained from very simple linear scaling relations (e.g., since there is a minimum size threshold for reproduction, the *percentage* of reproductive allocation of a plant will increase with size). In a more recent study, Thompson, Weiner, and Warwick (1991) present evidence that partially supports Weiner's hypothesis. Data from five species of agricultural weeds are reported to show a strong linear intraspecific relation between reproductive and vegetative biomass (sensu Samson and Werk 1986), while regression of data from many of the populations of these species evinced a negative Y_1-intercept, indicating a threshold size for reproductive biomass, particularly when size differences among conspecifics resulted from competition. Nonetheless, departures from linearity and positive intercepts were noted in a few instances (Thompson, Weiner, and Warwick 1991). Thus, the hypotheses of linear size-dependent reproductive effort and reproductive size thresholds warrant further research before they are categorically accepted.

Intraspecific comparisons often show that larger individuals of annual or nonwoody perennial species produce larger flowers and seeds than their smaller conspecifics (e.g., *Chelidonium majus;* see Kang and Primack 1991). A genetic explanation often suffices for this observation.

Pleiotropic gene effects are known to influence the size of reproductive and vegetative organs. Thus seed and plant size (which reflects the number and size of metameric modules) may be correlated because of parental genotype. In this regard Molau (1991) shows that seed size in the subarctic hemiparasitic angiosperm *Bartsia alpina* correlates significantly with maternal genotype, while Mazer (1987) reports strong maternal effects on seed size as well as the number of ovules per fruit in the wild radish (*Raphanus raphanistrum*). A positive intraspecific correlation between seed size and plant size, however, is not invariably obtained for all species because of the inverse relation between seed size and number. For example, the data set for *Capsella bursa-pastoris* shows that larger individuals (measured in terms of total vegetative biomass M_V) do not produce larger individual seeds. Rather, larger *Capsella* plants produce more seeds, presumably because the size of shoots increases by the developmental addition of more flowers, pedicles, and their subtending stem internodes (i.e., larger *Capsella* plants produce more seeds because they produce more flowers). In this regard, larger *Capsella* fruits contain either more seeds or larger seeds, while average seed size and number per fruit are inversely proportional. These intraspecific trends are reported with sufficient regularity for other species (e.g., see Mazer, Snow, and Stanton 1986, 507, fig. 5) to suggest that a biomass allocation rule may generally operate (the total biomass investment in seed biomass per fruit or inflorescence is developmentally partitioned among seeds reaching maturity; see Westoby, Jurado, and Leishman 1992). That reproductive structures themselves contribute a portion of the carbon required for seed development is now empirically established (see Bazzaz, Carlson, and Harper 1979; Watson and Casper 1984, 235–36). Based on a survey of fifteen species, estimates suggest that from 2.3% to 64.5% of the carbon required to fabricate flowers and fruits is directly supplied by these reproductive structures (Bazzaz, Carlson, and Harper 1979), although in most cases positive net photosynthesis is rarely observed for reproductive structures. Nonetheless, the photosynthesis of fruits may be a significant source of the assimilates each fruit invests in seed production. Since there is no reason to assume that unfertilized or aborted ovules impair the photosynthetic capacity of flowers and fruits, a reduction in the number of normally developing seeds within a fruit may permit viable seeds to receive a proportionately larger share of a limited metabolic resource. If it is reasonable to presume that larger fruits produce proportionately greater amounts of assimilates, then the observation that larger fruits contain either larger or more numerous seeds is not surprising.

The ability to allocate metabolic resources among the flowers, fruits, and seeds of an individual inflorescence appears to parallel the ability of an individual flower or fruit to partition photoassimilates among developing seeds. For example, as the number of developing fruits per inflorescence increases, the ability of an individual flower to set fruit tends to decrease along with seed size and viability (e.g., *Lavandula stoechas;* see Herrera 1991). Similarly, in a study of seed and pollen yield of *Lobelia cardinalis,* Devlin (1989, 210, fig. 2) reports that the relation between seed yield and flower number per plant complies with a quadratic regression formula in which the second term is negative—that is, seed yield begins to decline beyond a certain number of flowers per plant. The available data from several studies let us conclude that an inflorescence may operate to some degree as a physiologically autonomous module.

Finally, other factors may affect the allocation of resources within a developing reproductive structure, or possibly an entire reproductive module, and therefore may influence seed size and number (see Lee 1988 and references therein). The relative time of fertilization and ovule position may be important. The first ovules fertilized within a flower often develop into larger seeds than those produced by subsequently fertilized ovules (see Mazer, Snow, and Stanton 1986; Lee 1988, 188–90), while the pattern of embryo abortion and seed size within legume fruits is influenced by proximity to the nutrition provided by the sporophyte (Hossaert and Valéro 1988). In broad terms, the pattern of seed size within a fruit can reflect the countergradients of high probability of ovule fertilization (near the tip of the carpel) and proximity to sporophytic nutrition (near the base of the carpel). Likewise, the size of the reproductive structures of gymnosperms appear to be influenced by their relative position along the length of branches. Figure 4.6 shows a series of mature seed-producing (ovulate) cones removed from a single branch of *Larix,* a deciduous gymnosperm. For these cones, size tends to decrease toward the growing ends of the branches. The reason is unknown. It is possible that cones developing toward the base of the branch remove nutrients carried in xylem water and therefore impoverish the nutrition and growth of distal developing cones. Alternatively, the concentration of nutrients carried in the phloem sap supplied by exporting foliage leaves may increase toward the base of branches owing to the spatial distribution of leaves on branches. If either (or both) of these hypotheses is correct, then ovulate cones developing toward the base of branches occupy *preferred* positions regarding growth.

4.3 Flowers, Fruits, and Seeds

So far, size-dependent variations have largely been examined intraspecifically among the parts of individual plants characterized by determi-

Figure 4.6. Seed-producing (ovulate) cones removed from a single branch of *Larix*, a deciduous gymnosperm. Cones are arranged in a series as they were removed from the tip toward the base of the branch (A to F).

nate growth. One of the advantages of this approach is that we can deal with final "adult" size in an unambiguous way. In this section I continue with this approach and explore the scaling of the reproductive organ of angiosperms, the flower. As noted (see sec. 4.2), this reproductive unit is a determinate shoot. For those unfamiliar with the parts of the flower and the nomenclature for plant reproductive systems, the rest of this paragraph provides a brief review. The most prominent parts of the flower are highly (developmentally) modified leaves bearing sporangia, called sporophylls, and depending on the species, sterile appendicular organs that collectively comprise the perianth, composed of sepals and petals. There are two types of sporophylls: those that produce pollen, called microsporophylls (or more specifically in the case of angiosperms, stamens) and those that produce ovules, called megasporophylls (or more specifically carpels). The stamens of a flower are collectively referred to as the androecium. The carpels of a flower constitute the gynoecium. Most angiosperm species produce flowers with both the androecium and the gynoecium. Flowers of this sort are called perfect or hermaphroditic flowers. Those in which either the androecium or the gynoecium is lacking are called imperfect flowers. Depending on which is present, imperfect flowers are referred to as either staminate or carpellate. When staminate and carpellate flowers are produced on the same individual, the species is said to be monoecious. Species that are dioecious have individuals that bear either staminate or carpellate flowers. The following terms are used to describe the sex expression of an individual plant: hermaphroditic—bearing only hermaphroditic flowers; androecious—bearing only staminate flowers; gynoecious—bearing only carpellate flowers; andromonoecious—bearing hermaphroditic and staminate flowers; gynomonoecious—bearing hermaphroditic and carpellate flowers; and trimonoecious—bearing hermaphroditic, staminate, and carpellate flowers. Two other terms are needed to describe the sexual expression of a population of plants: androdioecious—consisting of hermaphroditic and androecious individuals; and gynodioecious—consisting of hermaphroditic and gynoecious individuals.

Much emphasis has been placed on the comparative energetics of different angiosperm breeding systems in terms of the allocation of biomass, particularly the allocation of biomass to the male and the female gender functions. Although all hypotheses relating to this topic assume that the resources to produce seeds are limited, a variety of criteria have been used to quantify the way these "limited" resources are allocated so as to maximize the efficiency of seed production. Pollen ovule ratios, the number of staminate versus carpellate flowers per individual or within a population, the relation between seed size and number, and such have been used in this regard. Hypothetically and only in the most general terms, selection pressures should favor a reduction in the ratio of pollen to ovules in perfect flowers and a reduction in staminate to ovulate flowers in species with imperfect flowers. Likewise, since smaller seeds can be produced in greater numbers, and since increased seed number should be advantageous in terms of colonization, selection pressures should favor species that produce smaller and more numerous seeds. A number of inter- and intraspecific comparisons lend credence to these speculations. For example, size and number are inversely proportional for pollen grains as well as seeds, while pollen/ovule ratios are higher for self-incompatible species than for self-compatible species as well as dioecious versus monoecious species (see, e.g., Cruden and Miller-Ward 1981; Mione and Anderson 1992; literature cited by Westoby, Jurado, and Leishman 1992). The relations between pollen grain size and number and between seed size and number have been used as circumstantial evidence for a *requirement* to allocate a limited nutritional resource among developing meiospores as well as ovules. It may just as well reflect nothing more than the simple geometric fact that a given volume can be partitioned only into either many small or a few large compartments—a fact, incidentally, that should cause us to expect an inverse relation between the ratio of pollen grains to ovules per flower and pollen grain size (see Cruden and Miller-Ward 1981).

Clearly, how hypotheses concerning biomass allocation patterns are constructed and subsequently tested depends on how biomass is measured. Consequently, much critical attention has been paid to its "currency" (see Goldman and Willson 1986). The issue is far from simple, from either a conceptual or a methodological point of view. Consider, for example, the biomass investment in the male gender function of flowers. Does the currency of investment pertain to the androecium relative to the gynoecium or the stamen relative to the anther sac? In the case of an andro-, gyno-, or trimonoecious species, this question requires information from a statistically representative sample of flowers from individuals that, in turn, are shown to be statistically representative for their populations. The issue becomes even more difficult to quantify in the case of an andro- or gynodioecious species. And in each case care must be taken to examine the allocation of biomass to the various "compartments" within each gender function. Even for hermaphroditic flowers this is far from simple. Based on the assumption that total stamen biomass is the appropriate currency with which to measure male effort, we might conclude that flowers with large androecia invest more in the male gender function than do flowers with small androecia. Although this may be true, each stamen consists of two compartments, the filament and the anther, into which the total stamenal biomass is partitioned. Clearly, therefore, the proportional allocation of biomass to the anther relative to the filament must be considered whenever we speculate on the reproductive market. Figure 4.7 plots the mass of individual anthers M_A versus the mass of individual filaments M_{fil} for fifty-six species of monocots and dicots with perfect flowers. Regression of these data shows that $M_A = 0.19\ M_{fil}^{0.852}$ ($r^2 = 0.97$, $N = 56$; $\alpha_{RMA} = 0.86 \pm 0.02$). The 95% confidence intervals for the RMA scaling exponent are 0.91–0.82. Thus the relation between anther and filament biomass is anisometric, indicating that, on the average, large filaments bear relatively smaller anthers than their smaller counterparts. Obviously a smaller anther mass does not imply a reduction in the number of pollen grains, since a reduction in average pollen grain volume permits equivalent pollen grain numbers among anthers differing in absolute volume. Nonetheless, a trade-off between pollen grain number and size is unavoidable for any given anther volume. For the data plotted in figure 4.7, measurements indicate that, on the average, the total volume of pollen grains per anther V_p is proportional to the 0.82 power of the mass of stamen filaments. Since $V_p \propto M_{fil}^{0.82}$, larger stamen filaments ap-

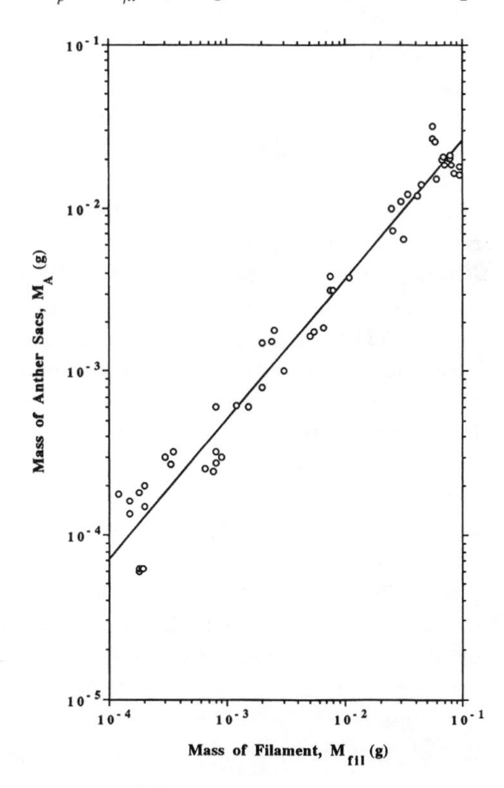

Figure 4.7. Log-log (base 10) plot of the mass of individual anthers M_A against the mass of individual filaments M_{fil} of stamens removed from monocot and dicot species with perfect flowers.

pear to subtend anther sacs that have either relatively smaller or fewer pollen grains ($V_p/M_A \approx 1.0$). The reasons for allocating more biomass to the filament relative to the anther are unknown, although Euler's formula shows that longer prismatic support members must be more robust in girth than shorter members in order to remain erect. This mechanical obligation may affect the differential allocation of biomass within the androecium. A curious footnote is that, for a number of angiosperm species with stamen filaments that mechanically operate primarily in tension rather than compression (e.g., grasses), M_{fil} is greatly reduced with respect to M_A (see Oka and Morishima 1967; Beri and Anand 1971; Vries 1974; Campbell 1982; and McKone 1989, who report strong correlations between pollen production and anther length among species of grasses). Regardless of the reason for the scaling demonstrated in figure 4.7, the relation between the biomass of the filament and the anther cautions against the supposition that the total dry matter invested in a floral organ type is invariably the best currency with which to measure gender function effort.

Just as the total biomass invested in the formation of the androecium is partitioned between stamen filaments and anther sacs, the total biomass invested in the formation of the hermaphroditic flower is partitioned among the various floral organ types (e.g., perianth, androecium, and gynoecium). The way biomass is allocated among the various parts of the flower has received much attention. By extension of Fisher's (1930) argument for progeny sex ratios, several theoretical models predict that the allocation biomass ratio should favor female gender function in proportion to the degree of selfing (Maynard Smith 1978; Charlesworth and Charlesworth 1981, 1987; Charnov 1982). Thus, obligate outcrossing (xenogamous) species are expected to produce flowers in which the ratio of gynoecial to androecial biomass is equal to or less than unity. By contrast, facultative self-crossing (autogamous) species are expected to have ratios above unity, presumably because the potential reduction in pollen loss during the pollination of autogamous flowers permits a greater allocation of metabolic resources to the female gender function. In all cases, selection is presumed to favor a shift toward the sexual role that produces the greater increase in parental fitness relative to investment.

Empirical data support the general predictions of these models. In a seminal paper, Lovett Doust and Cavers (1982) showed that the flowers of autogamous species have ratios of gynoecial to androecial biomass greater than unity, whereas the ratios for the flowers of xenogamous species are below unity. For example, the obligate xenogamous species *Lilium superbum* allocates six times as much biomass to the androecium as to the gynoecium, while the primarily autogamous flowers of *Nicotiana rustica* and *Amaryllis* spp. have high ratios of female to male gender biomass (see Lovett Doust and Cavers 1982). Similar results are reported by Cruden and Lyon (1985) and Ritland and Ritland (1989). An additional relevant observation is that obligate or facultative xenogamous species, on the average, tend to produce larger flowers than do autogamous species (Cruden and Lyon 1985).

Bearing in mind that flower size typically is measured by the size (biomass) of the perianth, of which the corolla is by far the largest compartment, the size-dependent variations in gynoecial and androecial biomass that figure prominently in empirical and theoretical attempts to evaluate the relation between allocation patterns and breeding systems may be the consequence of the physiological relations among developing floral organ types. Although the physiological effects of floral organ types on one another are highly complex, some evidence suggests that corolla expansion is under the physiological control of stamens. The reason for this is at present unknown. However, developing stamens appear to be a source of gibberellins. When exogenously applied to developing flowers, these plant hormones stimulate corolla development (see Lord 1979, 1980, 1981, 441). By the same token, externally applied plant hormones that stimulate stamen and corolla development appear to inhibit ovarian development (Lord 1981). Thus the physiological inhibitory and stimulatory interactions among developing floral organ types may account for the correlation between overall floral size and the relative size of the androecium and the gynoecium. Additional evidence, albeit circumstantial, supports this point of view. For example, male sterility is correlated with a reduction in corolla size and stimulated ovary growth (Heslop-Harrison 1957). Likewise, a common feature in the formation of flowers that normally fail to open and thus are largely self-pollinated (cleistogamous flowers) is a smaller corolla and androecium (Lord 1981). Reduction in corolla size also is observed for the carpellate flowers of many gynodioecious species (Baker 1948), as well as gynomonoecious and dioecious species (Darwin 1877). Indeed, numerous studies report a tendency for a high positive correlation between petal and stamen size, on the one hand, and an inverse correlation between the size of the androecium and the gynoecium, on the other (Lord 1981). Thus, provided the ability of floral organ types to produce or export plant hormones is size dependent, a developmental reduction in the size of one organ type could result in concomitant shifts in the size of other floral organ types, thereby prefacing a change in breeding system.

The developmental mechanisms responsible for regulating the patterns of biomass allocation to different floral organ types are largely unknown. Nonetheless, their morphometric consequences become apparent when we examine the scaling of gynoecial or androecial biomass by broad comparisons among species characterized by hermaphroditic flowers. Figure 4.8, for example, plots mean gynoecial and androecial biomass (M_G and M_A, respectively) versus mean perianth biomass M_P for ninety species with perfect flowers. These data include those reported by Lovett Doust and Cavers (1982, 2531, table 1; $N = 5$ species), Cruden and Lyon (1985 300, table 1; $N = 40$), Ritland and Ritland (1989 1735, table 3; $N = 8$). The two regression curves drawn on this figure were obtained from only 39 species (data from Niklas 1993c). The data points for these species are shown as small dots. The data from autogamous and obligate xenogamous species reported by other authors are denoted respectively by large open and closed symbols in figure 4.8. These data points were not used to determine the regression curves shown in this figure, because one of my objectives is to determine from the independent database of 39 species whether the flowers of auto- and xenogamous species differ in size. The regression curves shown in figure 4.8 have the formulas $M_G = 0.14\ M_P^{0.914}$ ($r^2 = 0.87$; $\alpha_{RMA} = 0.98 \pm 0.06$) and $M_A = 0.20\ M_P^{0.978}$ ($r^2 = 0.92$;

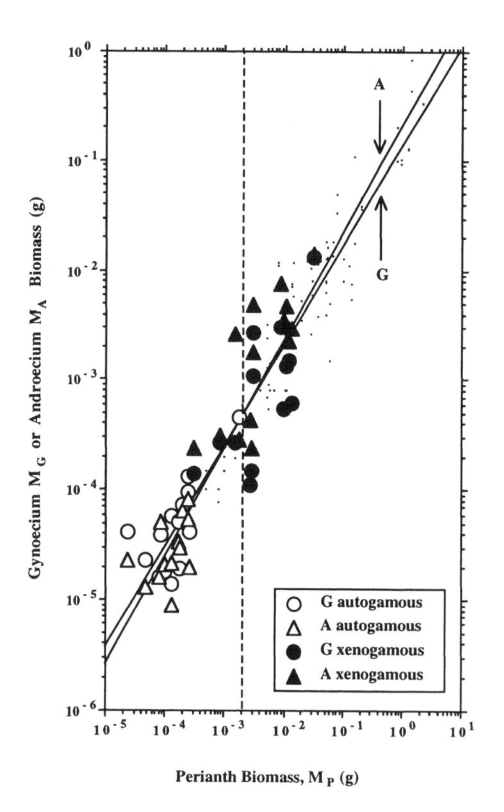

Figure 4.8. Log-log (base 10) plot of mean gynoecial and androecial biomass (M_G and M_A) against mean perianth biomass M_P of dicot and monocot species with perfect flowers. Data points for autogamous and xenogamous species are those reported by Lovett Doust and Cavers (1982, 2531, table 1; $N = 5$ species), Cruden and Lyon (1985, 300, table 1; $N = 40$), and Ritland and Ritland (1989, 1735, table 3; $N = 8$). Solid diagonal lines are the regression curves obtained from LS regression analysis of data points represented by small dots (data from Niklas 1993c). Vertical dashed line denotes $M_G = M_A$, which occurs when $M_P \approx 0.002$ g (i.e., $M_G < M_A$ when $M_P > 0.002$ g and $M_G > M_A$ when $M_P < 0.002$ g).

$\alpha_{RMA} = 1.02 \pm 0.04$). Given the standard errors associated with the scaling exponents in these two formulas, the allocation of biomass to the androecium as well as the gynoecium scales isometrically with regard to perianth size. Also, the two scaling exponents do not statistically differ from one another, as shown by their 95% confidence intervals (for the gynoecial α_{RMA}, the 95% confidence intervals are 1.1–0.86; those for the androecial α_{RMA} are 1.1–0.94).

To a certain extent, however, these statistics are misleading. Note that the regression curves in figure 4.8 for the scaling of andro- and gynoecial biomass (denoted as A and G) intersect approximately at the point where $M_P \approx 0.002$ g (shown by the dashed vertical line). Thus, on the average, $M_G = M_A$ when $M_P \approx 0.002$ g, and the ratio of gynoecial to androecial biomass is greater than unity when perianth size exceeds 0.002 g and vice versa. Further note that all the data from autogamous species plot to the left of the vertical dashed line, indicating that, on the average, autogamous species produce smaller flowers than the facultatively xenogamous species and that these small flowers have more massive gynoecia than androecia. This tendency agrees with the observations of Cruden and Lyon (1985), who report that obligate xenogamous species, on the average, tend to produce larger flowers with larger androecia than do autogamous species. To be sure, the statistics (summarized above) for the scaling exponents of the two regression curves shown in figure 4.8 would not have led us to conclude that gynoecial and androecial biomass evinces a size dependency or that flowers from the two types of breeding systems sort out as a matter of size. From a developmental perspective, however, the relations summarized in figure 4.8 are not surprising. Morphometrically, "larger flowers" translates into "flowers with larger corollas," because flower "size" is measured as perianth biomass, whose largest component is the biomass of the corolla. Developmentally, "larger flowers" is predicted to translate into "larger stamens" whenever the developing androecium provides a source of plant hormones that encourage the growth in size of petals. Finally, in terms of breeding systems, "larger flowers" translates into xenogamous species, which are predicted to require a proportionately greater allocation of biomass in the formation of stamens (and by inference pollen). Although not shown in figure 4.8, data points for facultatively xenogamous species plot out, on the average, between those shown for obligate xenogamous and autogamous species (they "hover" near the point where the two regression curves intersect).

The size dependency of the biomass allocation to the androecium and the gynoecium is more easily seen in figure 4.9, which plots the quotient of gynoecial and androecial biomass (denoted as Q) versus M_P for the xeno- and autogamous species examined by Cruden and Lyon (1985).

The dashed vertical line applies to $M_P = 0.002$ g, while the dashed horizontal line denotes the condition where $Q = 1.0$. Thus there are four quadrants (I–IV) in which data points may plot. Based on prior discussion, we predict that all autogamous species will plot in quadrant I, all obligate xenogamous species will plot in quadrant IV, and facultative xenogamous species will fall between the two extreme breeding systems. With few exceptions, these expectations are met in a reasonable manner. The data from most autogamous species plot in quadrant I, whereas the data from most obligate xenogamous species plot in quadrant IV (data from Cruden and Lyon 1985). With three exceptions, the data from facultative xenogamous species plot in quadrant III ($M_P < 0.002$ g and $Q < 1.0$). Thus, breeding system types among angiosperms appear to evince a size dependency that is predictable given our current understanding of the developmental mechanisms that affect the relative sizes of different floral organ types.

Thus far we have ignored the relation between floral organ biomass and shape, which in principle is just as important to pollination biology

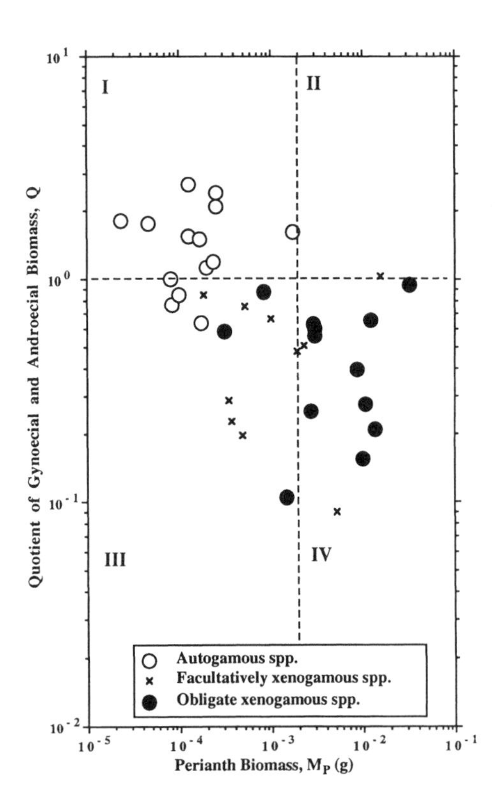

Figure 4.9. Log-log (base 10) plot of the quotient of gynoecial and androecial biomass (denoted as Q) against perianth biomass M_P. Data for xeno- and autogamous species examined by Cruden and Lyon 1985. The dashed vertical line denotes $M_P = 0.002$ g, while the dashed horizontal line identifies $Q = 1.0$. Four quadrants in which data points may plot are identified as I-IV.

as is the relative investment of biomass in the construction of the androecium and gynoecium. Unfortunately, this aspect of plant reproductive biology is often neglected when gender function allocation patterns are considered from a theoretical perspective. Consider the possibility that increased stamen biomass (which, as we have seen, correlates with filament length) is positively correlated with either the length or the width of the corolla or else that carpel biomass correlates with style exertion. If so, then size-dependent biomass allocations to male and female functions, however slight, could have significant effects on other floral traits that influence pollination and therefore breeding systems. In this regard, Stanton and Preston (1988) report a correlation between biomass allocation to male function and corolla diameter for *Raphanus sativus*, while Campbell (1989) reports a correlation between biomass allocation to carpels and stigma exertion for *Ipomopsis aggregata*. Finally, Plitmann and Levin (1983) show that, among closely related species, pollen grain diameter is positively correlated with style length. Although this correlation is absent among randomly selected species (Cruden and Miller-Ward 1981), larger pollen grains may be required to grow through longer styles within a species complex or family of plants. If so, then a redistribution of even a given amount of gynoecial biomass into the stylar compartment may require compensatory shifts in the allocation pattern of biomass within the conspecific androecium.

Other subtleties become evident when we consider the scaling of floral biomass allocation patterns in terms of the size of wind-pollinated plants. The pattern of female/male biomass allocation among anemophilous species has typically differed from theoretical expectations. For example, both Charnov (1982) and Charlesworth and Charlesworth (1981) adduce that male fitness in wind-pollinated plants will be linearly related to male reproductive investment because, unlike biotic pollination vectors, wind cannot be saturated as a pollen vector. Accordingly, these authors predict that the biomass allocation pattern between male and female gender functions for wind-pollinated species should be equal or nearly so. Empirically, however, this prediction is not borne out, perhaps because paternal fitness may not be a linear function of the capacity of plants to produce pollen but rather may be a nonlinear function of the capacity to disperse pollen. How thoroughly a wind-pollinated plant can saturate its immediate environment with pollen depends on the height at which pollen is released into the atmosphere. In theory a tall plant will have less local mating competition among sib pollen, and since it will have a wider range of reproductive influence owing to an enhanced pollen distribution range, it should have a less saturating male fitness curve than a shorter counterpart. Thus, although the traditional expectation is that female investment should increase with increasing plant size, this investment should decrease with increasing size for wind-pollinated plants. In their study of three subspecies of corn (*Zea mays* spp. *parviglumis, mexicana,* and *diploperennis*), Burd and Allen (1988) show that this prediction is borne out. These authors compared the ratios of staminate to ovulate flower spikelets produced by wind-pollinated plants differing in height. They found that this ratio increased with increasing plant height in four of their six study sites (see Burd and Allen 1988, 405, table 1). Indeed, the tallest plants had a ratio roughly ten times that of the smallest plants. True, it is not clear that the ratio of male to female spikelets is the most appropriate measure of gender function allocation, since the energy and nutrient costs of these reproductive structures have not yet been shown to be equivalent. Nonetheless, among conspecifics growing in the same locality, ratios of staminate to ovulate spikelets should provide a fair measure of reproductive biomass allocation patterns. Be that as it may, the results Burd and Allen reported suggest that allocation patterns to male and female functions among wind-pollinated plants may depend on the size and therefore the pollen dispersal range of individual plants. If so, then the theoretical expectation that female investment should increase with plant size needs to be adjusted. Without doubt, the reproductive ecology of anemophilous species requires closer inspection by empiricist and theorist alike.

Clearly, the problems of predicting and measuring relative paternal and maternal investment are numerous and potentially serious. Nonetheless, their resolution is important to a variety of questions regarding the evolution of plant reproduction. A gradual shift in the partitioning of biomass among types of floral organs or a sudden change in the identity of floral organs can result in plants with imperfect flowers. Either of these pathways can prefigure the evolutionary transition from monoecism to dioecism. Charlesworth and Charlesworth (1987) speculate that this transition results from the gradual redistribution of resources between staminate and carpellate flowers rather than from the complete sterilization of one gender function in a single episode. Accordingly, they regard the transition between a subdioecious species (one that regularly contains incompletely gender-differentiated individuals) and a dioecious species as a quantitative rather than a qualitative category. Alternatively, homeotic gene mutations altering floral organ identity may serve as a rapid evolutionary mechanism propelling a species along the pathway to imperfect flowers. The issue is not whether dioecism results from a gradual or sudden evolutionary transition (both likely have occurred, although the relative frequency of each is the important feature). Rather, it is that the allocation of biomass to developing primordia may remain indifferent to the morphogenetic effects of homeotic gene mutations, yet the observed pattern of resource partitioning might seem influenced simply as a consequence of the change of identity rather than the position of mature organs. Circumstantial evidence suggests that meristems observe independent developmental rules regarding the position, number, size, and identity of the parts they engender. For example, recent studies of the mouse-ear cress (*Arabidopsis*) and the snapdragon (*Antirrhinum*) have isolated homeotic gene mutations that alter floral organ identity. The effects of these genes, which can be obscured in flowers whose parts are arranged in spirals, are rendered by the formation of floral primordia in sequential whorls. The first whorl of primordia develops into sepals. Subsequently formed whorls develop into petals, stamens, and carpels. That some homeotic gene mutations can effect a sudden transition from perfect to imperfect flowers is shown by the *apetala3* and *pistillate* genes of *Arabidopsis* and the *dificiens* gene of *Antirrhinum.* Mutations in these genes yield flowers lacking stamens. By the same token, all whorls of primordia on triply mutant *Arabidopsis* flowers (i.e., *apetala2, apetala3,* and *agamous* mutant genes) develop into carpelloid leaves (Bowman, Smyth, and Meyerowitz 1989). Importantly, the number of floral primordia is indifferent to these homeotic mutations (see Chasan 1991 for a review). Primordia in the third whorl of *Arabidopsis* flowers normally develop into six stamens. In the *agamous* mutant, these primordia develop into six petals. Indeed, the number of parts of flowers is a highly conserved character. And, as we have seen from the limited data (fig. 4.7), the percentage of total biomass allocated to each type of floral part is essentially the same regardless of species and therefore independent of the number of parts. Specifically, analysis of the data plotted in figure 4.7 shows that, on the average, 67.8% of total biomass is invested in sepals and petals. The remaining fraction of biomass is more or less equally divided between stamens and carpels (16.6% and 15.5%, respectively), suggesting that the developmental mechanisms attending the formation of perfect flowers typically observe some common rules regarding the distribution of biomass to developing primordia regardless of their ultimate organ identity.

Clearly, data are essential to speculations regarding the role that shifts in biomass allocation play in the evolution of imperfect flowers. Fortunately, the plant world is resplendent with species performing natural experiments regarding the allocation of biomass to their floral organ types, although comparatively few have been carefully studied. One of these experiments is routinely performed by the melon *Cucumis melo.* This species is andromonoecious and therefore varieties typically produce staminate and perfect flowers on the same plant. Occasionally an individual plant of an andromonoecious variety may be gynomonoecious and therefore bears carpellate and perfect flowers. Thus, within a single population (or more fortuitously, on a single individual plant), the allocation of dry biomass to floral organs in all three types of flowers can be quantified. Figure 4.10 plots the dry mass M of the androecium and gynoecium versus the dry mass of the perianth M_P for thirteen perfect flowers of *Cucumis melo* variety Red Ace (Kyowo Seed Company). All thirteen flowers were removed from the same plant. Comparable data for fifteen staminate and five ovulate flowers are plotted in this figure. These imperfect flowers were removed from individuals growing in the same population as the individual from which perfect flowers were collected. Regression of the data from the perfect flowers yields $M_A = 0.40\ M_P^{1.47}$ ($r^2 = 0.90$; $\alpha_{RMA} = 1.55 \pm 0.02$) for the androecium and $M_G = 4.6\ M_P^{1.36}$ ($r^2 = 0.84$; $\alpha_{RMA} = 1.49 \pm 0.03$) for the gynoecium. These regression formulas would lead us to predict that $M_G/M_A \propto M_P^{-0.06}$, indicating that, on the average, the allocation of biomass to the gynoecium decreases relative to that of the androecium as the size of perfect flowers of this variety increases. Much more to the point, however, is the scaling relation between M_A or M_G and M_P for imperfect flowers. Regression of the data for imperfect flowers

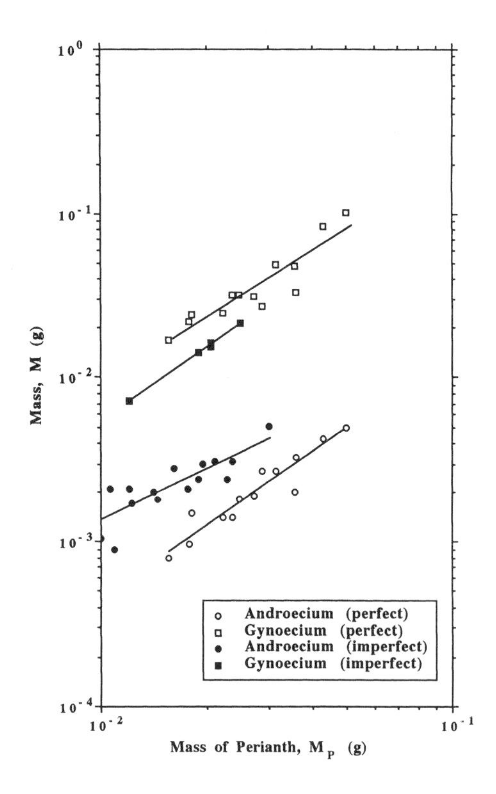

Figure 4.10. Log-log (base 10) plot of dry mass M of the androecium and gynoecium against dry mass of the perianth M_P for perfect and imperfect flowers of Red Ace (a variety of melon, *Cucumis melo*), Solid diagonal lines are the regression curves obtained from LS regression analysis.

yields $M_A = 0.15\ M_P^{1.02}$ ($r^2 = 0.68$; $\alpha_{RMA} = 1.23 \pm 0.02$) and $M_G = 4.7\ M_P^{1.47}$ ($r^2 = 0.99$; $\alpha_{RMA} = 1.48 \pm 0.03$). Clearly, the scaling relations for the androecium and the gynoecium are anisometric such that larger imperfect flowers have proportionately larger male and female reproductive biomass. In this regard, the scaling exponents for the gynoecium of perfect and imperfect flowers are statistically indistinguishable. By contrast, the scaling exponent for androecial biomass of imperfect flowers is significantly reduced compared with that for perfect flowers. The loss of male-gender function does not lead to a proportional reallocation of the conserved biomass to female gender function. That patterns of biomass reallocation in other species are likely to be complex and far from the simple cost-profit economy envisioned by some should warn us against any generalized theory. Incidentally, the genetic control of gender expression in *Cucumis melo,* which appears to be basically the same as in the cucumber (the related species *Cucumis sativus*), involves two allele pairs: G/g controls the length of the male phase (gg gives no male phase), and M/m controls unisexuality (dominant) and bisexuality (recessive). Additional modifying genes are known to influence flower shape and the pattern of flowering. No heteromorphic sex chromosomes are known.

Thus far we have examined the way plants allocate biomass to male and female gender functions in terms of individual flowers, whereas the best way to explore biomass allocation patterns is at the level of the whole organism. For example, in the case of *Cucumis melo* just reviewed, the loss of male gender by a flower was not reciprocated by a gain in gynoecial biomass. But this tells us nothing about how conserved male biomass may be utilized by other portions of the same plant. That is, could the loss of male gender at the level of individual flowers result in the production of more perfect flowers by the same individual? Data for *C. melo* are not available to answer this question, but a study of *Cucurbita foetidissima* plants suggests that biomass allocation patterns to gender function can be complex when examined at the whole-plant level. Individual plants of *C. foetidissima* produce either ovulate flowers or separate ovulate and staminate flowers (the species is gynodioecious), and therefore the allocation of biomass to each gender function is comparatively easy to measure for each individual plant. Based on a study of eight populations, Kohn (1989) reports that female plants produced more female biomass than hermaphrodites. Since the two types of plants did not differ in vegetative size (e.g., number and size of leaves and stem internodes), and since the total sexual biomass did not differ between female and hermaphroditic plants, the biomass of male flowers produced by hermaphrodites was about equal to the extra female biomass produced by females. Accordingly, a trade-off appears to exist between male and female reproductive effort when the biomass allocation pattern of this species is compared at the whole-plant level. The methods of analysis Kohn employed are exemplary, particularly since a comparison was made between female and hermaphroditic plants in terms of seed production, which revealed that the former produced 1.5 times as many seeds as the latter. Many more studies of this sort are required to gauge the full import of gender function reallocation patterns, however.

The scaling relations among the parts of perfect and imperfect flowers do not exhaust the scaling of angiosperm reproductive structures. The gynoecium develops into a fruit in which fertilized ovules develop into mature seeds. Consequently, the relation between the size of the flower and either the mass of the fruit or the total mass of seeds per fruit developing per flower deserves attention, as may be illustrated by the data from Ibarra-Manríquez and Oyama (1992, app. 1, 390–94), who measured the size of flowers, fruits, and seeds for 139 Mexican rain forest tree species. Although the most extensive, their data do not permit as rigorous an analysis as desirable, since size was not always measured in the same manner: the size of flowers was determined by the average floral length L_F; fruit size was measured in terms of the average length as well as the fresh mass of individual fruits M_{IF}; and the size of seeds was measured as the average fresh mass of individual seeds M_{IS}. Fortunately, data on the average seed number per fruit N are given, and therefore the total seed mass per fruit can be computed as well as the mass of the pericarp (the mass of the fruit sans total seed mass). Pericarp biomass is an important variable, since the total fruit mass and total seed mass are obviously autocorrelated.

Figure 4.11 plots the mass of the pericarp (= actual fruit tissue biomass) and the total mass of seeds per fruit versus length of flower. No correlation is evident among these three variables in terms of an interspecific comparison. Curiously, however, comparatively strong correlations between petal length and fruit length are frequently observed when closely related species are considered. For example, Primack (1987) reports significant correlations between these two variables among species belonging to the same genus. He suggests that evolutionary changes in the size of flowers are closely related to changes in fruit size. Clearly, the data sets considered when intraspecific or congeneric comparisons are drawn are expected to evince some degree of a "phyletic effect." Thus what may be true among closely related species need not be true at the level of broad intergeneric comparisons, which provide a better measure of general evolutionary "trends." And in any circumstances we must be sensitive that "fruit length" is but one measure of "fruit size." Whether it is an *adequate* measure of fruit size depends on whether fruit length scales isometrically with respect to fruit diameter such that length provides a good indirect measure of fruit volume, hence mass. An anisometric scaling relation between these two variables could easily lead to failure to detect a correlation between pericarp mass and flower length, as shown in figure 4.11.

The mass of individual fruits M_{IF}, mass of individual seeds M_{IS}, and seed number N are significantly correlated, as shown by the data provided by Ibarra-Manríquez and Oyama (1992). Figure 4.12 plots average individual fruit mass versus average individual seed mass. Regression of these data reveals that $M_{IF} = 4.65\ M_{IS}^{0.78}$ ($r^2 = 0.55$; $\alpha_{RMA} = 1.05 \pm 0.06$). Thus we see that the biomass of the individual fruit scales nearly in an

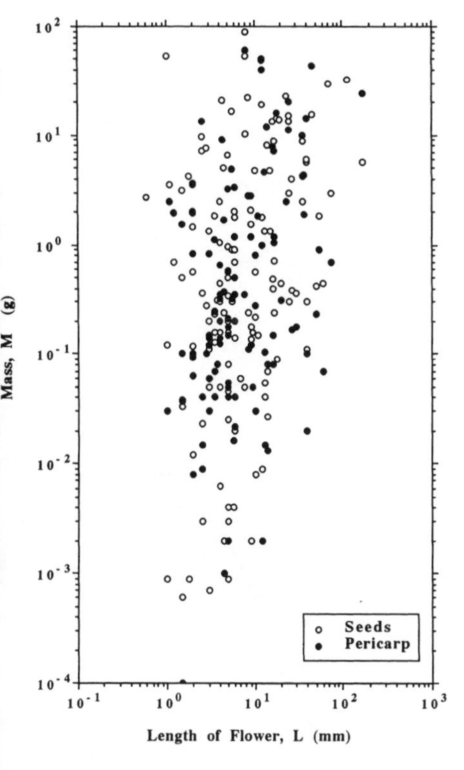

Figure 4.11. Log-log (base 10) plot of pericarp mass M (= actual fruit-tissue biomass) and the total mass M of seeds per fruit plotted against length of flower L. Data from Ibarra-Manríquez and Oyama 1992, app. 1, 390–94.

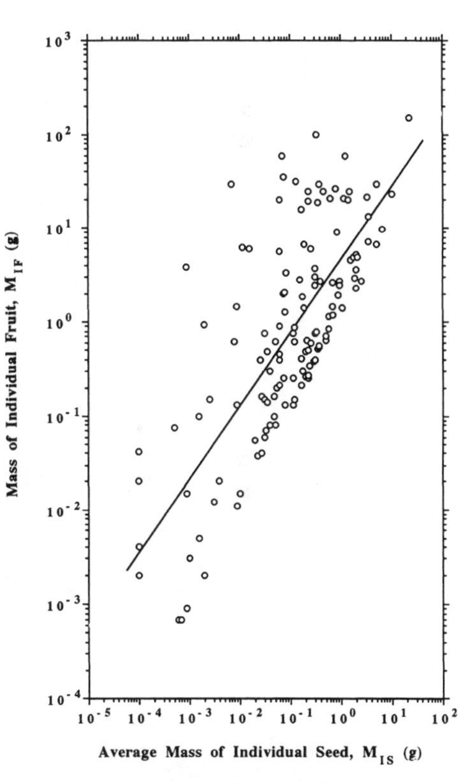

Figure 4.12. Log-log (base 10) plot of average individual fruit mass M_{IF} against the average mass of individual seed M_{IS}. Solid diagonal line is the regression curve obtained from LS regression analysis. Data from Ibarra-Manríuez and Oyama 1992, app. 1, 390–94.

isometric function of the mass of individual seeds. This is hardly surprising, since the bulk of most mature fruits consists of mature seeds. This is reflected by the sharp boundary in the distribution of data points below the regression curve drawn in figure 4.12. As we might anticipate, the data points falling just along the boundary are those for fruits bearing a single large seed that occupies most of the volume of its fruit. Nonetheless, it is conceivable that many different scalings, each defined by a class of fruit with an equivalent number of seeds, could be submerged within the scatter of data points in this figure. This possibility is quickly examined and rejected when the data from fruits with one, two, three, and four seeds are treated separately. Figure 4.13 plots M_{IF} versus M_{IS} for fruits like these. Regression reveals an isometric relation between M_{IF} and M_{IS} for each of the four classes of fruits defined by number of seeds per fruit: for one-seeded fruits, $M_{IF} = 2.0\ M_{IS}^{1.0}$ ($r^2 = 0.93$; $\alpha_{RMA} = 1.04 \pm 0.02$); for two-seeded fruits, $M_{IF} = 2.9\ M_{IS}^{0.91}$ ($r^2 = 0.88$; $\alpha_{RMA} = 0.97 \pm 0.03$); for three-seeded fruits, $M_{IF} = 3.8\ M_{IS}^{0.91}$ ($r^2 = 0.86$; $\alpha_{RMA} = 0.98 \pm 0.05$); and for four-seeded fruits, $M_{IF} = 4.3\ M_{IS}^{0.91}$ ($r^2 = 0.85$; $\alpha_{RMA} = 0.99 \pm 0.04$). Thus, for each class of fruit defined by seed number per

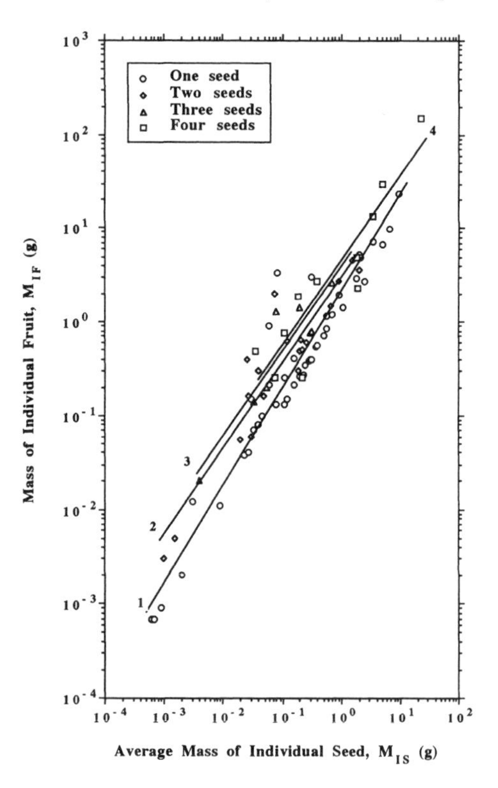

Figure 4.13. Log-log (base 10) plot of average individual fruit mass M_{IF} against average mass of individual seed M_{IS} for fruits containing different numbers of seeds (see inset). Solid diagonal lines are the regression curves obtained from LS regression analysis. Data from Ibarra-Manríquez and Oyama 1992, app. 1, 390–94.

fruit, the relation between fruit and total seed size is isometric.

Turning to seed size and number, the total average seed mass per fruit M_{TS} can be calculated from the product of the average number of seeds per fruit N and the average individual seed mass M_{IS}. Figure 4.14 plots M_{TS} versus M_{IF}. Regression of these data yields $M_{TS} = 0.67\ M_{IF}^{0.93}$ (r^2 = 0.84, N = 139; $\alpha_{RMA} = 1.01 \pm 0.04$), indicating a near isometric relation between the total seed mass per fruit and individual fruit mass. Perhaps it is belaboring the rather obvious to say that bigger fruits produce more numerous or more massive seeds, but this is emphasized by considering the relation between the ratio of M_{IF} and M_{IS} and the number of seeds per fruit N. Figure 4.15 plots M_{IF}/M_{IS} versus N. Regression of these data yields $M_{IF}/M_{IS} = 1.9\ N^{0.84}$ ($r^2 = 0.73$; $\alpha_{RMA} = 0.98 \pm 0.06$). That is, plants with more seeds per fruit and larger individual seeds have larger individual fruits. The scaling of fruits and seeds complies with the notion that there are advantages to bearing large fruits, either in terms of the number of potential progeny (seeds) each fruit can produce or in terms of enhancing the survival potential of each seed by investing proportionally more assimilates (see Westoby, Jurado, and Leishman 1992). As we shall

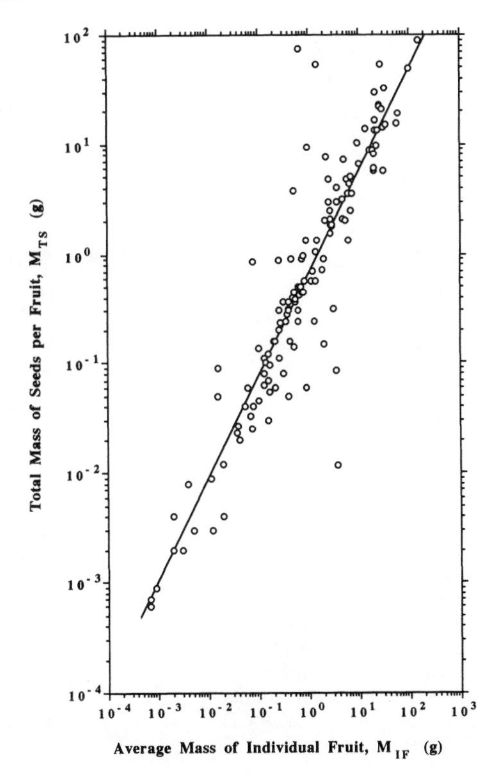

Figure 4.14. Log-log (base 10) plot of total seed mass M_{TS} against average mass of individual fruit M_{IF}. Solid diagonal line is the regression curve obtained from LS regression analysis. Data from Ibarra-Manríquez and Oyama 1992, app. 1, 390–94.

see, these advantages require mechanically more robust stems and branches to support reproductive organs (see sec. 4.6).

Before leaving the relation between fruit and seed size and number, it is appropriate to examine the reproductive organs of gymnosperms, specifically conifers. Obviously these seed plants do not produce fruits, since they lack ovaries. Nonetheless, the seeds of conifers develop and mature within ovulate cones that fulfill some of the functional roles of the fruit (nutrition of developing ovules and protection and dissemination of ripened seeds). Figure 4.16 plots the number of seeds per ovulate cone

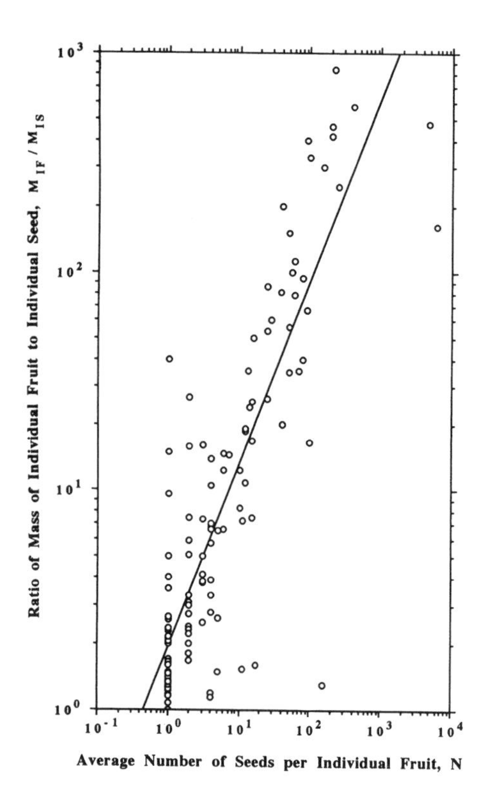

Figure 4.15. Log-log (base 10) plot of the ratio of individual fruit biomass to individual seed biomass M_{IF}/M_{IS} versus the number of seeds per fruit N. Solid diagonal line is the regression curve obtained from LS regression analysis. Data from Ibarra-Manríquez and Oyama 1992, app. 1, 390–94.

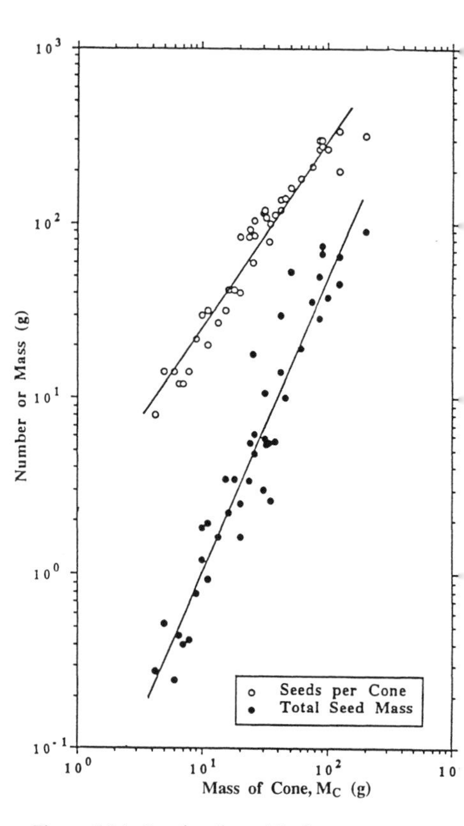

Figure 4.16. Log-log (base 10) plot of seed number per ovulate cone N and total seed mass per cone M_{TC} (see inset) against the mass of the ovulate cone M_C for conifer species. Solid diagonal lines are the regression curves obtained from LS regression analysis.

N and the total seed mass per cone M_{TC} versus the mass of the ovulate cone M_C for ten conifer species from a total of five genera. The scaling exponents determined from this data set naturally will be idiosyncratic, owing to the small number of species and genera examined. But my purpose is to consider conifers, of which there are comparatively few extant genera. Mature ovulate cones were collected from trees before ovulate scales had opened to release seeds. The ovulate scales of each cone were pried apart, and the number and weight of individual seeds were recorded. Regression of these data yields $N = 2.20\ M_C^{1.06}$ ($r^2 = 0.94$; $\alpha_{RMA} = 1.09 \pm 0.04$) and $M_{TC} = 0.022\ M_C^{1.67}$ ($r^2 = 0.91$; $\alpha_{RMA} = 1.75 \pm 0.03$). Accordingly, on the average larger cones bear a disproportionately larger total seed mass. Further, on the average the ratio of total seed mass per cone to number of seeds per cone increases roughly as the 2/3 power of ovulate cone mass ($M_{TC}/N \propto M_C^{0.66}$). But larger cones tend to produce disproportionately smaller seeds. The correlation between the average individual seed mass and the mass of the cone, however, is not particularly robust ($r^2 = 0.53$), although it is statistically significant. Similar results have been reported for intraspecific comparisons among cone size and seed number and size (e.g., *Pinus albicaulis*; see Lanner 1982).

Thus far we have not considered the joint effects of vegetative plant size and average individual seed weight on the total number of seeds produced by plants differing in absolute size. As noted, developing seeds are presumed to be strong physiological sinks for carbon and mineral nutrients that are provided by the photosynthetic biomass of the plant on which they develop. In theory, therefore, we might expect a positive correlation between total seed N and vegetative biomass M_V and a negative correlation between seed number and average individual seed mass M_{IS}. Such appears to be the case, although the paucity of empirical studies makes any general claim concerning these correlations premature. Based on a fairly robust interspecific comparison among fifty-seven herbaceous angiosperm species, however, Shipley and Dion (1992) report that $N = 1.3\ M_V^{0.93} M_{IS}^{-0.79}$. This regression formula indicates that the number of seeds produced per plant scales roughly isometrically with respect to the biomass of aboveground vegetative tissues (the scaling exponent for M_V does not statistically differ from unity) and that an increase in average seed size leads to a less than proportional decrease in seed number (the scaling exponent of M_{IS} is significantly less than unity). Accordingly, simple size-related scalings appear to be the primary determinants of seed production among herbaceous angiosperms. Shipley and Dion note that the large number of seeds produced by species occupying disturbed habitats is not primarily a consequence of producing small seeds but rather results from a particular combination of comparatively small seed size and vegetative biomass. These authors further claim that the quotient of reproductive and vegetative biomass increases with increasing seed size and either remains the same or decreases slightly with increasing plant size. This claim, which contradicts the hypothesis of Samson and Werk (1986), that reproductive effort will change with plant size, is not based on straightforward data because vegetative biomass appears in both x variables regressed by Shipley and Dion (1992, 479). Given the presumed physiological source-sink relation between photosynthetic biomass and seed number and average size, it should not escape our attention that the scaling formula reported by Shipley and Dion which is based on an interspecific comparison among herbaceous species, will likely overestimate the number of seeds produced by plants whose aboveground biomass includes a significant volume fraction of nonphotosynthetic tissues (e.g., wood). Finally, the interspecific scalings reported by Shipley and Dion do not necessarily reflect intraspecific scalings (e.g., the number of seeds and total seed biomass increase proportionally with increasing plant size for the monocarpic perennial plant *Cynoglossum officinale;* Klinkhamer and De Jong 1987).

A curious footnote is that the isometric interspecific scaling exponent reported by Shipley and Dion (1992) for the relation between total seed number per plant and aboveground vegetative biomass per plant is very similar to that reported by Blueweiss et al. (1978) for the scaling relation between poikilotherm litter biomass and body size for animals (α_{LS} = 0.93 and 0.92, respectively). Although no obvious mechanistic explanation exists for this similarity in the two scaling exponents, a tenuous sink-source analogy can be drawn between developing seeds and photosynthetic tissue biomass, on the one hand, and developing animal embryos and maternal body size, on the other.

4.4 Branches and Reproductive Structures

As the size or number of reproductive organs increases, the loading on a vertical stem or branch that mechanically supports these structures also increases. Provided there are selective advantages to keeping fruits and seeds aloft (see sec. 4.5), it seems reasonable to suppose that the scaling of reproductive effort must in some fashion correlate with the scaling of mechanical support, although the logic claiming that plants producing larger or more numerous fruits must be proportionally more mechanically robust than those that bear smaller or fewer fruits is quickly deflated when we compare the size of fruits borne on squash vines and on orange trees. With this caveat in mind, for plants whose fruits impose mechanical stresses on stems, the relation between the total biomass invested in fruits per branch M_{TF} and the branch or stem biomass M_S becomes biologically important. Fortunately, the investment in biomass for either function can be measured empirically, although for large arborescent plants, particularly those of commercial interest, direct measurements of M_S are not practical. For these plants, however, the biomass of a stem or branch can be estimated by employing Murray's scaling equation, which states that the proportionality between branch diameter D and M_S measured distal to D is $M_S \propto D^{2.49}$. Accordingly, the ratio of M_S to M_{TF} provides a reasonable estimate to the trade-off between mechanical and reproductive effort. The data and analyses of Peters et al. (1988) are particularly interesting in this regard. These authors measured D (in units of cm) and M_{TF} (in units of g) distal to D for twenty-two tree and shrub species from Barbados. For garden, feral, and native species, they report that $M_{TF} = 47\ D^{1.9}$ ($r^2 = 0.78$; see Peters et al. 1988, 613, table 1). Thus $\alpha_{RMA} = 2.15$, indicating that total fruit biomass scales roughly as the square of stem diameter.[2] In other words, bigger (older) branches bear disproportionately more fruit mass. This is hardly surprising, given that the number of twigs producing flowers and therefore ultimately fruits increases as a woody branch increases in length and girth. That D and M_{TF} are correlated is not the point, however. Rather, the object of inquiry is the scaling exponent between these two parameters, since this exponent cannot be disclosed by a theoretical approach.

Recall that Murray's empirically determined equation accords with the stress similarity model. From this model, total fruit mass is predicted to increase almost as the fourth power of plant height H (in terms of LS regression scaling exponents, $M_{TF} \propto D^{1.9}$ and $D \propto H^2$, and thus $M_{TF} \propto H^{3.8}$). One of the advantages to reproduction resulting from increasing overall plant height therefore appears to be the capacity to produce more or larger fruits. However, since $M_{TF} \propto D^{1.9}$ and $M_S \propto D^{2.49}$, the ratio of mass invested in reproductive to mass invested in mechanical effort theoretically is proportional to $D^{-0.59}$. That is, the investment in reproduction relative to structural support decreases roughly as the 1/2 power of stem size. From the empirically determined relation between tree height H and trunk diameter (see fig. 3.23), the relation $H \propto D^{1/2}$ should hold for the arborescent species studied by Peters et al. (1988). If so, then the ratio of reproductive to structural investment is predicted to decrease anisomet-

2. $M_{TF} \propto D^{1.9}$ does not imply that bigger branches produce bigger fruits. M_{TF} is the product of average individual fruit biomass M_{IF} and the number of fruits N_F per branch. Accordingly, the relation between M_{IF} and N_F is also of interest, and Peters et al. (1988, 615) found that $N_F \propto D^{1.2}\ M_{IF}^{-0.43}$. That is, plants with larger branches and smaller individual fruits produce more fruits. Once again, we see circumstantial evidence for a trade-off between the size and number of reproductive structures.

rically with increasing branch diameter or total plant height. Thus arborescence appears to come at a cost in terms of reproduction. Needless to say, however, this is pure supposition, since the scaling exponents of proportionalities derived from mathematical manipulations of empirically determined scaling exponents (e.g., $M_{TF} \propto H^{3.8}$) are highly suspect on statistical grounds (see appendix). Unfortunately, therefore, much of the quantitative speculation entertained by Peters et al. (1988) regarding the trade-off between mechanical and reproductive effort must be discounted.

It should be noted further that the biomass of the branch or an entire tree largely reflects the amortized investment made in wood. Old growth layers of wood do not draw on the metabolic resources of the plant. Thus, in terms of annual costs, a better reflection of the ratio of reproductive to mechanical biomass would be the ratio of total fruit biomass to the biomass of the woody growth layer deposited in branches during the concurrent year of fruit production. This form of analysis rarely has been undertaken, in large part because it is extremely tedious to gather data and because it is sample destructive. However, Popp and Reinartz (1988) examined the relation of biomass allocation and sexual dimorphism in northern prickly ash (*Zanthoxylum americanum*) in terms of the new wood production in first- and second-year twigs. They report that clones that produce only staminate flowers invest a greater proportion of their metabolic resources (measured in terms of biomass) into new wood than do clones that produce only carpellate flowers. Although the results reported by Popp and Reinartz suggest that the high resource allocation to fruit production can reduce the vigor of vegetative growth relative to *male* plants in some species, it is premature to draw this conclusion. Studies of other species that follow a similar protocol are badly needed if we are really to understand the quantitative relation between reproductive and vegetative biomass allocation.

Returning to the previously criticized scaling relation $M_{TF}/M_S \propto D^{-0.59}$ adduced by Peters et al. (1988), it is noteworthy that this relation differs from its analogue reported for annual or perennial nonwoody species, perhaps because the scaling of growth of nonwoody species does not involve the amortization of secondary xylem and therefore complies with the geometric similitude model rather than with either the stress similarity or elastic similarity model (see fig. 3.32). Unfortunately, the empirically determined scaling factor between nonwoody or shrub plant height and stem diameter varies depending on the plants considered. Since the choice of the scaling exponent for the relation of biomass to plant height is critical to any hypothesis about the relation between reproductive and structural investment, the range of empirically determined scaling exponents for nonwoody plants presents a problem. In terms of nonwoody species, Peters et al. (1988) speculate that the biomass invested in leaves and twigs M_{L+T} provides an analogue to M_S for woody species. Drawing on Whittaker and Woodwell (1968), who report that $M_{L+T} \propto D^{1.65}$ for nonwoody species and presuming that $M_{TF} \propto D^{1.9}$ also holds true for these as well as woody species, Peters et al. adduce that $M_{TF}/M_{L+T} \propto D^{0.25,}$ which predicts that the relative investment in reproduction should increase with the amount of photosynthetic tissue. Noting that Whittaker and Woodwell also report $H \propto D^{1.02}$ for nonwoody or shrub species, Peters et al. further argue that $M_{TF}/M_{L+T} \propto H^{0.24.}$ As noted, all the scaling relations derived by Peters et al. are statistically unreliable. But as hypotheses, both $M_{TF}/M_{L+T} \propto D^{0.25}$ and $M_{TF}/M_{L+T} \propto H^{0.24}$ suggest that the mass invested in reproductive organs relative to the mass invested in photosynthetic organs should decrease as a function of the size of nonwoody or shrub species. This is easily shown by a simple graph. Figure 4.17 plots the hypothetical size-dependent variations of $M_{TF,}$ MS, and M_{L+T} in terms of D, from which we may hypothesize that small plants with slender branches will invest more biomass in leaves and fruits and less in stems, whereas large plants

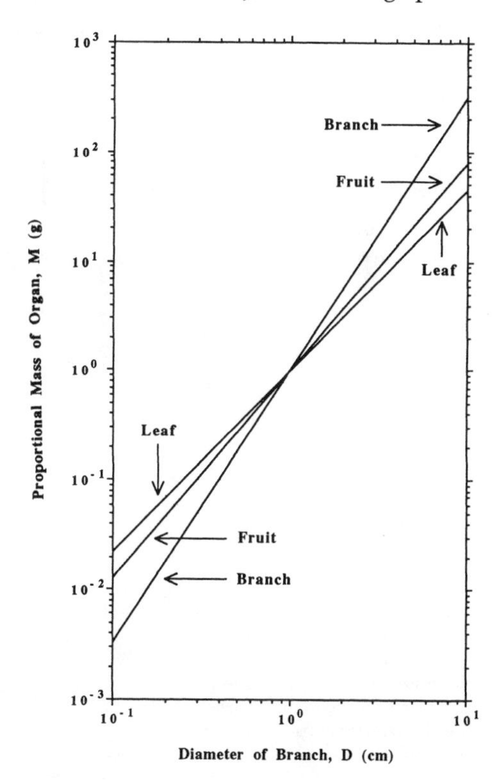

Figure 4.17. Hypothetical regression curves showing the scaling relation of the proportional log-transformed biomass of organ type *M* with respect to log-transformed branch diameter *D*. See text for further details.

are expected to invest more mass in branches and fruits and less in leaves. Figure 4.18 plots these size dependencies in terms of the ratios $\Re$ of organ mass (M_{TF}/M_S and M_{TF}/M_{L+T}) versus D. It cannot be overemphasized, however, that the relations shown in figures 4.17 and 4.18 were derived from mathematically valid, albeit statistically inappropriate, manipulations of three empirically determined scaling relations, $M_{TF} \propto D^{1.9}$, $M_S \propto D^{2.49}$, and $M_{L+T} \propto D^{1.65}$, which obtain the following hypotheses:

Woody species	Nonwoody/shrub species
$\frac{M_{TF}}{M_S} \propto D^{-0.59}$	$\frac{M_{TF}}{M_{L+T}} \propto D^{0.25}$

A variety of other highly speculative hypotheses can be developed relating plant size to the average mass of individual fruits and seeds based on the proportionalities adduced by Peters et al. (1988) and those determined from the data reported by Ibarra-Manríquez and Oyama (1992) as well as others. Recall from section 4.4 that, based on LS regression analyses, the proportionalities $M_{TS} \propto M_{IF}{}^{0.93}$ and $M_{IF}/M_{IS} \propto N^{0.84}$ were ob-

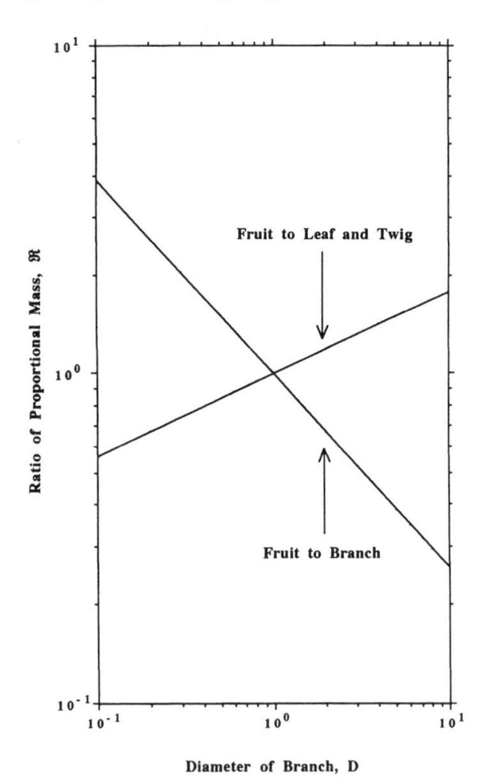

Figure 4.18. Hypothetical regression curves showing the scaling relation of the ratio $\Re$ of the log-transformed proportional biomass of organ type with respect to log-transformed branch diameter D. See text for further details.

served. Accordingly, larger plants that produce fruits with smaller average seed size and number are predicted to produce more fruits per branch ($N_F \propto D^{1.2} M_{IS}^{-0.43} N^{-0.36}$), while the ratio of the number of seeds per fruit and the number of fruits per branch is predicted to decrease with increasing plant size or increasing average seed size ($N_S/N_F \propto D^{-1.84} M_{IS}^{-1.19}$). Additionally, the ratio of total seed mass per branch M_{SB} to mass of branch M_S is predicted to decrease with plant size ($M_{SB}/M_S \propto D^{-1.75}$). Figure 4.19 summarizes these hypothetical relations by plotting the ratios of total seed mass to branch mass, total fruit mass to total leaf and twig mass, and total fruit mass to branch mass as a function of branch diameter. Obviously this figure is not meant to present the actual scaling relations among these variables. This is why the scaling exponents determined from LS regressions were purposely used. Rather, the objective of figure 4.19 is to expose these hypotheses to future empirical inquiry and criticism.

I conclude this discussion about the relation between branch biomechanics and reproductive effort with an empirical comparison among mosses, pteridophytes, gymnosperms, and angiosperms. This comparison

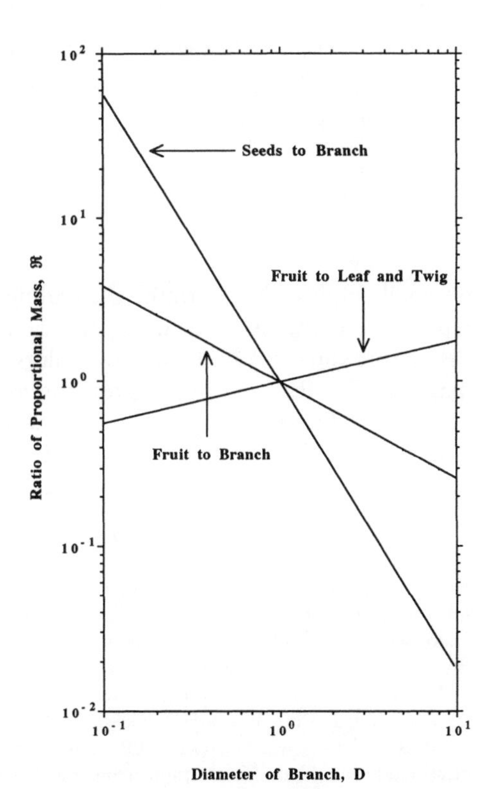

Figure 4.19. Hypothetical regression curves showing the scaling relation of the ratio $\Re$ of the log-transformed proportional biomass of organ type with respect to log-transformed branch diameter D. See text for further details.

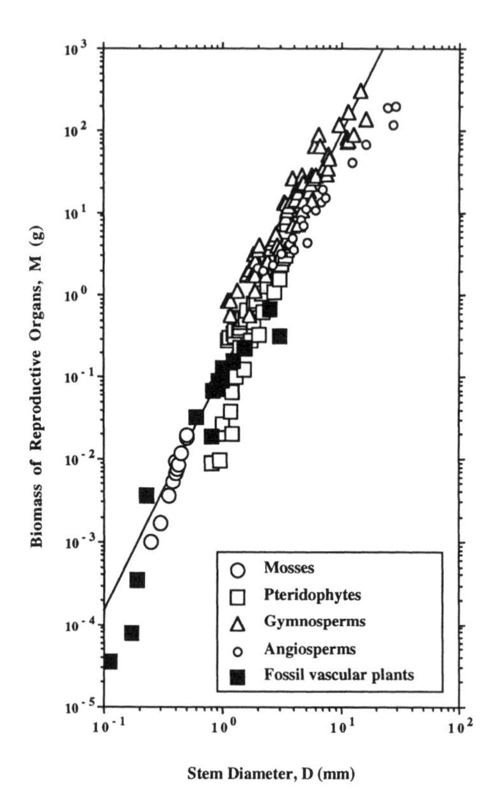

Figure 4.20. Log-log (base 10) plot of reproductive biomass M against stem diameter D measured proximal to M for living and fossil plants (see inset). Solid diagonal line is the regression curve obtained from LS regression analysis.

follows the experimental protocol used by Peters et al. (1988). That is, we shall empirically determine the scaling exponents for the biomass of reproductive organs M with respect to stem diameter D measured just below (proximal to) M. Figure 4.20 plots M (measured in g) versus stem diameter D (measured in mm) for one moss species (*Polytrichum commune*) and fifteen vascular plant species, two of them fossil vascular plants (*Cooksonia pertoni* and *Psilophyton princeps*) (data from Niklas 1993d).[3] The data set is heavily biased toward vascular plants, among

3. Data from Edwards, Davies, and Axe (1992, 684, table 1) were used to calculate the volume V of three *Cooksonia pertoni* sporangia. Based on the values of sporangial height H and width W reported by Edwards, Davis, and Axe, each sporangium has an oblate spheroidal geometry. Thus, $V = (4/3)a^2b$, where a is the major semiaxis or $W/2$ and b is the minor semiaxis or $H/2$. Sporangium mass M was estimated by assuming tissues had a density of 1,100 kg m^{-3}. The width of the proximal end of the subtending axis was used for D. For *Psilophyton princeps*, a single branched specimen bearing numerous prolate sporangia was selected. Thus, $V = 4ab^2/3$, where a is the major semiaxis and b is the minor semiaxis. Tissue density was taken as 1,100 kg m^{-3}. The diameter D of reproductive axes was measured proximal to M in the same manner as for living plant materials.

which gymnosperms (*Larix, Pseudotsuga, Tsuga, Pinus, Picea,* and *Taxus*) and pteridophytes (*Cooksonia, Psilophyton, Psilotum, Lycopodium clavatum,* and *L. lucidulum*) are more or less equally divided. The remaining taxa from which data were gathered are two angiosperms (*Quercus* and *Crategus*). Accordingly, we anticipate that the scaling exponents revealed by the data set will reflect a pronounced "vascular effect." Also, because we need numerous data points from each species, each species is represented by a series of measurements of M and D. A total of 195 data points are plotted in figure 4.20, and on the average each species is represented by 12 data points. Note that a "mean" or "weighted mean" value for M and D for each species has no biological meaning because there is no "representative" intraspecific reproductive biomass or stem diameter. Although the representation of each species by numerous data points and the unequal representation of nonvascular and vascular plants appear to present problems in terms of "data point inflation" and the "phyletic effect," the latter is an inescapable feature of determining the scaling of M with respect to D, whereas the phyletic effect on scaling exponents is precisely what we are interested in.

Regression of the data shown in figure 4.20 reveals that $M = 0.12\ D^{2.90}$ ($r^2 = 0.94$, $N = 16$ species; $\alpha_{RMA} = 2.99 \pm 0.05$, 95% confidence intervals = 3.10–2.88). Thus, for these organisms reproductive biomass scales roughly as the cube power of stem diameter. The phyletic effect on the scaling exponent of reproductive biomass is very pronounced, however. For example, regression of the data from fossil and extant pteridophyte species shows that $M = 0.07\ D^{3.20}$ ($r^2 = 0.90$, $N = 5$ species; $\alpha_{RMA} = 3.37 \pm 0.13$, 95% confidence intervals = 3.63–3.11), while regression of the data from gymnosperm species shows that $M = 0.56\ D^{2.21}$ ($r^2 = 0.92$, $N = 6$ species; $\alpha_{RMA} = 2.29 \pm 0.08$, 95% confidence intervals = 2.45–2.13). For angiosperms, $M = 0.44\ D^{1.79}$ ($r^2 = 0.97$, $N = 2$ species; $\alpha_{RMA} = 1.82 \pm 0.07$, 95% confidence intervals = 1.96–1.68). It is noteworthy that the α_{RMA} values for pteridophytes, gymnosperms, and angiosperms differ at the 95% confidence level and assume the descending numerical sequence $3.37 > 2.29 > 1.82$ that by and large mirrors the chronology of evolutionary appearance of these three plant groups: pteridophytes → gymnosperms → angiosperms. This is curiously reinforced by the scaling relation for the only nonvascular plant in the data set (the moss *Polytrichum commune;* $\alpha_{RMA} = 4.60 \pm 0.06$, $r^2 = 0.99$). Qualitatively, this numerical sequence of α_{RMA} values demonstrates that larger, more robust vertical stems are required to support the comparatively more massive reproductive organs of pteridophytes (compared with those of mosses), while in turn the still more massive reproductive organs of gymnosperms require thicker stems than those of pteridophytes.

The interspecific scaling shown in figure 4.20 represents the statistical summation of sixteen intraspecific relations, each with its own scaling relation, and each appearing to fall into one of only two general categories distinguishable by the numerical value of $\alpha_{RMA.}$ This can be seen by focusing on a few representative species. Figure 4.21 plots M versus D for seven species, only one a conifer (*Larix decidua*). Unfortunately, only three data points are available for the vascular fossil remains called *Cooksonia pertoni*. Consequently, only six regression curves are statistically legitimate. When these curves are drawn, a crosshatched pattern is created by the two sets of diagonally intersecting regression curves. Each set has nearly equivalent α_{RMA} values. Within and between these two sets, the values for the Y_1-intercept differ, sometimes significantly. I shall return to the significance of variations in β shortly. For the time, simply note that the average α_{RMA} value for the regression curves of the fossil pteridophyte *Psilophyton princeps*, the extant pteridophyte *Lycopodium clavatum*, and the conifer *Larix decidua* is 2.06 ± 0.04. By contrast, the average scaling exponent of the regression curves of the moss *Polytrichum com-*

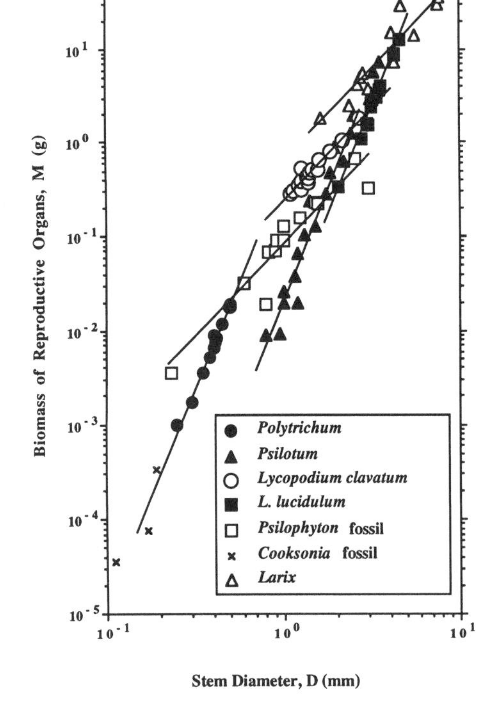

Figure 4.21. Log-log (base 10) plot of reproductive biomass M against stem diameter D measured proximal to M for seven plants (see inset; data from fig. 4.20). Solid diagonal lines are the regression curves obtained from LS regression analysis. Three curves are shown for plants with low scaling exponents (*Lycopodium clavatum*, *Psilophyton*, and *Larix*, $\overline{\alpha}_1$ = 2.06); three curves are shown for plants with high scaling exponents (*Polytrichum*, *Psilotum*, and *L. lucidulum*, $\overline{\alpha}_1$ = 4.63).

mune and the pteridophytes *Psilotum nudum* and *Lycopodium lucidulum* is 4.63 ± 0.04. The intraspecific scaling exponents of the species excluded from figure 4.21 have values very near the average for *Psilophyton princeps, Lycopodium clavatum,* and *Larix decidua* ($\bar{\alpha}_1$ = 2.08 ± 0.04). The regression curves for these species were not plotted in figure 4.21 because they would make the figure too hard to read.

Those already familiar with plant systematics and evolution may be perplexed by the somewhat bizarre taxonomic associations implied by the two sets of regression curves shown in figure 4.21. Each set of curves contains one species of *Lycopodium,* while the data for the vascular plant remains of *Cooksonia* fall along the regression curve for the nonvascular moss *Polytrichum.* The grouping of the fossil pteridophyte *Psilophyton* with a lycopod and a conifer defies every proposed phyletic scheme that warrants serious attention. Likewise, in terms of morphology and anatomy, the two sets of curves defined by essentially equivalent scaling exponents make little apparent sense. For example, *Cooksonia* sporophytes are vascular and branched; *Polytrichum* sporophytes lack vascular tissues and are unbranched, although anomalous bifurcate setae are known. The principal factor responsible for sorting taxonomically unrelated species into the two categories of regression curves appears to be the slenderness factor of stems. All the plants with very slender, untapered stems have high scaling exponents; those with stems thickened by secondary xylem have low scaling exponents. Additionally, plants that fall into one or the other set of regression curves share similar reproductive morphologies. Plants with slender reproductive stems for which $\bar{\alpha}_1$ = 4.63 bear reproductive structures more or less flush to stem surfaces (*Cooksonia, Polytrichum, Psilotum,* and *Lycopodium lucidulum*). By contrast, the plants with comparatively thickened stems for which $\bar{\alpha}_1$ = 2.06 bear clustered or aggregated sporangia subtended by comparatively slender, twiglike lateral branching systems (*Psilophyton, Lycopodium clavatum,* and *Larix*). The difference in the average scaling exponent between the two sets of regression curves therefore appears to relate to stem morphology and to how much reproductive organs are clustered.

Shifts in the Y_1-intercepts (the scaling coefficient β) of the regression curves shown in figure 4.21 seem to relate to the comparative stiffness of stems supporting reproductive organs. From chapter 3 we know that the flexural stiffness of a beamlike stem is the product of Young's modulus *E* and the second moment of area *I.* We also know that, for a stem with a terete cross section, *I* increases as the fourth power of stem diameter. For any value of stem diameter, therefore, the mass that can be mechanically supported by the stem is proportional to *E.* Thus, for any given stem diameter an increase in β requires an increase in this elastic modulus.

Accordingly, it is reasonable to speculate that the Young's modulus of sporangia-bearing axes of the fossil *Psilophyton* is lower than that of *Lycopodium clavatum* axes and that the stems subtending reproductive organs of this lycopod have lower values of E than the stems of *Larix*. Although a direct comparison between the fossil and either of these two extant plants is not possible, direct measurements of elastic moduli verify that the Young's modulus of *Lycopodium clavatum* stems is one order of magnitude less than that of *Larix* twigs.[4] In terms of the regression curves for *Cooksonia, Polytrichum, Psilotum,* and *Lycopodium lucidulum,* the available data suggest that differences in E are in part responsible for shifts in the regression parameter β. Within the size range appropriate for the diameter of sporangia-bearing axes (2 mm $< D <$ 3 mm), the elastic moduli of the stems of *Psilotum* and *Lycopodium lucidulum* are nearly comparable in magnitude, although the numerical values for the E of *Psilotum* stems are slightly higher than for the lycopod species. A comparison between the intraspecific scalings of the moss *Polytrichum* and the pteridophyte *Psilotum* is more difficult. The sporangia-bearing axes of these two plants do not overlap in size, and we cannot legitimately (statistically) extend a regression curve beyond the actual distribution of data points. This would not matter if we were dealing with engineered materials, because the values of elastic moduli determined for one sample of material can be used for any other sample of the same material (the values are size independent). However, the values of the elastic moduli of plant tissues depend on the cross-sectional areas of tissue samples (see Niklas 1992). Thus we cannot argue that the tissues of *Polytrichum* setae have higher E than those of *Psilotum* for comparable stem diameters, even though extension of the regression curves for these two species suggests this is so. But empirical measurements show that the elastic moduli of moss setae are comparable to those of sclerenchyma. Since this tissue is poorly developed in the distalmost axes of *Psilotum,* it is not unreasonable to argue that the vertical axes of *Polytrichum* sporophytes are relatively stiffer than the most distal axes of *Psilotum.* Under any conditions, the regression curves for *Psilotum* and *Lycopodium clavatum,* which are shifted to the right of the curve for *Polytrichum,* demonstrate that the

4. That shifts in β relate to differences in the Young's moduli of stems does not imply that the magnitude of Young's modulus is uniform throughout an entire plant body. The data plotted in figure 4.21 come from measurements taken on the most distal portions of peripheral plant stems, not from the main plant stems or trunks. Thus the larger axes found at the base of *Psilophyton* may have had elastic moduli whose magnitudes equal or exceed those of the distal portions of *Lycopodium lucidulum* stems. In other words, the shifts in β values evident among the regression curves within each set of regressions relate only to those levels of branching that bear reproductive organs.

vascular reproductive axes of pteridophytes mechanically sustain relatively smaller loadings than do the setae of the moss *Polytrichum.*

The situation for the fossil remains of *Cooksonia* is ambiguous at best because we have so few data, although they fall near the regression curve for *Polytrichum,* suggesting that the material properties of the axes of *Cooksonia* were nearly the same as those of moss sporophytes. Three data points, however, should inspire no confidence regarding this inference. Indeed, a variety of competing regression curves exists for these few data, and their coefficients of correlation are not helpful in resolving which if any curve is most appropriate.[5] Be that as it may, even for a comparatively large data set a regression curve serves simply as a hypothesis. Thus the regression curve for *Polytrichum* sporophytes may be cautiously used as a crude model for the scaling relation between sporangial mass and axial diameter for *Cooksonia pertoni.* Since the sporangial mass of *Cooksonia* is an order of magnitude less than that of the moss, and since the scaling exponent for the regression of sporangial mass versus axial diameter is 4.59, the setae of *Polytrichum* theoretically can mechanically sustain relatively much more reproductive biomass than the axes of the currently known most ancient vascular land plant. Keep in mind, however, that a sizable volume fraction of the typical moss sporangium consists of sterile tissues. Thus the actual spore biomass within each *Cooksonia* sporangium may have equaled or exceeded that within a *Polytrichum* capsule. That the moss capsule has an elaborate spore-dispersal design whose additional weight may come at a cost in terms of requiring proportionally more mechanically robust subtending axes speaks volumes for the possible selective advantages to releasing spores when atmospheric conditions are ripe for long-distance dispersal and spore survival.

In summary, a comparison between figures 4.20 and 4.21 emphasizes that intra- and interspecific scaling relations yield very different kinds of information. It also highlights how scaling analyses may be used to speculate on evolutionary patterns. More important, an interspecific scaling relation may give the appearance that a very basic limitation or constraint exists ($M \propto D^{2.99}$). Typically, however, no single plant taxon occupies the entire range of size considered in the previous interspecific comparison. Consequently the relation $M \propto D^{2.99}$ does not apply to a "particular" plant or even a particular group of plants, so attempts to deduce the raison d'être for $M \propto D^{2.99}$ from biomechanical "first principles" may be largely misplaced. Indeed, figure 4.21 shows that evolution attains two

5. Likewise, the values for *Cooksonia* sporangial biomass are estimates, rather than the result of direct measurements, while the data for axial diameter are suspect owing to compressional distortions that likely attended the fossilization of these plant remains.

very different scaling relations that appear to result from dissimilar ontogenetic patterns (see figs. 5.10 and 5.11). Indeed, the interspecific scaling relation $M \propto D^{2.99}$ is a statistical consequence of the superimposition of very different evolutionary patterns.

4.5 The Scaling of Long-Distance Dispersal

I begin the discussion of long-distance dispersal by considering its consequences on gene flow. Among embryophytes, two broad categories of interpopulational gene flow are distinguishable and relate either directly or indirectly to how far meiospores are dispersed from their parent plants. These categories are the homosporous and heterosporous plants. The former produce highly dispersible meiospores that grow to become free-living, potentially bisexual gametophytes. For these plants, three types of fertilization are possible (Klekowski 1979; Lloyd 1974; Soltis, Soltis, and Noyes 1988): (1) self-fertilization of the same gametophyte (intragametophytic selfing); (2) cross-fertilization between gametophytes derived from the sporophyte (intergametophytic selfing); and (3) cross-fertilization between gametophytes derived from different sporophytes (intergametophytic crossing). Intragametophytic selfing has been argued to be the predominant mode of reproduction in natural populations of homosporous species, particularly ferns (see Klekowski and Baker 1966; Klekowski 1979), although this type of fertilization has severe genetic consequences (e.g., inbreeding depression can be reached in one generation rather than many). Intergametophytic selfing and intergametophytic crossing, which respectively are equivalent to selfing and outcrossing in seed plants, have been considered less frequent modes of gene flow for homosporous plants, although these types of fertilization confer genetic advantages. If intergametophytic selfing is the dominant mode of fertilization among homosporous species, then natural populations of these species should evince little heterozygosity, indicative of outcrossing, since homozygous sporophytes can be produced in a single generation as a result of intragametophytic selfing. For some homosporous fern species (e.g., *Botrychium* and *Asplenium*), intragametophytic selfing appears to dominate. For others (e.g., *Bommeria, Pellaea,* and *Polystichum*), however, substantial levels of heterozygosity are evident (see Soltis and Soltis 1992 and references therein). Clearly, the issue is not "exclusivity" but rather "relative frequency" of occurrence of selfing versus outcrossing. Along these lines, Soltis and Soltis (1992) examined intragametophytic selfing rates of twenty species of homosporous ferns. They report a pronounced bimodal distribution with most species (80%) characterized as extreme outcrossers. Likewise, Soltis, Soltis, and Noyes (1988) and Soltis and Soltis (1988) show that intragametophytic selfing is rare or totally absent for natural populations of a common horsetail (*Equisetum arvense*) and three species of lycopods (*Lycopodium clavatum, L. annotinum,* and *Huperzia miyoshiana*). For these homosporous pteridophytes, intergametophytic fertilization predominates.

The rarity of highly inbreeding homosporous species is permitted by three mechanisms: (1) inbreeding depression with prominent elimination of genotypes characterized by intragametophytic selfing, (2) asynchrony of antheridia and archegonia formation on the same thallus, and (3) long-distance dispersal leading to intergametophytic crossing among sporophytes. These three mechanisms are not mutually exclusive, but I list them separately to emphasize that genetics, development, and the physical environment play equally important roles in determining gene flow patterns in natural populations. My emphasis here is on long-distance dispersal, which permits plants to colonize new sites, increase gene flow, avoid possible density-dependent negative effects, and escape competition from siblings or parents.

Long-distance dispersal can be effected by biotic or abiotic vectors that disperse diaspores (i.e., meiospores, seeds, and fruits). For example, the production of edible fruits containing inedible, toxic, or indigestible seeds is common among taxonomically diverse angiosperms. Just as common and perhaps as efficient in terms of dispersal and metabolic investment is the production of wind-dispersed diaspores. Although a number of ecological and morphological traits correlate with this mode of abiotic dispersal, two are easily addressed in terms of scaling analysis: a tendency for convergence in diaspore size and shape on efficient aerodynamic design (e.g., the autogyroscopic winged seeds of *Pinus* and the samaras of *Acer*) and a tendency to locate diaspores as far aboveground as vegetative mechanical architecture permits. The most elementary (and therefore naive) ballistic model of dispersal quickly demonstrates that these generalized traits have predictable consequence on the horizontal distance x traveled by an individual pollen grain, seed, or fruit. This model contains only three variables: the height H at which a diaspore is released from its parent, the diaspore settling velocity U_T, and the mean horizontal wind speed U_h (which must be averaged between H and the deposition site on the ground because wind speed tends to diminish parabolically from the tip to the base of the plant). Intuitively, we expect x to proportionally increase in respect to H and U_h. Indeed, these two variables are interrelated since, all other things being equal, U_h will increase as a function of H. Likewise, the higher the point of diaspore release, the greater the potential dispersal distance. By the same token, we expect x to decrease proportionally with respect to U_T. The faster something falls downward, the less likely it is to be carried away by horizontal gusts of wind. Combining these proportional relations yields $x \propto (HU_h)(U_T)^{-1.}$ Application of the tool of dimensional analysis shows that the quotient of HU_h and U_T has the same units as dispersal distance (length). Accordingly, the most basic ballistic model for seed and fruit wind dispersal is given by the formula

$$x = \frac{HU_h}{U_T} \tag{4.1}$$

It must not escape our attention that plant size dictates H and indirectly affects $U_{h.}$ From prior discussions, recall that plant size and shape influence U_T (see table 2.8).

The variables contained in eq. (4.1) can operate in complex ways. For example, the influence of vertical stature can be compensated for by a change in U_T. A short plant may be as effective in dispersing its diaspores as a tall counterpart if the settling velocity of its seeds or fruits is slow enough. Also, regardless of how slowly a diaspore descends through the air, its lateral transport depends on the meteorological variable $U_{h.}$ The complex interrelations among wind speed and plant size and shape are simplified by relating two dimensionless expressions obtained from the variables in eq. (4.1):

$$\frac{x}{H} = \frac{U_h}{U_T} \tag{4.2}$$

The expression to the left of eq. (4.2) is the normalized dispersal distance. This parameter is useful when comparisons among plants differing in height are required in terms of the dispersal distance of diaspores. The expression to the right of eq. (4.2) is average wind speed normalized with respect to diaspore settling velocity. This parameter is useful when comparing plants with different habitat preferences. In any circumstances, the juxtaposition of the two expressions in eq. (4.2) manifests the biological strategy of wind dispersal. The "objective" is to maximize the normalized dispersal distance x/H. This can be achieved by either increasing U_h or decreasing $U_{T.}$ In this regard we may ask, How do diaspores (pollen, seeds, and fruits) differ? Figure 4.22 plots U_h/U_T versus x/H for a number of wind-dispersed seeds, real and model fruits, and meiospores.[6] Regression of these data indicates that U_h/U_T and x/H are anisometrically related such that, on the average, $x/H \propto (U_h/U_T)^{1.18}$ ($r^2 = 0.75$, $N = 65$; $\alpha_{RMA} = 1.36$). This scaling relation indicates that a slight increase in the

6. Data from Okubo and Levin (1989, 334–35, table 1) who provide a summary of data on x, H, U_h, and U_T from other authors.

averaged horizontal wind speed or a slight decrease in settling velocity realizes a disproportionate increase in normalized dispersal distance. Since U_h is indirectly dependent on H, an increase in plant size is evidently beneficial to dispersal. Likewise, any modification in the physical properties of a diaspore that reduces U_T is beneficial. Indeed, even a small decrease in settling velocity has a significant affect on dispersal distance. For example, regression of x/H versus U_T for the data plotted in figure 4.22 shows that $x/H \propto U_T^{-1.72}$. Thus, a comparatively small reduction in U_T can result in a dramatic increase in normalized dispersal distance.

Clearly, the settling velocity U_T can be reduced by decreasing the mass M or increasing the maximum cross-sectional area A of a diaspore, since in general $U_T \propto M/A$. Inserting this proportional relation into eq. (4.2) relates diaspore size (measured in terms of M) and shape (measured in terms of A) to the normalized dispersal distance:

$$\frac{x}{H} \propto \frac{A}{M}. \tag{4.3}$$

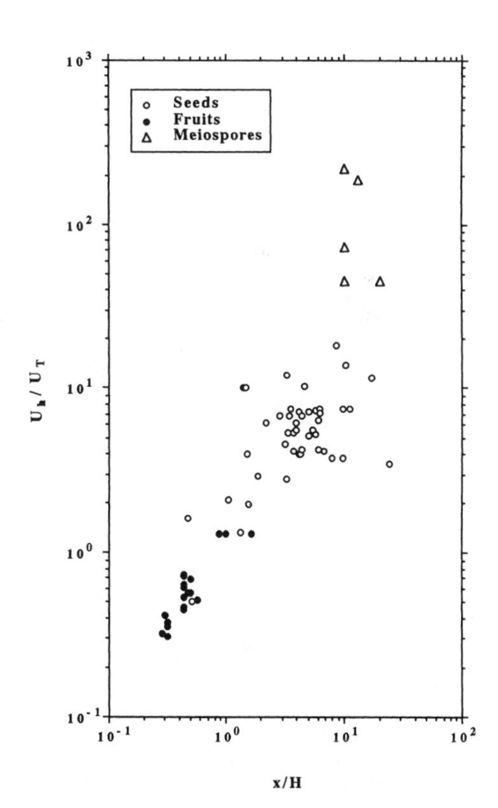

Figure 4.22. Log-log (base 10) plot of the quotient of horizontal wind speed U_h and settling velocity U_T against the quotient of lateral distance of transport x and release height H. Data from Okubo and Levin 1989, 334–35, table 1.

Thus, in terms of diaspore dispersal and metabolic investment in seedling establishment, an obvious tactic is a scaling relation such that $A \propto M^{\alpha}$, where $\alpha > 2$. In this circumstance, x/H would increase disproportionately with respect to diaspore biomass. Aerodynamic theory asserts that any diaspore mass can be conveyed at any settling velocity provided the shape (i.e., cross-sectional area) of the falling object is sufficiently modified. There are biological limitations to the relation between size and shape, however. Figure 4.23 plots A versus M for forty-four species of wind-dispersed fruits and twenty-four species of wind-dispersed seeds (data from Augspurger 1986, 356–57, table 1) and Matlack 1987, 1152–53, table 1). Regression of the data for fruits indicates that $A \propto M^{0.47}$ ($r^2 = 0.60$; $\alpha_{RMA} = 0.61$), while regression of the data for seeds gives $A \propto M^{0.78}$ ($r^2 = 0.72$; $\alpha_{RMA} = 0.92$). Inserting these proportionalities into eq. (4.3) yields $x/H \propto M^{-0.39}$ for fruits and $x/H \propto M^{-0.08}$ for seeds. These proportionalities are purely hypothetical, but they suggest that x/H is less dependent on seed size than on fruit size (seeds are "better"). When both types of diaspores are considered, regression of A versus M shows that $A \propto$

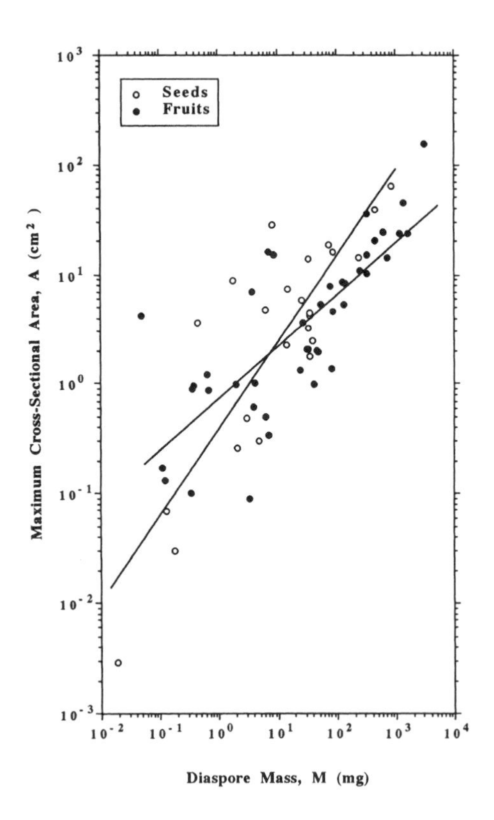

Figure 4.23. Log-log (base 10) plot of maximum cross-sectional area A of diaspore against diaspore mass M. Solid diagonal lines are the regression curves obtained from LS regression analysis. Data from Augspurger 1986, 356–57, table 1, and Matlack 1987, 1152–53, table 1.

$M^{0.59}$ ($r^2 = 0.62$, $N = 68$; $\alpha_{RMA} = 0.75$). Thus, on the whole, the capacity to modify diaspore shape in response to increasing seed or fruit biomass hypothetically is rather dismal, falling far short of the theoretically optimal condition suggested earlier ($A \propto M^{\alpha>2}$).

The distinction between fruits and seeds as effective wind dispersal agents can be empirically tested based on an independent data set. Figure 4.24 plots M versus x/H (once again, data from Okubo and Levin 1989). The distinct segregation based on M versus x/H suggests that, on the average, fruits tend to fall proportionally closer to their parent plants than do seeds, presumably because the scaling relation between fruit mass and maximum surface area is less efficient than for seeds. Indeed, at first glance the evolution of wind-dispersed fruits seems counterintuitive. The mature ovarian tissues of a fruit confer an added mass, and the aggregation of seeds into a single dispersal package potentially can lead to sibling rivalry at the deposition site. Wind-dispersed fruits may represent an alternative route, however, in terms of the developmental capacity to alter the shape of botanical structures in accordance with aerodynamic design.

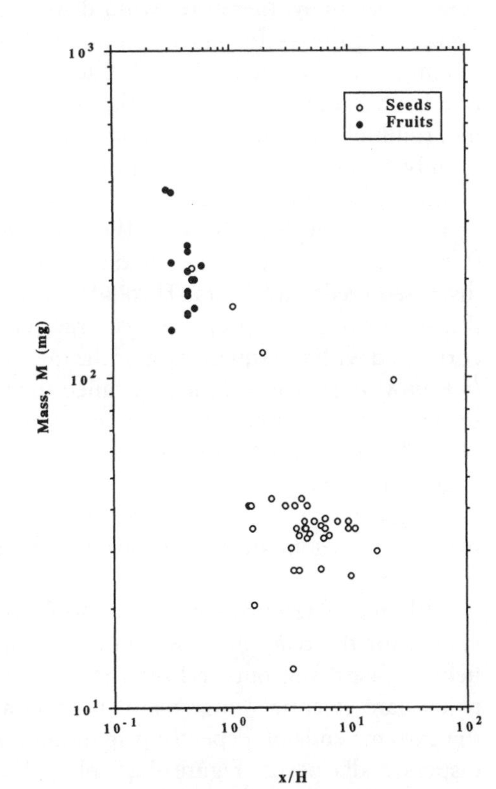

Figure 4.24. Log-log (base 10) plot of diaspore mass M against the quotient of the quotient of lateral distance of transport x and release height H. Data from Okubo and Levin 1989.

Presuming that selection pressures in some habitats favor wind dispersal and that some species are developmentally incapable of modifying seed morphologies to obtain effective diaspore dispersal mechanisms, modification of fruit morphology represents a functional compromise. Likewise, fruit tissues can offer an additional barrier to seed predation (Casper 1987). Consequently, developmental limitations as well as seed protection may explain the evolution of wind-dispersed fruits in some taxa. Finally, it is noteworthy that ovule abortion of wind-dispersed fruits is not uncommon (see Augspurger and Hogan 1983; Augspurger 1986), perhaps because of natural selection operating at the level of seedling mortality.

In terms of morphological convergence in diaspore morphology, it is important to note that two fundamental mechanisms reliably oppose the influence of gravity: intertial drag and rotation-induced lift. Thus seeds and fruits either can slow their descent through air by parachute-like structures that generate drag forces over and around their surface (e.g., the pappus of a *Tragopogon* fruit) or can rotate and generate lift by means of airfoil-like extensions (e.g., the blade of an *Acer* samara or a *Pinus* seed). Essentially, therefore, wind-dispersed diaspores fall into two extreme categories—*floaters* and *rotators*. Rotation about an axis converts potential into kinetic energy. This additional component of free-fall motion further reduces U_T. Hence diaspore morphologies that produce rotating, tumbling, or some other motion about one or more of their axes are an advantage in terms of long-distance dispersal. This has been demonstrated by Carol Augspurger, who provides perhaps the most comprehensive empirical analysis of the consequences of size (measured in terms of M) and shape (measured in terms of A) on the behavior of wind-dispersed seeds and fruits. Through a detailed study of the diaspores of thirty-four tropical species, Augspurger (1986) found that U_T is linearly correlated with the square root of the quotient of M and A—that is, $U_T = b + m(M/A)^{1/2}$. This quotient is called the wing loading when applied to seeds and fruits, although not all diaspores possess *wings*. Also note that U_T and wing loading are necessarily autocorrelated. Nonetheless, Augspurger reports that the slope m of the regression for floaters is significantly greater than that for rotators, demonstrating that for comparable mass and area floaters have faster settling velocities, on the average, than rotators.

Although Augspurger presents a sophisticated and useful classification system for the relation between diaspore morphology and aerodynamic behavior (see Augspurger 1986, 355, fig. 1), this system can be reduced to its bare essentials in terms of floaters and rotators, which represent the extreme ends of a spectrum spanning the behavior of all other wind-dispersed diaspores. Figure 4.25 plots U_T versus [wing loading]$^{1/2}$ for twenty-three floaters, fifteen rotators, and twenty-seven other morphologically diverse diaspores (whose aerodynamic behavior involves tumbling, undulation, rolling, etc.). These data are from Augspurger (1986) and Matlack (1987). As expected, regression shows that the relation between U_T and [wing loading]$^{1/2}$ is linear for each category of diaspore: floaters, $U_T = -0.12 + 0.61$ [wing loading]$^{1/2}$($r^2 = 0.98$); rotators, $U_T = -0.07 + 0.32$ [wing loading]$^{1/2}$($r^2 = 0.75$); and others, $U_T = 0.31 + 0.15$ [wing loading]$^{1/2}$($r^2 = 0.80$). Moreover, the slopes of these regressions qualitatively reveal that floaters descend faster than rotators of comparable mass and area. And as predicted, the data points for other diaspores plot between the regression curves drawn for floaters and rotators.

Since the shape dependency of wing loading obscures the relation between M and U_T, it is instructive to relate U_T directly to M, particularly since the dependent variable is a crude measure of seedling fitness. Figure 4.26 plots U_T versus M for floaters, rotators, and other types of seeds and fruits (once again based on data from Augspurger 1986 and Matlack 1987). Regression of the entire data set shows that $U_T \propto M^{0.18}$ ($r^2 = 0.50$,

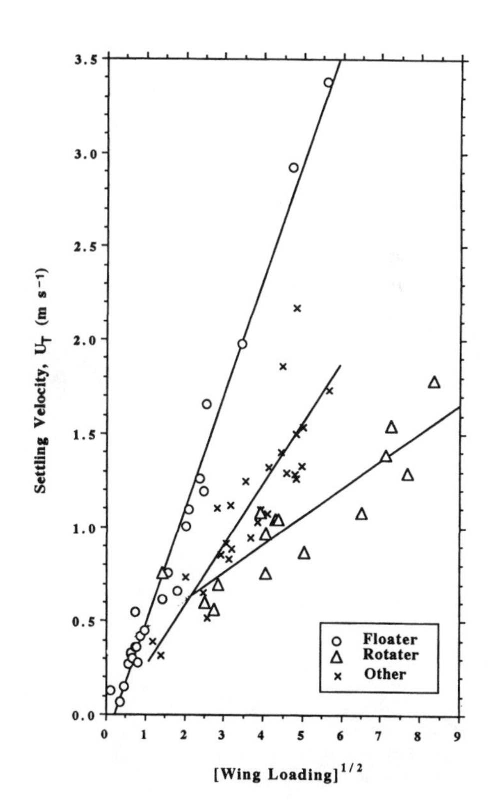

Figure 4.25. Log-log (base 10) plot of settling velocity U_T against [wing loading]$^{1/2}$ for morphologically diverse diaspores (whose aerodynamic behavior involves tumbling, undulation, rolling, etc.). Solid diagonal lines are the regression curves obtained from LS regression analysis. Data from Augspurger 1986 and Matlack 1987.

$N = 66$). Inspection of this figure shows that the smallest diaspores ($M < 1$ mg), which are predominantly (but not exclusively) seeds, tend to be floaters, whereas most of the more massive diaspores ($M > 100$ mg), which are all fruits, are nonfloaters. Although the data are scanty, this interspecific comparison suggests that floaters may be a preferred seed-dispersal mechanism, whereas rotators may be a preferred fruit-dispersal mechanism in terms of reconciling mass with settling velocity. The data shown in figure 4.26, however, have little bearing on the relation between U_T and M at the species level, since the natural variation in diaspore mass and therefore in the settling velocity of the diaspores produced even on a single inflorescence can be considerable. Accordingly, it is useful to illustrate the influence of diaspore mass on settling velocity for a particular species. Figure 4.27 plots the settling velocity versus fruit mass measured for eighty-five fruits taken from a single inflorescence of *Tragopogon dubius*. The fruits of this species are achenes; each is wind dispersed by means of a modified parachute-like calyx, called a pappus. These fruits therefore are floaters, relying on inertial drag to slow their

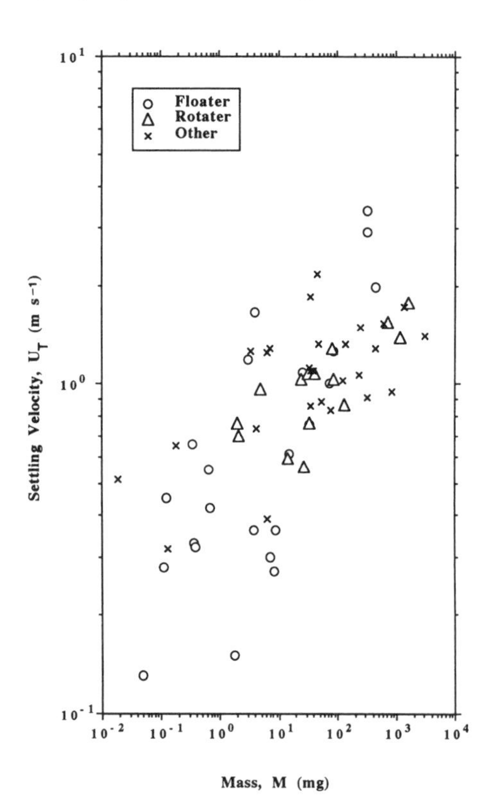

Figure 4.26. Log-log (base 10) plot of settling velocity U_T against diaspore biomass M. Data from Augspurger 1986 and Matlack 1987.

descent through the air. Regression of the data shows that $U_T \propto M^{0.33}$ ($r^2 = 0.79$; $\alpha_{RMA} = 0.37$). That is, on the average, the settling velocity is approximately proportional to the cube root of achene mass. Although the anisometric scaling of *Tragopogon* is better than that evinced by an interspecific comparison, it is nonetheless in keeping with the conclusion that larger diaspores disproportionately settle in the air column at faster rates than do smaller seeds or fruits. Incidentally, the natural variation in fruit or seed mass, even for an inflorescence that releases its diaspores at a single source point (e.g., *Tragopogon*), is rarely taken into account by those who attempt to mathematically predict the number of propagules per ground surface area as a function of distance from the parent plant. This is unfortunate, since the shape of the dispersal curve will be affected by the scaling relation between U_T and M.

The influence of parental plant height on long-distance dispersal, which has been alluded to, can be assessed crudely by taking the quotient of the height H of diaspore release to settling velocity U_T, which has the unit of second and therefore permits a rough estimate of the time required

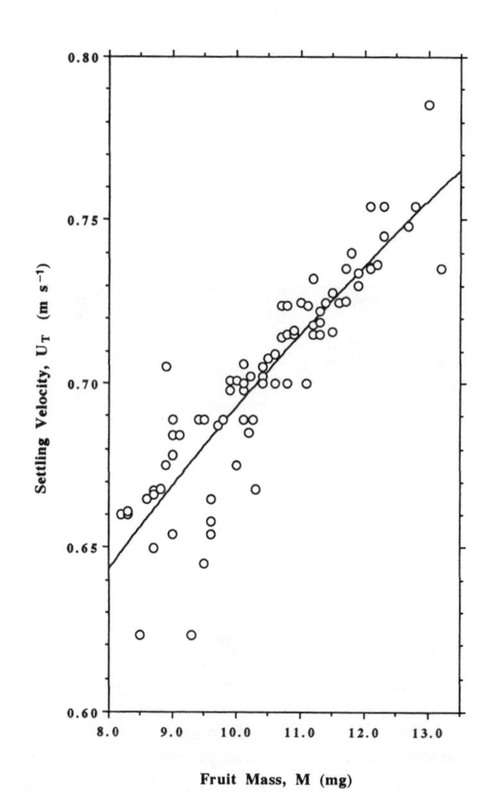

Figure 4.27. Log-log (base 10) plot of settling velocity U_T against mass M of fruits removed from a single inflorescence of *Tragopogon dubius*. Solid diagonal line is the regression curve obtained from LS regression analysis.

for a seed or fruit with a given settling rate to fall to the ground. The longer the time required for vertical descent, the greater the probability that the diaspore will be swept away from its parent plant by wind currents. Although data are scanty in this regard, those accumulated by Augspurger (1986) suffice. Figure 4.28, which plots H/U_T versus [wing loading]$^{1/2}$ for seven floaters, six rotators, and twenty other diaspores, shows that, for any given diaspore size, rotators tend to take longest to fall to the ground, even though the smallest floaters have the slowest rates of descent (see figs. 4.25 and 4.26). This is because, on the average, rotators are produced by taller plants than floaters. In summary, therefore, floaters tend to be small seeds borne on short plants, while rotators tend to be large fruits borne on tall plants.

4.6 Size and Gender Expression

In this section we examine the correlation between plant size and gender expression. This topic is important because the ratio of the pollen- to

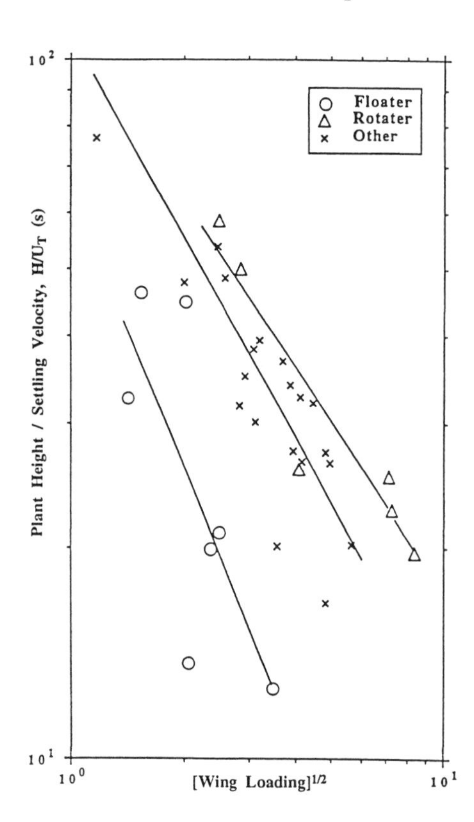

Figure 4.28. Log-log (base 10) plot of quotient of plant height H and settling velocity U_T against [wing loading]$^{1/2}$ of morphologically diverse diaspore (dispersal) types. Solid diagonal lines are the regression curves obtained from LS regression analysis. Data from Augspurger 1986.

ovule-bearing reproductive structures borne on an individual of a monoecious species may change, so much so that the gender expression of an individual may be unalloyed for a particular season. Likewise, the sexual expression of the individual of a dioecious species is not always genetically fixed or independent of size or local environmental conditions. Hermaphoroditic individuals may be common in a species traditionally viewed as dioecious. Significantly, several of the species known to occasionally produce hermaphroditic individuals are reported to have sex chromosomes (e.g., *Cannabis sativa, Humulus japonicus,* and *Rumex hastatulus;* see Freeman, Harper, and Charnov 1980). To be sure, some of these individuals may be the consequence of genetic mosaics, but instances of true gender expression plasticity in individual plants of dioecious as well as a monoecious species are well documented. Significantly, the frequency of production of pollen- or ovule-bearing reproductive structures often correlates with total plant size and therefore may legitimately be viewed as a scaling phenomenon (in the crudest sense). As I will show, an important conclusion resulting from an examination of these species is that an increase in plant size is not unidirectional—older plants can be smaller than in their immediate past. Thus, the size-dependent trend in gender expression is reversible. This is permissible provided plant "size" is measured in terms of metabolic reserves that can increase or decrease depending on transient or persistent changes in the local environmental conditions attending the growth of the individual. The difficulty with this definition of *size* is that taking measurements is extremely troublesome or, in some cases, impractical.

Obviously, the first task is to quantify gender expression. A variety of equations have been proposed for this purpose, but in each case these equations attempt to index the allocation of energy (or some other expression of resource) to male versus female function. The most forthright index in this regard is the caloric content of pollen- and ovule-bearing reproductive structures. But this is sample destructive and largely unworkable, particularly when changes in the gender expression of an individual plant over many years are the focus of attention, because the removal of reproductive structures may alter the pattern of energy allocation in subsequent years. Perhaps for this reason, most of the equations used to quantify gender expression employ phenotypic variables presumed to reflect energy allocation patterns. For example, Lloyd and Bawa (1984, 258–59) introduce a standardized measure of phenotypic gender for each individual in a population. This measure is weighted as a function of the population average. In terms of the standardized phenotypic female (male) gender of an individual plant, their equation takes the form

(4.4a) $$F_i = \frac{c_i}{c_i + s_i E}$$ (phenotypic female gender)

(4.4b) $$M_i = \frac{s_i E}{c_i + s_i E},$$ (phenotypic male gender)

where c_i is a measure of the allocation to female gender (e.g., number of carpels or carpellate flowers), s_i is a measure of the allocation to male gender (e.g., number of stamens or staminate flowers), and E is the equivalency factor that standardizes the gender allocation of the individual in relation to the entire population of plants. Specifically, E is the quotient of the sum of c_i and the sum of s_i for the entire population ($E = \Sigma c_i / \Sigma s_i$). Equation (4.4) yields $F_i = 1.0$ and $M_i = 0.0$ for an individual plant that produces only carpellate flowers and $F_i = 0.0$ and $M_i = 1.0$ for an individual that produces only staminate flowers. A hermaphroditic individual would have a value of F_i less than unity but greater than zero ($0 < F_i < 1$).

The equation proposed by Lloyd and Bawa has been used with great success in a variety of studies comparing populations growing under different environmental conditions. It is evident, however, that the magnitudes of F_i and M_i depend on the equivalency factor E, which can change from year to year (Delesalle 1989, 32). Equation (4.4) therefore can present problems when we compare the gender expression of an individual over many years. This is easily illustrated by a simple numerical example. Consider an individual that invariably produces carpellate and staminate flowers such that $c_i/s_i = 10$. If in two successive years the population in which the individual grows is such that $E = 0.5$ and $E = 10$, then eq. (4.4a) yields $F_i = 0.95$ and $F_i = 0.50$, respectively, thereby indicating that the *female*ness (or *male*ness) of an individual depends on the gender dynamics of the population even though the individual evinces a constant floral ratio. Removing E from eq. (4.4) resolves this dependency:

(4.5a) $$f_i = \frac{c_i}{c_i + s_i}$$

(4.5b) $$m_i = \frac{s_i}{c_i + s_i}.$$

For this reason, an investigator may prefer to use eq. (4.5) when comparing year-to-year phenotypic gender expressions of a population whose equivalency factor fluctuates from year to year (Delesalle 1989).

A great deal has been written about the factors that influence gender expression. Among them are injury, disease, temperature, soil fertility, moisture, light availability, and the external application of growth hormones. When they are surveyed, however, no clear pattern emerges, although in the broadest terms male gender expression tends to be induced by environmental stress (e.g., dry soil, extreme temperature, low light intensity, or low nitrogen), whereas under optimal growth conditions the tendency is to favor femaleness. This has led some to speculate that labile gender expression is an adaptive function that has survival value in patchy environments. Although this is an attractive hypothesis for some species, the growth rates of female individuals are known to decline in comparison with conspecific male individuals (e.g., *Acer negundo* and *Fraxinus lanceolata;* see Lysova and Khizhnyak 1975). Likewise, a shift to female gender may be induced by environmental stress in other species (e.g., *Atriplex* spp. and *Simmondsia chinensis;* see Stark 1970; Hikmat, Strain, and Mooney 1972; and Freeman, Harper, and Charnov 1980, 226n). Thus the *adaptiveness* of labile gender expression in terms of stressful or patchy environments must be viewed with some skepticism.

Leaving aside the issue of adaptiveness, the influence of environmental factors on gender expression is important because the size of the individual plant is correlated with changes in gender expression in woody and nonwoody species and because plant size may be influenced by growth conditions (e.g., *Arisaema* spp., *Castillea clastica, Ilex opaca,* and *Metasequoia glyptostroboides;* see Freeman, Harper, and Charnov 1980, 224, table 2). In general, female gender expression tends to increase with plant size because individuals undergo a transient male transition as they increase in size before they manifest phenotypic female gender. That this feature is size dependent rather than age dependent is revealed by the reversibility of gender expression with a reduction in plant size, as when the amount of photoassimilates stored in corms, taproots, and the like is reduced owing to poor growing conditions. For example, Condon and Gilbert (1988) compiled observations taken over seventeen years on the gender expression of the cucurbits *Gurania spinulosa* and *Psiguria bignoniacea.* Long believed to be dioecious, these species were found to be monoecious. Small plants either fail to produce vegetative structures or produce only staminate flowers, whereas Condon and Gilbert found that large vines of both types of cucurbits tend to produce carpellate flowers. Under poor growth conditions and a concomitant diminution in size, however, the same genotypes were found to revert to producing only staminate flowers, presumably because stressful environmental conditions diminish plant size measured in terms of metabolic resources.

Likewise, Delesalle (1989) recorded variations in phenotypic gender in monoecious cucurbit *Apodanthera undulata* plants over a period of two years. Individuals of this species may persist as long as thirty years by growing from a perennial tuberous root. As in *Gurania spinulosa* and

Psiguria bignoniacea, small *Apodanthera undulata* plants produce no flowers, and larger plants produce only staminate flowers. Delesalle reports that large plants are cosexual. Of the total number of blooming plants, 85% did not change gender between successive years, presumably because in most cases the growth conditions of the previous year were found to be correlated with those of the second year. Beyond a certain threshold plant size, Delesalle reports that female and male allocation may increase in either similar or dissimilar ways depending on growth conditions, and values for female gender expression f_i (see eq. 4.5a) were higher for plants growing under less crowded conditions, presumably because plants had more access to sunlight.

The scaling of gender expression is illustrated further by species of jack-in-the-pulpit (*Arisaema triphyllum, A. dracontium,* and *A. japonica;* see Atkinson 1898; Maekawa 1924; Camp 1932; and Bierzychudek cited in Lloyd and Bawa 1984, 300). For these plants, gender expression is highly correlated with the size of the individual's long-lived corm. Individual plants with large corms growing in sunny microhabitats or in rich soils produce carpellate flowers. Conversely, individuals with small corms growing in shady microhabitats or in poor soils either fail to flower or produce only staminate flowers. In the case of one species (*A. dracontium*), the alternative of either all-male or male-female flowers exists. Maekawa (1924), who explicitly related gender expression to photoassimilate reserves, showed that gender expression in *A. japonica* is primarily unidirectional with increasing corm size (nonflowering → male → bisexual → female). This trend is reversed when corm size decreases.

Collectively, therefore, the data reported for a variety of botanical systems support the hypothesis that the allocation to female function increases as the size of an individual exceeds a certain threshold and that this allocation pattern can be reversed under stressful growth conditions (see also Bazzaz and Harper 1977; Clay and Shaw 1981; Freeman, McArthur, and Harper 1984; Allison 1991). It cannot be overemphasized, however, that patterns of gender expression can be extremely complex because of the interplay of genotype and growth conditions. This complexity is most evident when we are working with perennial large plants whose metabolic reserves have been accumulated over many years and therefore permit the buffering of year-to-year variations in growing conditions. For these plants, individuals may "assess" their immediate growth conditions in terms of assimilates accumulated from previous years and respond by shifting gender expression accordingly. If so, then even large plants may "opt" to allocate resources to the male rather than the female function when poor growth conditions persist for a few years.

Thus, depending on the interplay between immediate growth conditions and plant size, size-dependent variations in gender expression are often obscured when differences among populations are emphasized at the expense of the differences of a single population over many years of growth and reproduction. This emphasizes the need to work with large data sets on gender expression accumulated over many years (see Condon and Gilbert 1988).

4.7 Summary

The question, Do larger plants produce larger parts? or Do larger plants produce larger reproductive structures? was considered and shown to be largely meaningless unless framed in the context of plant architecture and growth pattern and also in terms of the distinction between size-dependent variations among parts of modular (meristic) reproductive units and size-dependent variations among comparable modular units on the same plant. The partitioning of reproductive biomass M_R (measured in terms of the dry matter of flowers, fruits, and seeds) and vegetative biomass M_V (measured in terms of the dry matter of leaves, stems, and roots) among annual species of angiosperms was shown to comply with $M_R = mM_V + b$. When b is positive, M_R/M_V will monotonically decrease with size. When b is negative, M_R/M_V will monotonically increase with size. The latter was illustrated by a data set for *Capsella bursa-pastoris*, for which $M_R = 0.025 + 0.176\ M_V$ ($r^2 = 0.97$). M_R/M_V was seen to decrease monotonically with increasing size. By contrast, $b < 0$ for many perennial plants, presumably because the biomass allocated to storage organs disproportionately increases with plant size, shifting the ratio of M_R to M_V accordingly. The tendency for M_R/M_V to decrease with the size of woody species was explained in terms of the amortization of secondary xylem in both root and stem. In general, average seed size and number per fruit are inversely proportional, suggesting that an allocation rule may generally operate (that the total biomass investment in seed biomass per fruit or inflorescence is developmentally partitioned among seeds reaching maturity). The inverse relation between seed size and number, however, was found not to be invariable. For *Capsella bursa-pastoris*, larger plants were found to produce more but not larger individual seeds, because larger plants produce more flowers. Other factors, which in turn may affect the allocation of resources within a developing fruit or possibly an entire inflorescence, may play a role in influencing seed size and number (e.g., the relative time of fertilization and ovule position are often important).

In terms of biomass allocation to the gynoecium M_G, androecium M_A, and perianth M_P of perfect flowers, data from ninety species yielded $M_G \propto M_P^{0.914}$ and $M_A \propto M_P^{0.978}$, indicating that the biomass investments to both floral organ types scale interspecifically in an isometric manner with respect to flower size. On the average, however, we saw that the allocation ratio M_G/M_A is size dependent. The mass of individual fruits M_{IF} and mass of individual seeds M_{IS} evince an isometric relation, and seed number per fruit and individual seed size are inversely correlated. For gymnosperms, data on the number N of seeds per ovulate cone, the total seed mass per cone M_{TC}, and the mass of the ovulate cone M_C for ten conifer species showed that, on the average, larger cones bear a disproportionately greater total seed mass. Also, larger cones tend to produce larger seeds. For angiosperms, the total fruit biomass is roughly proportional to the square of branch diameter, and the mass invested in photosynthetic organs decreases as a function of the size of nonwoody or shrub species. Based on a broad interspecific comparison, total reproductive biomass was found to scale roughly as the cube power of stem diameter. The scaling exponents for different groups of plants (mosses, pteridophytes, gymnosperms, and angiosperms), however, differed significantly. In general terms, intraspecific scaling exponents fell into one of two broad categories ($\alpha_{RMA} \approx 2.0$ or 4.0), each of them explicable in terms of the slenderness factor of stems and whether or not reproductive organs were aggregated.

In terms of dispersal, seeds were shown, on the average, to be far better agents of wind dispersal than fruits. Diaspores that rely on inertial drag tend to be small seeds borne on short plants, whereas diaspores that rely on a gyroscopic aerodynamic behavior tend to be large fruits borne on tall plants.

The size-dependent trend in gender expression of an individual plant was seen to be reversible when size was measured in terms of metabolic reserves, which can increase or decrease depending on transient or persistent changes in the local environmental conditions attending the growth of the individual. Injury, disease, temperature, soil fertility, moisture, light availability, and the external application of growth hormones are among the factors known to influence gender expression. When surveyed, however, no clear pattern emerged, although male gender expression tends to be induced by environmental stress (e.g., dry soil, extreme temperature, low light intensity, or low nitrogen), whereas optimal growth conditions tend to favor femaleness. In general, female gender expression increases with plant size because individuals undergo a transient male transition as they increase in size before they manifest phenotypic female gender. That this feature is size rather than age dependent was revealed by the reversibility of gender expression with a reduction in plant size. Collectively, the data reported for a variety of systems supported the hypothesis that the allocation to female function increases as the size of an individual plant exceeds a certain threshold and that this allocation pattern can be reversed under stressful growth conditions.

5 Evolution

5.1 Introduction

This chapter will examine the consequences of size on plant evolution. Even a cursory treatment of this topic, however, is daunting in terms of both the time-span involved and the tremendous diversity in fossil and extant plant morphology, anatomy, and reproduction. The first known unicellular photoautotrophs made their appearance roughly 3.1 billion years ago, and the earliest currently known vascular plant remains are from late Silurian sediments dated at roughly 420 million years. Many morphological, anatomical, and physiological modifications occurred between the advent of the first photosynthetic organisms and the first vascular plants (tracheophytes), and many more made their appearance after the late Silurian. Consequently, this chapter provides only a cursory treatment of the evolutionary significance of scaling relations by drawing on a few examples calculated to emphasize the relations among plant size, form, and process.

We shall specifically examine three propositions: (1) that the domain of size and form in which a particular scaling relation attains functional similitude is limited; (2) that scaling relations are transformed by alterations in the way plants grow and develop; and (3) that these transformations prefigure or attend the appearance of evolutionarily new groups of plants. In a variety of guises, these propositions were explored in previous chapters, where we saw (*a*) that form-function relations evince size-dependent behavior, (*b*) that the procession of ontogenetic and developmental events must be charted so as to permit the identification of when and how modifications affect the adult phenotype, and (*c*) that phyletic hypotheses regarding precise ancestor-descendant relations among major plant taxa are required whenever our concern is with whether the shifts from ancestral scaling patterns coincide with the appearance of a new lineage (cladogenesis). Previous discussions stressed that whether viewed in isolation or juxtaposition, propositions (1) to (3) do not presume that scaling relations are the consequence of adaptive evolution. Ontogenetic and developmental modifications are required a priori to change a scaling relation, since every intraspecific scaling relation is merely the statistical consequence of the way organisms sharing the same ontogenetic and developmental repertoire grow in size, shape, or structure. Likewise, there is no a priori reason to presume that an organism's ontogenetic and developmental repertoire is the result of natural selection, although this may be the case. And finally, the maintenance of functional similitude among closely related organisms can not be taken as prima facie evidence for adaptive evolutionary change, because closely related organisms differing in size share many features owing to common ancestry and because natural selection is envisioned to come into play only after a descendant taxon departs from its ancestral scaling relation. Indeed, dramatic departures from previously established scaling relations may produce less efficient derived taxa, particularly if we believe that ancestral scaling relations are adaptive in the sense that they have stood the test of time by attaining workable, albeit perhaps not perfect, form-function relations. In this respect it is reasonable to think that natural selection may "work as hard" to maintain the status quo as to effect change.

The proposition that scaling relations are adaptive will be evaluated by the comparative method (see Alexander 1979; Maynard Smith and Holliday 1979; Pagel and Harvey 1988; Harvey and Pagel 1991), which differs from the experimental method in using information obtained from broad intertaxonomic comparisons rather than from observations gleaned from one or a few species manipulated in the laboratory or in the field. The experimental method, which obtains greater resolution of the interplay of biological variables and thereby attempts to resolve cause and effect relations, must abstain from generalizing results to other taxa, because its results may reflect the idiosyncrasies of the taxa placed under observation. By contrast, the comparative method adduces capacious generalizations across a broad spectrum of ecologically or taxonomically diverse organisms. By so doing, it can detect patterns of convergent or parallel evolution. Since the resemblance among the features of phyletically and ecologically dissimilar organisms in conjunction with similar selective forces is unlikely to occur by chance alone, it provides strong circumstantial evidence for adaptive evolutionary change.

The successful implementation of the comparative method requires accurate phyletic hypotheses regarding the relations among taxa. We need to know the subordinate taxa within each higher taxon being compared as well as the phyletic relations among the lineages across which intertaxonomic comparisons are drawn. The most accurate method of constructing these phylogenetic hypotheses is cladistic analysis, which groups organisms according to shared, derived characters under the criterion of parsimony. The phyletic hypotheses obtained from this approach are depicted as a bifurcating "tree" whose branching topology establishes taxa sharing a common ancestry. Although a number of preliminary topologies for nonvascular and vascular extant and fossil embryophytes have been constructed based on cytological, biochemical, and morphological characters (e.g., Mishler and Churchill 1984, 1985; Doyle and Donoghue

1987; Gensel 1992) as well as on ribosomal RNA, chloroplast DNA, and other molecular data sets (e.g., Zimmer et al. 1989), the phylogenetic relations among plants as well as animals are still far from resolved, and there is a paucity of "fine-grained" taxonomic analyses, which are required for detailed comparisons among the scaling relations attained by different albeit related clades. Additionally, discrepancies exist between morphological and "molecular" topologies, suggesting that evolutionary changes in morphological and biochemical traits may not invariably be concordant.

These short-term limitations of cladistic analysis, however, do not detract from the general importance of considering phylogenetic effects in scaling analysis. Nor do they pose a problem when our objective is to examine "coarse-grained" phylogenetic relations in terms of the scalings of major plant groups. Figure 5.1 provides a very coarse-grained topology for the group of plants called embryophytes. More will be said about these relations later, when particular scaling relations are discussed. But for now it is sufficient to note that the embryophytes are viewed as a monophyletic group nested in turn within a larger monophyletic group that includes the charophytes, a group of green algae whose subordinate taxa have a number of important cytological and physiological features in common with the embryophytes. Perhaps more important, a number of evolutionary transformations are hypothesized to have occurred within the embryophytic clade. Pivotal among these are the retention of the zygote and the subsequent development of the embryonic diploid generation (the sporophyte) within the archegonium produced by the haploid generation (the gametophyte). This is the sine qua non of the embryophytes. Subsequent elaboration of an elongate prismatic sporophyte is hypothesized to have occurred independently at least four times—three times in the nonvascular embryophytes (liverworts, hornworts, and mosses), which constitute the grade referred to as bryophytes, and at least once within the vascular embryophytes, which compose the grade called tracheophytes. If this hypothesis is true, then the evolution of prismatic sporophytes likely was the consequence of similar selection pressures favoring long-distance spore dispersal as well as the interception of sunlight by means of elevated photosynthetic organs (Niklas 1992). In terms of the latter, "leaves" appear to have evolved more than once within the embryophytes (e.g., on the gametophytes of some liverworts and mosses as well as on the sporophytes of lycopods, horsetails, ferns, and seed plants), as the arborescent growth habit has done in virtually every major vascular plant lineage.

The phyletic hypotheses summarized in figure 5.1 let us compare numerous scaling relations in an effort to determine whether they are a con-

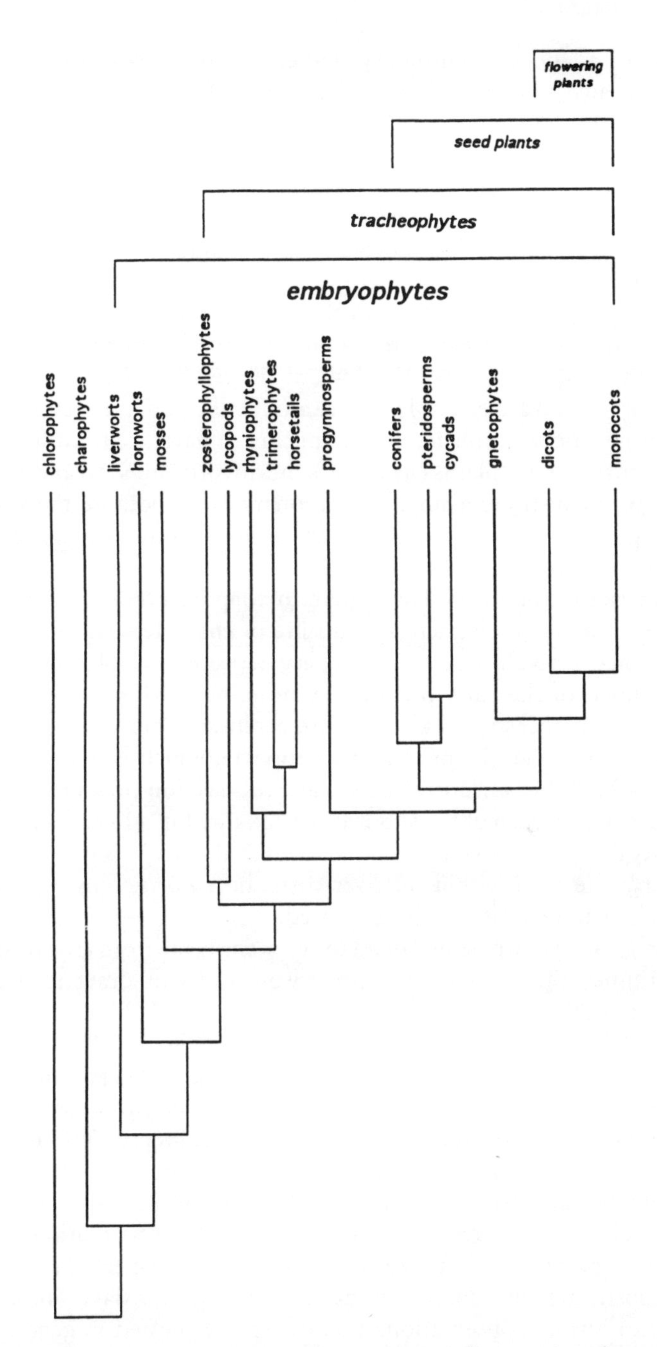

Figure 5.1. Cladistic relations among major plant groups. Data from a variety of sources.

sequence of adaptive evolutionary change. Yet this assessment is contingent on the answer to a fundamental question: What kind of evolutionary event constitutes an adaptation? There are two highly polarized answers. One is that an "adaptive event" is one where a change in function accompanies a shift in morphology or a particular scaling relation. Logically, this outlook must take the stance that the persistence of any combination of functional change with a morphological character within a lineage is not an "adaptive event" but rather the consequence of common ancestry. Accordingly, the scaling relation for the relevant biological variables must be weighted by the number of times the scaling relation has arisen rather than by the numbers of taxa that inherit the relation (Pagel and Harvey 1988; Harvey and Pagel 1991). For example, the "adaptive event" of an axially prismatic sporophyte, which appears to have occurred four times in the history of land plants (liverworts, hornworts, mosses, and tracheophytes), may constitute a single evolutionary event because the subordinate taxa within these four lineages share this feature as a consequence of common ancestry.

An alternative point of view is that an adaptive event is the "maintenance" of functional characters or a suite of characters within a lineage by selection over evolutionary time. This perspective, which is not uncommon in much of the early literature treating adaptive evolution, argues that "common ancestry" and "adaptive evolution" are not mutually exclusive. Features like the prismatic sporophyte, which is undeniably underlain by heritable variation, are subject to selection pressure operating on species within each of the subordinate taxa within a lineage. Logically, therefore, this point of view would argue that the scalings of sporophytes of subordinate taxa within the liverworts, hornworts, mosses, and tracheophytes can be legitimately compared.

Clearly, the way we define an adaptive event is far from trivial, since it predetermines the number of comparisons we can draw to evaluate whether a scaling relation is the consequence of adaptive evolutionary changes. Additionally, at a very basic level the definition of an "adaptive event" predisposes the numerical values of the regression parameters that characterize the mathematical form of a scaling relation. A basic assumption of every regression analysis is that each datum represents an *individual observation* obtained from the *smallest sampling unit.* This assumption militates against the likelihood that the results from regression analysis reflect the overrepresentation of a single "observation," a bias called data point inflation (see sec. A.3). In this respect the question, What constitutes an adaptive event? statistically translates into, What constitutes a single observation? because species nested in genera share features as a result of common ancestry as well as convergent or parallel evolution. This is true for every taxonomic level subsumed by higher levels of classification (genera nested in families, families nested in orders, etc.). If we take the view that an adaptive event is the circumstance leading to the first appearance of a functional suite of characters, then the smallest sampling unit is the taxon in which these functional characters made their first appearance. Alternatively, if we take the view that an adaptive event is evinced by the persistence of functional characters that obtain a particular scaling relation in subordinate taxa, then each subordinate taxon may be a legitimate single observation. The dilemma posed by the question, What constitutes a single observation? is aggravated by the fact that there is no a priori method of adducing the smallest taxonomic sampling unit for regression analysis. The smallest sampling unit may vary depending on the objective of the regression analysis. Taxonomic levels below the species—for example, subspecies, varieties, and demes as well as the individual plant—may be appropriate sampling units when the concern is with phenotypic or genotypic variation among individuals within a population, across an ecological gradient, or among individuals occupying the same particular microhabitat. The smallest sampling unit may even be organs drawn from an individual plant, provided our interest focuses on organographic variability. True, in the context of phylogeny and adaptation the smallest sampling unit is a taxon, but once again we have no reason to adduce a priori which taxon is the smallest sampling unit. Depending on the variance of the functional characters within and among different taxonomic levels, the smallest sampling unit may be the species or a higher level of classification.

A number of methods have been developed to determine the smallest taxonomic level appropriate for regression analysis (see Pagel and Harvey 1988; Harvey and Pagel 1991). The very best of these methods rely on cladistic phylogenetic hypotheses. Many, however, are useful only when the functional character under study is a discrete biological variable (e.g., Ridley 1983). Although continuous variables can always be made discrete, the rules for categorizing continuous variables tend to be arbitrary or are adduced from statistical rather than biological criteria. Additionally, "parsimonious" methods of determining the smallest taxonomic sampling unit are subject to well-reasoned criticism (see Felsenstein 1983). In the absence of a rigorous phylogenetic hypothesis, alternative and admittedly less reliable methods must be used. Among these is nested analysis of variance. This "fallback" method, which identifies the smallest taxonomic sampling unit for scaling analysis, may be used for continuous biological variables. It is discussed in the appendix, where it is illustrated by a small data set (see sec. A.6).

Nested analysis of variance permits us to determine the percentage distribution of the variance in a character among subordinate and higher taxonomic levels. The taxon providing independent data points for regression analysis is elected using the percentage distribution of variance for the particular character. Table 5.1 provides nested analyses of variance for some of the biological variables considered in this chapter and shows that roughly 70% of their total variability is accounted for by differences among families within orders and orders within classes. Accordingly, plant species and genera typically appear not to add substantial variance that is independent of phyletic association. A corollary of this generalization is that, in terms of taxonomic structure, little or no evolution has taken place below the level of the family for most of these particular variables. Consequently, family mean values for them should be used as the data points for regression analysis.[1]

Regardless of its usefulness, we must be very sensitive to a fundamental assumption whenever we use nested analysis of variance to determine the smallest taxonomic sampling unit for regression analysis. Nested analysis of variance presumes that the taxonomic structure of a data set accurately reflects a natural classification system. That is, it assumes that we have accurately grouped species within genera, genera within families, and so forth so that our "taxonomy" reflects a true phylogenetic relation. Clearly, if the taxonomic structure of our data set does not adequately reflect the phylogenetic relations among the taxa being considered, nested analysis of variance obtains unreliable and potentially completely erroneous results concerning the distribution of variance among data classification levels. At a fundamental level, therefore, the use of nested analysis of variance does not circumvent the logical requirement for rigorous phylogenetic hypotheses.

Finally, it is a curious feature of plants that regression of species mean values or genus mean values often yields scaling parameters statistically indistinguishable from those obtained by regression of family (or higher) mean values. One might argue, therefore, that an important role that nested analysis of variance plays in determining botanical scaling relations is to establish which taxonomic level is most appropriate for regres-

1. It is worth noting in this regard that efforts to identify and quantify mass extinction events (statistical outliers from background extinction levels) would benefit from nested analysis of variance. By definition, a mass extinction event involves the extirpation of many closely related subtaxa. Logically, therefore, each mass extinction event to some degree involves selection against features shared among taxa by virtue of common ancestry. Since the extinction of a higher taxonomic level is a cumulative variable whose magnitude reflects the summation of the extinction of subordinate taxa (see Raup and Sepkoski 1984), the distribution of variance for biological features among taxonomic levels should never be neglected.

Table 5.1. Nested Analyses of Variance of the Percentage Variance Distributions of Some Vegetative and Reproductive Biological Variables among Taxonomic Levels for Angiosperms, Gymnosperms, and Pteridophytes

	Species within Genera	Genera within Families	Families within Orders	Orders within Classes
Angiosperms				
Vegetative variables				
Shoot height	21.0%	15.5%	21.1%	42.4%
Stem diameter	25.6	23.0	25.6	25.8
Shoot biomass	16.1	20.4	26.6	36.9
Leaf biomass	26.5	18.9	20.4	34.2
average	22.3	19.5	23.4	34.8 = 100
Reproductive variables (biomass)				
Gynoecium	15.0	6.0	21.0	58.0
Androecium	10.7	7.5	26.4	55.4
Perianth	20.5	13.5	26.6	39.4
Seed	8.4	19.4	21.1	48.9
Average	13.7	11.6	23.8	50.4 = 99.5
Gymnosperms				
Vegetative variables				
Shoot height	18.1	15.5	31.1	35.3
Stem diameter	20.6	20.0	20.6	38.8
Shoot biomass	14.8	18.1	29.4	37.7
Leaf biomass	9.9	10.1	29.9	50.1
Average	15.9	15.8	27.8	40.5 = 100
Reproductive variables (biomass)				
Oculate cone	10.4	18.2	20.1	51.3
Pollen cone	9.7	10.1	15.6	64.6
Seed	10.2	11.1	26.7	52.0
Average	10.1	13.1	20.8	55.9 = 99.9
Pteridophytes				
Vegetative variables				
Shoot height	8.1	13.5	30.4	48.0
Stem diameter	16.6	18.0	20.6	44.8
Shoot biomass	12.8	17.1	19.4	50.7
Leaf biomass	9.3	13.1	18.3	59.3
Average	11.7	15.4	22.2	50.7 = 100
Reproductive variables (biomass)				
Sporangia	8.4	12.6	24.1	51.3

sion analysis when the scaling exponents and scaling coefficients obtained by regressions of subordinate and higher taxonomic levels differ statistically. Since nested analysis of variance requires mean values at the species, genus, family, order, and class levels, the scaling parameters obtained for these different mean values may easily be compared. Nonetheless, it cannot be overemphasized that regardless of the similarity in the scaling parameters obtained between species mean values and the mean values of higher taxa, nested analysis of variance does not eliminate the problem of the nonindependence of data points whenever our operative phylogenetic hypotheses are faulty.

5.2 Heterochrony

I begin the treatment of the effects of size on plant evolution by discussing the proposition that modification of adult size, shape, or internal structure involves alterations of growth and development. It has been fashionable to describe these alterations in terms of heterochrony. In this section, the definition and application of heterochrony are reviewed in the context of plant growth, development, and evolution.

Heterochrony has been defined either as a change in an ontogenetic sequence of events or as a change in the time of appearance of ontogenetic events. For example, Raff and Wray (1989) refer to heterochrony as "a shift in the relative timing between two developmental processes in a descendant ontogeny" (410) or "any change in the relative order of events in a developmental sequence" (411). Thus, relative *order* may substitute for relative *timing*. Likewise, in his classic book *Embryos and Ancestors,* De Beer (1958) alternatively defines heterochrony as any "alteration and reversal of the sequence of [ontogenetic] stages" (8) and as any "shifting along the time-scale [of the time of appearance of a structure]" (34). Once again, the *sequence* of events may substitute for the *timing* of events. Although a distinction may be drawn between a change in the sequence of ontogenetic events and a change in the relative time of appearance of these events, it is undeniable that whereas a change in the order of ontogenetic events will produce a change in the relative times of appearance of these events, the reverse is not invariably true. Accordingly, the term heterochrony has two different guises and therefore shares with the word allometry an unnecessary ambiguity.

That sequence and timing are different and yet interrelated is quickly made apparent when we consider a simple hypothetical example. Figure 5.2 diagrams an ancestral ontogenetic system as well as three hypothetical descendant ontogenies, denoted as types I–III. The ancestral ontogenetic sequence consists of three phenotypic events *A, B,* and *C,* which are the consequence of a sequence of unsuspected physiochemical, genetic, or some other manner of events, denoted as $c \rightarrow b \rightarrow a$. The relative rate of action for each causal event is indicated by an arrow whose slope has the dimensions of some unspecified quantity per unit time. For illustrative purposes only, *quantity* may refer to size or shape in terms of morphogenesis, histogenesis, or the like, whereas *time* refers specifically to develop-

ANCESTRAL ONTOGENY

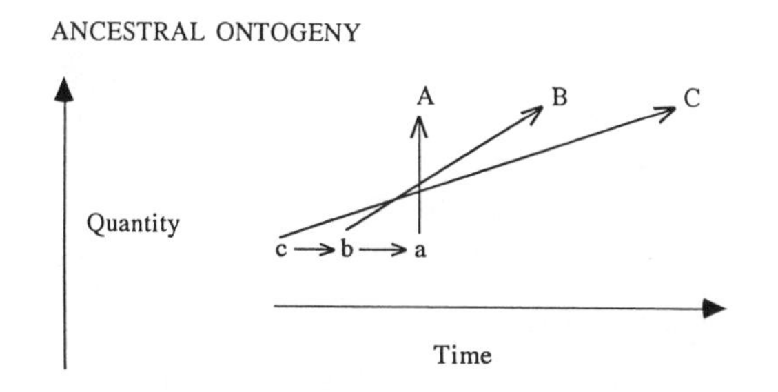

DESCENDANT ONTOGENY: TYPE I

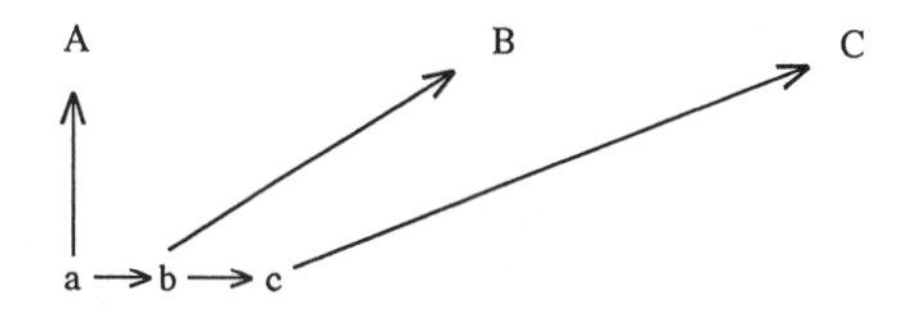

DESCENDANT ONTOGENY: TYPE II

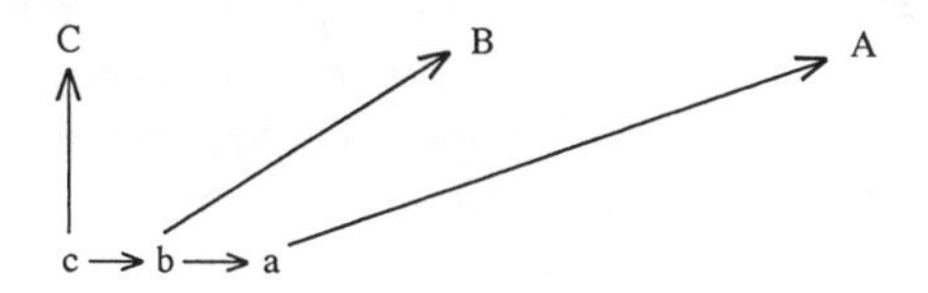

DESCENDANT ONTOGENY: TYPE III

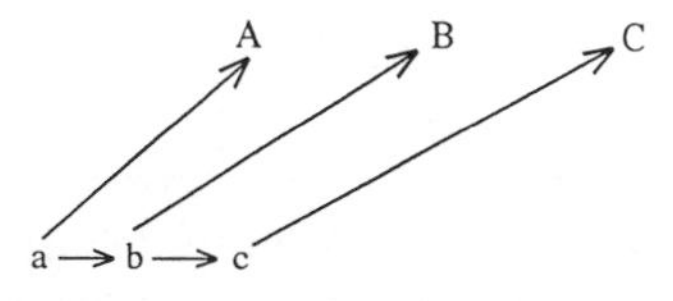

Figure 5.2. Hypothetical ancestral ontogenetic pattern and three hypothetical descendant ontogenies, denoted as types I–III. The ancestral ontogeny consists of three phenotypic events (*A*, *B*, and *C*) resulting from underlying events ($c \rightarrow b \rightarrow a$). The relative rate of action for each event is indicated by an arrow whose slope has the dimensions of quantity per unit time.

mental time. These hypothetical ontogenies were constructed such that the last event *a* in the causal chain induces the first observed phenotypic event *A* by its high rate of influence. By contrast, because of its slow rate of influence, the first event *c* in the causal chain is contrived to induce the last observed phenotypic event *C* in the ancestral ontogeny. In this way the temporal sequence of events *A*, *B*, and *C* in the ancestral ontogeny intentionally does not map onto the temporal sequence of causal events $c \rightarrow b \rightarrow a$. The consequences are that (1) a change in the pattern of causal events ($c \rightarrow b \rightarrow a \Rightarrow a \rightarrow b \rightarrow c$) but no change in their rates of influence produces no change in the pattern of observed events, although the relative times of appearance of these events are shifted (type I descendant ontogeny); (2) a change in the rates of influence of causal events but no change in the pattern of causal events ($c \rightarrow b \rightarrow a$) gives rise to a change in the pattern of observed events (type II descendant ontogeny); and finally (3) a change in both the sequence and rates of influence of a causal chain of events can yield a sequence of phenotypic events identical to that of the ancestral ontogeny (type III descendant ontogeny). The last point emphasizes that complex ontogenetic modifications can occur at a submerged level of cause and effect yet obtain identical descendant ontogenies in terms of both the pattern and the tempo of appearance of phenotypic events. Although "cryptic" ontogenetic modifications are distressing, particularly since the paleontologist can only hope to observe a phenotypic ontogenetic pattern and since the neontologist is rarely cognizant of causal mechanisms, the immediate concern here is not with cause and effect, although this is an important aspect of relating ontogenetic to phylogenetic modifications. Rather, the present focus is the sequence and tempo of phenotypic events. The hypothetical ontogenies diagrammed in figure 5.2 show that shifts in the order of phenotypic events and shifts in the relative time of appearance of phenotypic events are very different aspects of ontogeny and that there is no a priori reason to accept the often-held notion that modifications of early ontogenetic events are less common because they are more intrusive on subsequent development than alterations occurring later in the ontogenetic sequence. Nor is there reason to accept the corollary of this point of view—that a descendant ontogeny most frequently results from a terminal change (addition, deletion, rearrangement, etc.) in the ancestral ontogenetic pattern of events.

The application of the term heterochrony leads to yet another type of ambiguity, regardless of the precision of definition. For example, heterochronic processes often are inferred from differences in the adult conditions of a descendant and its presumed ancestor. Diagnosis based purely on external symptoms can be a very dangerous affair—the legitimate application of heterochrony requires direct observation of ontogeny, as is seen when we examine the terminology of heterochrony. As originally proposed by De Beer (1958), heterochrony refers to eight categories of presumed mechanism, each resulting in some manner of temporal or spatial displacement capable of transforming the size or shape of a descendant relative into that of the presumed ancestor (Gould 1977; Albrech et al. 1979; McNamara 1986; Raff and Wray 1989). These categories are often referred to as *pure* heterochronies, since each can be combined in a variety of ways to yield more complex heterochronic phenomena. Table 5.2, which summarizes the eight categories of pure heterochrony, is constructed to show that heterochrony falls into three general classes at the level of describing the morphological expression of a descendant adult in terms of the ancestral adult. The first of these three categories is called paedomorphosis (from Greek *pais, paidos,* "child," and *morphe,* "form, shape"). Paedomorphosis results in the reduction of adult size but obtains a morphometric similitude between the derived adult and the juvenile ancestral condition. The second class of heterochrony is peramorphosis (from Greek *pera,* "beyond, across," or *peraiteros,* "further"), in which both adult shape and size extend beyond those of the adult ancestral condition. The third class of heterochrony has not previously been draped with the respectable mantle of the ancient Greek language. I presume to call it akratomorphosis (from *akratos,* "pure, unmixed"). Akratomorphosis is distinguishable from paedomorphosis and peramorphosis

Table 5.2. Three Classes of Morphological Expression and Eight Categories of Heterochrony

Class of Morphological Expression	Category	Alteration		Growth	
		Sexual	Vegetative	Rate	Onset
Paedomorphosis					
	Postdisplacement			Same	Later
	Neoteny	Same	Delayed		
	Progenesis	Accelerated	Same		
Peramorphosis					
	Acceleration	Same	Accelerated		
	Hypermorphosis	Delayed	Same		
	Predisplacement			Same	Earlier
Akratomorphosis					
	Dwarfism			Slower	Same
	Gigas			Faster	Same

Note: Peramorphosis = shape and size altered, adults with morphologies "beyond" those of the ancestor; paedomorphosis = shape and size altered, adults resemble juvenile forms of ancestor; akratomorphosis = shape unaltered, descendant differs from ancestor in size only.

because the size of the descendant adult differs from the adult ancestral condition but shape is conserved.

I shall return to these three class of heterochrony later on. For now simply note that authors tend to affirm a class (or worse, a category) of heterochrony without directly observing ontogeny. This is most apparent in the paleontological literature, but treating living organisms in this cavalier fashion is regrettably not uncommon. Direct observation of ontogeny is absolutely essential whenever the claim is made that a descendant has evolved as a consequence of some heterochronic phenomenon. At a minimum, three principal ontogenetic variables influence the time of appearance of phenotypic events and therefore must be established before the terminology of heterochrony may legitimately be used. These variables are the time when each ontogenetic event is initiated, the time it is terminated, and the rate at which the event proceeds. Although the duration of the ontogenetic process leading to a particular event may be under explicit control, we have no a priori basis to assume that process initiation and termination are ontogenetically coupled. In the simplest case, additionally, each variable has three conditions relative to that of the ancestral ontogeny: *same, early,* or *late* in terms of initiation or termination, and *same, fast,* or *slow* in terms of the rate of progress. Consequently, there exists a $3 \times 3 \times 3$ matrix containing all possible permutations. Figure 5.3 displays this matrix. Each permutation is identified by a lowercase letter, while the ancestral condition is denoted by *A*. From figure 5.3, a descendant ontogeny can come to differ from that of an ancestor in one of twenty-six possible ways—the number of permutations $= (\text{variables})^{\text{conditions}} = 3^3 = 27$, but one of these (initiation, termination, rate)^{same}, gives rise to *A*. Since the number of possible permutations in the $3 \times 3 \times 3$ matrix far exceeds the current terminology employed by heterochrony, it is likely that present terminology conflates different routes by which modified ontogeny can achieve the same phenotypic result.

It is comparatively simple to show that different combinations of process initiation, termination, and rate produce the same adult form and size. Figure 5.4 diagrams an ancestral ontogeny that achieves the adult condition *A* and three descendant ontogenies that converge on the same adult condition, denoted as *D*. In this diagram the abscissa is developmental time t and the ordinate is some morphological or other kind of phenotypic quantity Q. The diagram assumes that the difficult task of unambiguously specifying the adult condition of both the ancestor and the descendant has been accomplished. That of the ancestor is shown by a dashed horizontal line. The vertical dashed line denotes the time at which maturity is achieved. In this example the descendant ontogeny takes longer to achieve the adult condition, and the adult phenotypic condition extends beyond that of *A*. In reference to figure 5.3, it is important to note that *D* can be achieved by three very different ontogenetic routes: (1) an ontogeny that has the same process initiation and same rate of growth as the ancestral ontogeny but a late process termination (type *f*); (2) an ontogeny that has a late process initiation and termination and a faster rate of growth (type *q*); or (3) an ontogeny that has an early process initiation, a late process termination, and a slow rate of growth (type *y*). Without direct observation of ontogeny, a mere phenotypic comparison between the adult conditions of *A* and *D* simply tells us that the adult

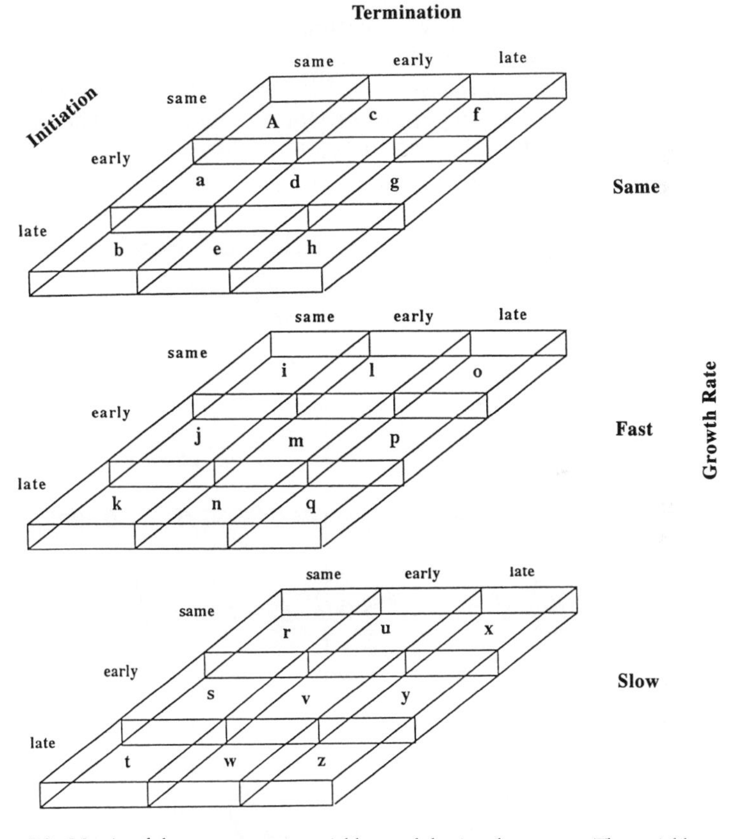

Figure 5.3. Matrix of three ontogenetic variables, each having three states. The variables are (1) the time when each ontogenetic event is initiated, (2) the time the event is terminated, and (3) the rate at which the event proceeds. The three states (relative to the ancestral ontogeny) are same, early, or late in terms of event initiation or termination or same, fast, or slow in term of the rate of progress. Each permutation is identified by a lowercase letter; the ancestral adult condition is denoted as *A*. Descendant ontogeny can differ from *A* in one of twenty-six possible ways: number of permutations = $(\text{variables})^{\text{conditions}} = 3^3 = 27$, one of which, $(\text{initiation, termination, rate})^{\text{same}}$, produces *A*.

condition of D is a giant version of A, which form table 5.2 may be achieved by the class of heterochrony called akratomorphosis.

Akratomorphosis can have important taxonomic and phylogenetic implications because the extension of a scaling relationship into a larger or smaller size range can account for the appearance or disappearance of morphological features that may serve to distinguish taxa, even though the novel shape may not be a legitimate independent morphological or taxonomic character. A classic example of akratomorphosis of the "gigas" type (see fig. 5.4, *f*) was published by Edmund Sinnott (1936) for the ontogeny of the fruits of the cucurbit *Lagenaria*. These fruits have an upper sterile lobe and a lower seed-bearing lobe separated by a contracted region or isthmus, hence the common name for the genus (bottle gourd). Sinnott examined two varieties of the bottle gourd, "miniature" and "giant," whose mature fruits differ in shape as well as size. The original data for fruit length L and width W were reconstructed from his data. Figure 5.5 shows that the difference in the shape of "giant" and "miniature" fruits is not the result of differences in the relative rates of growth in L and W. Least squares regression of the data for "miniature" indicates

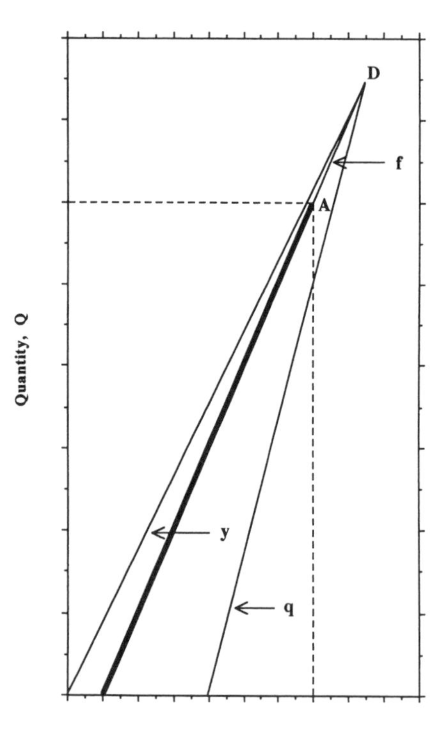

Figure 5.4. Three ontogenies (defined in terms of permutations, *f*, *q*, and *y*, described in fig. 5.3) converging on the adult condition *D*. Ancestral adult condition denoted as *A*.

that $L = 1.96 \ W^{0.807}$ ($r^2 = 0.99$, $N = 42$; $\alpha_{RMA} = 0.81 \pm 0.002$), whereas LS regression of the data for "giant" shows that $L = 1.87 \ W^{0.839}$ ($r^2 = 0.99$, $N = 16$; $\alpha_{RMA} = 0.84 \pm 0.003$). The scaling exponents for these two regression curves do not differ statistically, showing "giant" fruits result from the continuation of the "miniature" scaling relationship into a new, larger range of fruit size. Sinnott showed that the scaling of the growth of fruit length with respect to width, which results in what superficially appear to be two different fruit shapes, segregates in a simple Mendelian fashion, suggesting that it is under direct genetic control (Sinnott 1936).

Further examples of how different permutations of "initiation," "termination," and "growth rate" can achieve the same phenotypic end result are informative. Figure 5.6 illustrates how a descendant adult condition achieving a lower level of phenotypic expression can result from very different permutations of process initiation, termination, and growth rate. In general terms, the descendant condition achieved by these different routes would be diagnosed as the consequence of paedomorphosis. And if the

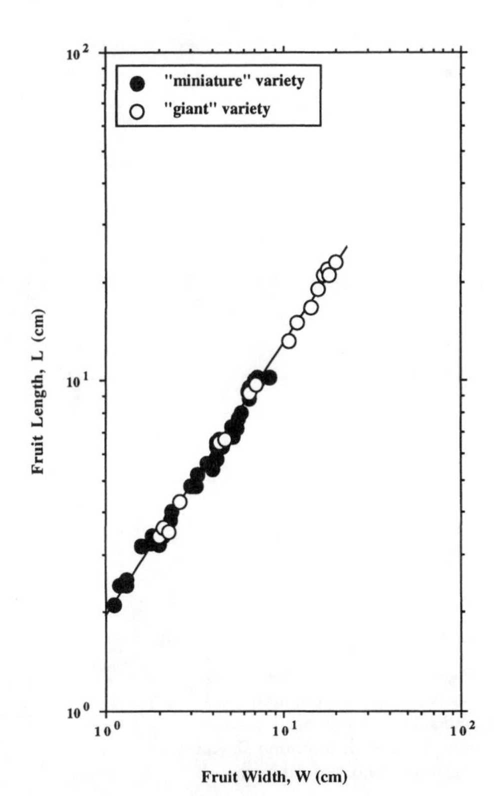

Figure 5.5. Log-log (base 10) plot of fruit length L against fruit width W of "miniature" and "giant" varieties of the bottle gourd, *Lagenaria*. Solid diagonal line is the regression curve obtained from LS regression analysis of the pooled data (reconstructed from Sinnott 1936).

adult condition was defined by sexual maturity, then the descendant would likely be considered an example of neoteny. An interesting example of botanical neoteny was shown by Edward Guerrant (1982), who studied two species of *Delphinium*, one pollinated by hummingbirds (*D. nudicaule*) and another by bumblebees (*D. decorum*). The flowers of the hummingbird-pollinated *D. nudicaule* have strongly tubular corollas that bear a striking similarity to the buds of the bumblebee-pollinated *D. decorum*. By means of scaling analysis, Guerrant was able to show that the growth rates of *D. nudicaule* sepal length and petal length are slower than those of *D. decorum* (Guerrant 1982, figs. 8 and 11). Consequently the floral morphology of the hummingbird-pollinated species is a neontogenic derivative of that of bumblebee-pollinated species. Apparently, however, the heterochrony evinced by a comparison of these two species is not "pure." The nectariferous petal that is critical to the hummingbird-pollination syndrome is not "juvenilized" but rather grows in a manner that recapitulates (by acceleration and hypermorphosis) the morphogenesis of the presumed bumblebee-pollinated ancestral species.

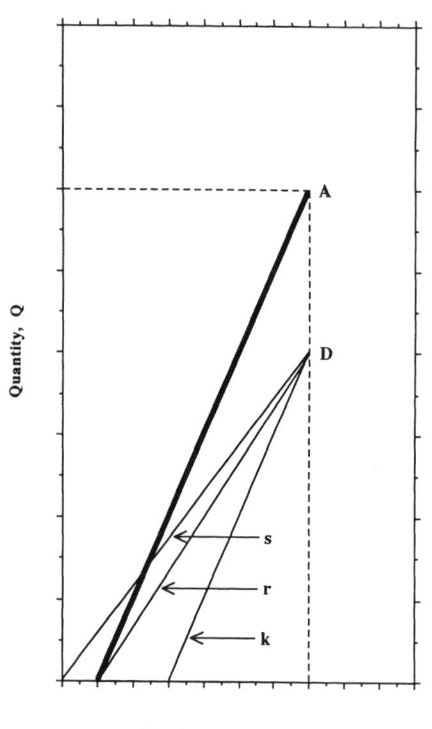

Figure 5.6. Three ontogenies (defined in terms of permutations, *k*, *r*, and *s*, described in fig. 5.3) converging on the adult condition *D*. Ancestral adult condition denoted as *A*.

This study of *Delphinium* illustrates yet another aspect of the study of heterochrony—the role of phyletic hypotheses. Guerrant made two phyletic hypotheses, neither one directly tested in the context of his scaling analysis: (1) that hummingbird-pollinated species are evolutionarily derived from more generalized bumblebee-pollinated species, and (2) that the floral morphogenesis of *D. decorum* is representative of the ancestral condition. The first hypothesis established "polarity" of the presumed heterochronic ancestral to derived transition, while the second isolated which among the numerous *Delphinium* species was to serve as the "exemplar" of the ancestral condition. These hypotheses were not tested by cladistic analysis. However, both *Delphinium decorum* and *D. nudicaule* are members of the tuberiform series of *Delphinium,* which contains an additional twelve bumblebee-pollinated species, only one of which (*D. nudicaule*) is pollinated by hummingbirds. It is reasonable to conclude, therefore, that *D. nudicaule* is "unique" because it is an evolutionarily "derived taxon." In terms of the second phyletic hypothesis, hybrid swarms of *D. decorum* and *D. nudicaule,* which produce viable hybrid seeds, occur naturally, suggesting that the genetic distance between these two species is not great.[2] Consequently, *D. decorum* appears to have been a reasonable choice as the "representative" ancestral condition. The important point is not whether the phyletic hypotheses underlying Guerrant's study of *Delphinium* are "correct" but that hypotheses of this sort are implicit to any analysis of heterochrony.

Returning to the main focus of this discussion, figures 5.4 and 5.6 make the implicit assumption that we can unambiguously specify the beginning and the closure of an ontogenetic sequence. A variety of features may be elected. In the case of floral development, meiosis and anthesis can serve as temporal referents for the initiation and closure of ontogeny (e.g., Guerrant 1982). The tradition in zoology is to use the onset of sexual maturity as the temporal referent for comparing ancestral and descendant ontogenies (see De Beer 1958; Gould 1977; Albrech et al. 1979; McNamara 1986). As pointed out in chapter 4, "sexual maturity" is rarely a practical temporal referent in terms of plants, and as noted by Raff and Wray (1989), small shifts in the time of appearance of early ontogenetic events may go undetected when sexual maturity is attained very late in development. Additionally, there is no reason to suppose that sexual maturity is independent of heterochronic shifts. The issue of temporal referent is far from trivial, but it is fair to say that too much has

2. A comparison among the floral morphogenesis of *D. decorum, D. nudicaule,* and their F_1 hybrid would be extremely interesting because it might shed light on the inheritance of developmental heterochronic features.

been made of the problem (see, e.g., Raff and Wray 1989). Although we may not be able to use absolute ontogenetic time to compare ancestral and descendant ontogenies, since few ancestor-descendant organisms grow and develop at the same overall rate even when reared under identical environmental conditions, we can use the absolute times of appearance of many or all ontogenetic events and normalize these times with respect to one or more temporal referents in comparing any pair of ontogenies that contain much the same events. By so doing, the case of heterochrony becomes more convincing because the times of appearance of events are considered in terms of a network or system of interconnected ontogenetic events. One such method, based on a mathematical technique derived from systems analysis, is presented in the appendix. To my knowledge it has never been employed and therefore remains untested.

Many developmental phenomena affect the pattern as well as the tempo of ontogenetic patterns. Three that appear to dominate phyletic trends in plant morphology are reduction, fusion, and asymmetry. Reduction in size results from a decrease in either the rate or the duration of growth. For example, the difference between the foliage leaves and bud scales on a branch typically results from differences in the rates of growth of parts of the primordia given rise to these appendicular organs. The rates of growth of the seedling pinelike foliage leaves on *Chamaecyparis* are nearly identical to that of the scalelike foliage leaves on the mature branches of this gymnosperm. The difference in the size and shape of these two types of foliage leaves is due exclusively to the duration of leaf growth. The difference between foliage leaves, bud scales, and floral parts also depends largely on differences in growth rate between parts of their primordia (see Foster and Barkley 1933). The rosette growth habit is the consequence of early termination of growth of successively produced stem internodes. This growth habit has been achieved repeatedly in very different lineages (e.g., ferns, cycads, monocotyledons, and dicotyledons). The acquisition of determinate shoot growth is another example of how reduction has figured in the evolutionary history of plants. The short shoots of many conifers, such as the fascicles of needlelike leaves in the genus *Pinus,* is the consequence of the early termination of shoot growth. Reproductive structures also have undergone reduction. The ovulate cones of pine species and the strobili of cycads and horsetails are examples of reproductive organs that evince reduction compared with their presumed antecedent condition. These examples of reduction appear to result from genetic mutations that have slowed or prematurely terminated the growth of primordia, resulting in a diminution of the parts of organs. Reduction in organ size can also result from the extension of the entire organism into a larger size range when different organs or parts of organs grow at different rates. Consider the case of an organ whose growth rate is half that of another. The relative size of these two organs would be dictated by the absolute size of the entire organism as a simple consequence of relative growth rates.

The fusion or union of plant parts is accomplished by changes in the rate, duration, and location of growth centers of neighboring primordia. Note that location is a spatial rather than a temporal variable and therefore affects the pattern of an ontogenetic process. That temporal and spatial variables operate in biological networks is obvious. For example, in the case of a deeply lobed sympetalous corolla, the primordia of the apical lobes, which correspond to the separate petals on the apopetalous ancestral form, are the first to appear. The rate of growth of these primordia is typically high at first but then declines with the appearance of a basal zone of tissues whose growth rate below the sinuses of apical lobes is as rapid as that directly below each apical lobe. The basal zone of growth produces the floral tube. It is a novel meristematic region in terms of location. Initially the tube is short relative to the apical lobes, but over the course of development its length equals and then exceeds that of the lobes. Among species with sympetalous corollas, differences in the final length of apical lobes relative to that of the floral tubes depend upon the relative times at which apical lobes and the basal zone of tissue initiate growth, the ratio between the rates of growth of these two parts of the corolla, and the absolute size of the mature corolla.

Changes in plant symmetry, though less common than reduction or fusion, occasionally characterize long-term phyletic trends. A transition from radial to bilateral symmetry can be accomplished by a change in the relative rates of growth in the same organ or different organs. For example, how fully foliage leaves assume a dorsiventral morphology depends on the relative growth rates of the leaf apical meristem (which dictates the mature length of the leaf) and the marginal meristem (which dictates the lateral expansion of the lamina). The transition from radial to bilateral floral symmetry results from the differences in the relative growth rates of perianth floral parts. The tendency to form flattened systems of leaves and branches is the consequence of conspicuous dissimilarities in growth rates of primordia. For example, leaf and branch primordia of the lycopod *Selaginella* are produced in a spiral pattern. Primordia confined to the two sides of horizontally oriented shoots grow faster than those on the dorsal and ventral surfaces of stems. Laterally located leaves are accordingly larger than leaves on the upper and lower surfaces of stems, while branching appears to be confined to the horizontal direction.

Thus, the shoots of *Selaginella* have a planated appearance that belies the radial arrangement of branch and leaf primordia. A similar pattern of growth occurs in many conifers (e.g., *Thuja*).

A number of evolutionary innovations have resulted from permutations of reduction, fusion, and change in the symmetry of organs. The morphology of the ovulate cones of pine species and other conifers evinces reduction and fusion as well as changes in the symmetry of parts. But when the individual components of these innovations are considered separately, major evolutionary innovations in external form typically are seen to result from a change in the location as well as the rate and absolute time of meristematic activity. An evolutionarily successful alteration of ontogeny and development requires some independence among the processes that give rise to the final form of the organism. Evidence that some degree of independence among ontogenetic processes exists in terms of the genetic information regulating external plant form (morphology) and internal structure (anatomy) was reviewed in section 1.2. I also noted that, although they are somewhat independent, ontogenetic and developmental processes are under the direction of integrated genetic systems. That is, an alteration in one aspect of the system is likely to have multiple ramifications and consequences. In this regard, apparently large differences in plant morphology can result from comparatively small genetic changes (Kaiser 1935; Bachmann 1983; Gottlieb 1984). For example, the Norin 10–derived *Rht* genes (*Rht1* and *Rht2*), which are used to reduce the height of many wheat cultivars (Gale and Youssefian 1985), reduce the elongation of individual organs (coleoptiles, leaves, and culms; see Lenton, Heddon, and Gale 1987) by conferring an insensitivity to gibberellic acid (GA). The *Rht* genes therefore produce plants that behave as if they were GA saturated. At the cellular level, the principal effect of *Rht* is a reduction in cell length rather than cell number (Keyes, Paolillo, and Sorrells 1989). Consequently *Rht* depresses the rate of growth, reduces the plasticity of growing leaves, reduces the slenderness ratios of coleoptiles (thereby influencing seedling establishment), and alters a variety of other morphological features (see Niklas and Paolillo 1990).

5.3 Heterauxesis and Allomorphosis

As we have seen in previous chapters, scaling relations may be drawn from ontogenetic, intraspecific, and interspecific (or higher taxonomic) comparisons. The first is referred to as heterauxesis, while the latter two are called allomorphosis (Huxley, Needham, and Lerner 1941). The imposition of this erudite and somewhat imposing terminology does not detract from the fact that heterauxesis cannot be used as a model for intraspecific allomorphosis, nor can intraspecific comparisons be used to predict the scaling of an interspecific comparison. As we saw in earlier chapters, heterauxesis charts the progress of a particular ontogenetic pattern in response to the particular external environmental conditions attending the growth and development of the individual organism. It is rendered intelligible by plotting some phenotypic, biochemical, or other biological variable against some measure of the individual's instantaneous absolute size. The scaling of ontogeny therefore is a mathematical distillation of growth and development, ordered in terms of increasing size measurements that provide a spatial surrogate for the dimensional scale of time, leading to a very specific adult form and size. The obviously counterfeit nature of this practice is evident when we call to mind the following five points: (1) The mathematical convention of using size measurements as the abscissa does not imply that size is the cause of the biological variable plotted against the ordinate. Clearly, a change in organic proportion may extend the functionally permissible range in size—to be more precise, size is not an independent variable. (2) Ontogenetic time is expressed crudely by the progressive increase in the magnitude of some operationally useful but typically otherwise completely arbitrary measurement of size. The procession of size is not smooth and stately but rather erratic, in fits and starts. (3) The relation between time and the measurement of size may not be adequately expressed by even the most complex mathematical function. Additionally, more than one mathematical function can compete as the predictive model for the relation between size and form. (4) Although it is preferable to determine the scaling of ontogeny directly from observations made as an individual grows ("longitudinal" data), mass samples of individuals in various stages of ontogeny ("cross-sectional" samples) are often used to infer heterauxesis, particularly in the case of fossil species, for which direct measurements on growing individuals often are not practical. Unfortunately, the scaling of a mass sample of juveniles often fails to obtain the *longitudinal* heterauxesis derived from an individual as it grows. (5) The distinction between heterauxesis and intraspecific allomorphosis is much less precise for organisms characterized by indeterminate growth.

We shall soon see that each of these points bears on how ontogenetic modifications affect intra- and interspecific scaling relations. But for now we must turn to allomorphosis, which records the size-correlated variations in the organic proportions of conspecific adults or representative adults drawn from different species or higher taxa. The former, intraspecific allomorphosis, treats adult organisms that share the same general ontogenetic and developmental patterns but nonetheless differ in size and proportion owing to natural variation, whereas the latter, interspecific (intergeneric, etc.) allomorphosis, treats adult organisms that share a phyletic legacy but differ in size and proportion because of often profound differences in ontogenetic and developmental patterns. Interspecific (or higher taxa) allomorphosis may be used to identify shifts or transpositions in scaling parameters that may attend the expansion or contraction of the size range of different types of organisms. This level of scaling analysis helps identify the range of size over which a particular relation is functional. It can be used simply as a method of description, or it may serve as a basis to test an analytical scaling relation whose objective is to predict from first principles when alterations in organic proportions must attend a phyletic change in size in order to maintain equivalent levels of functional performance.

Traditional definitions of allomorphosis invariably emphasize that size-correlated variations must be based solely on a comparison among adult individuals (see Gould 1966, 601–2). Thus, in terms of what is generally understood as standard practice, allomorphosis is not legitimately determined from a mass sample containing juveniles as well as adults because the size-correlated variations such a comparison evinces will necessarily contain elements of heterauxesis. For a crude analogy, consider the mechanical scaling of twigs and branches drawn from a mature tree, which represent juveniles and adults. From previous treatments in chapters 3 and 4, we know that the scaling of twig length or mass versus girth differs significantly from that of branches, and the scaling obtained from a collection of twigs and branches differs from that of either twigs or branches. For those who study organisms characterized by determinate growth in size, the construction of a legitimate intraspecific allomorphosis is comparatively easy. The adult organism is defined simply as one that has reached its final maximum size or sexual maturity. By contrast, for those who study organisms that grow indefinitely in size, the distinction between heterauxesis and intraspecific allomorphosis is somewhat ambiguous, since an end point in final size and proportions may not exist, except in the trivial case of decedent size. By the same token, sexual maturity, which is commonly used as the temporal referent for defining the adult animal and for comparing ancestor-descendant pairs of ontogeny to evaluate for heterochrony, may change owing to environmental stress for plants. In consequence, an individual plant can appear to vacillate between the juvenile and adult conditions. This only goes to show that the choice of the criterion used to define the adult condition is sometimes arbitrary in the face of a much more complex biological reality.

In general practice, the heterauxesis of an organism that grows indefinitely in size is dealt with by decomposing the individual into its metameric (modular) units and considering each unit as analogous to an individual organism. As I noted in chapter 1, leaves, stem internodes, and other types of plant organs (= modules) are typically determinate in growth. Thus, we often speak of the *ontogeny* of a leaf, stem, or root. This is practical because each module of a plant has a distinct final size and organic proportions, but it runs the risk of conflating terminology and therefore obscuring communication with those who would reserve the term heterauxesis to cover size-correlated phenotypic variations attending the growth and development of an entire organism. The latter is reasonable since organisms, rather than organs, evolve. Also, if alterations in heterauxesis are the basis of evolutionary innovations or, at the least, if changes in the mathematical parameters that describe the scaling of ontogeny sensu stricto help to identify when or how these innovations come into being, then the distinction between heterauxesis and the scaling of organogenesis is biologically real and should be emphasized by our terminology. If, on the other hand, the distinction between ontogeny and organogenesis is inappropriate in terms of organisms with a modular construction and indeterminate pattern of growth by reiterative organogenesis, then it may be legitimate to apply the term heterauxesis to the scaling of individual plant organs.

Whatever our terminological predilections, the consequences of continued growth in size by adding new organs are intrinsic features of the biology of most plant lineages and therefore cannot be ignored. In this regard previous discussions have strongly reinforced the notion that plant organs can physiologically, developmentally, and structurally operate at least as quasi-independent units. In this sense organogenesis is dissociated to a degree from heterauxesis sensu stricto. That morphogenesis can yield specialized or different types of modules, that each module is not a wholly independent living thing, that its activities are coordinated with those of other modules at higher levels of organization, and that each is derived from a meristem rather than another module of the same type and therefore cannot give rise directly to others of the same or modified type are not viewed as compelling arguments against this point of view. As a crude analogy, consider social insects. Different types of individuals (workers, guards, etc.) operate within the community of the hive or nest. Each insect within the community is derived from a single fertile individual. Each is sterile and therefore cannot give rise to other individuals. Each is a component of a larger biological entity (the hive or nest): its activities are to a certain degree independent of the activities of others, yet the collective activities of individual insects are coordinated at higher levels of organization, thereby perpetuating a social contract that ensures the survival of the species. The zoologist does not hesitate to apply the term heterauxesis to the ontogenetic allometry of the individual social insect even though this ontogeny is not a property independent from the *ontogeny* of the hive or nest. By the same token, the botanist, with due caution, can employ the same terminological freedom provided it does not lead to confusing the ontogeny of an organ and the ontogeny of the individual plant as homologous processes.

The view that each organ has its own ontogeny offers some insight into the distribution of the variance of vegetative and reproductive variables among different levels of taxonomic classification. Table 5.1 offers nested analyses of variance for some vegetative and reproductive variables among subordinate and higher taxonomic levels of classification for angiosperms, gymnosperms, and pteridophytes. Within each of these three major vascular plant groups, roughly 70% of the total variance of these variables is accounted for at family and ordinal levels, suggesting that comparatively little evolution has occurred in these variables below the level of the family. Within each of the three plant groups, however, the percentage distribution of variance for vegetative variables at the species and genus levels, differs, often significantly. In general the variance of vegetative variables tends to be higher than that of reproductive variables; that is, comparatively more evolution of vegetative variables than of reproductive variables has occurred at the level of the species and genus. One way to look at this is that reproductive variables are evolutionarily more conservative than vegetative features. In general terms this tendency has been noted by many workers in comparative plant morphology, who have offered a variety of explanations, among which the most generally accepted is that any significant variation in the reproductive features of a plant is likely to diminish fitness (reproductive efficiency), and therefore that natural selection will favor reproductive organs with highly conservative ontogenies. By contrast, a comparatively greater level of variation for vegetative plant organs may be permissible because these organs influence reproductive success only indirectly and because the functional obligations of these organs are not as sensitive to differences in size, shape, and internal structure. To be sure, the size, shape, and structure of leaves drawn even from the same individual often vary markedly compared with the size, shape, and structure of the reproductive organs borne by the same individual. The extent to which the reproductive and vegetative structures produced by the same plant may differ evidently relates to the latitude permitted in the ontogeny of these organs. And it is in this context that the extent to which the ontogeny of an organ may be dissociated from the ontogeny of the individual plant assumes its greatest importance.

From prior discussions on the modular growth of plants, which have emphasized that juvenile organs are metabolically and mechanically supported by older, more mature ones, we can appreciate that the scaling of the ontogeny of an organ need not be under the direct or immediate influence of natural selection. For vegetative organs, heterauxesis is the scaling of *preparatory growth* (*Vorbereitungs-wachstum;* see Kramer 1959; Gould 1966). Natural variation in the scaling of preparatory growth is permissible provided the organ fulfills its functional obligations toward or at the conclusion of ontogeny. Theoretically, ontogenetic latitude can serve as the basis for evolutionary innovation because the end product, not the intermediate organic expressions of ontogeny, experiences selection. At the level of the whole plant, the growth and development of plant meristems can be evolutionarily modified, perhaps radically, provided they give rise to primordia that develop into functionally mature plant organs. At the level of the developing plant organ, how far ontogenetic modifications are permitted depends on how far these modifications influence an organ's functional obligations. For reproductive organs such as ovules, the intermediary products of ontogeny often assume functional obligations and therefore influence the survival and performance of later ontogenetic stages. The growth of the ovule into a seed or the gynoecium into a fruit illustrates transitions from one phase of *functional growth* to another. Alterations in the ontogeny of the ovule will affect the functional performance of the seed, as alterations in gynoecial development will affect the performance of the fruit. The scaling of structures such as these therefore is likely to be much more conservative than that of vegetative structures. The percentage distributions of variance for reproductive and vegetative variables appear to support this notion, since these distributions are in large part a consequence of how far ontogenetic patterns are permitted to vary at the level of the species, genus, and so on. For example, at the species level the variance of the biomass of foliage leaves tends to be greater than that of seed biomass for angiosperms, gymnosperms, and pteridophytes, showing that evolution at the subordinate taxonomic level in these three phyletically different vascular plant groups has progressed to significantly different degrees for these vegetative and reproductive organs.

The really important issue in terms of evolutionary transformations of one taxon into another is the way heterauxesis and allomorphosis are interrelated. To be sure, each datum in an intraspecific allomorphosis is the terminus of the former, and therefore the scalings of ontogeny and intraspecific allomorphosis are interrelated at a basic mathematical level. Curiously, ontogenetic scaling exponents frequently are negatively correlated with the mature size of organs and organisms, whereas the scaling exponents obtained from intraspecific comparisons tend to be negatively correlated with the mean adult size of different species or the mean size of subordinate taxa within higher levels of taxonomic classification. This tendency can be illustrated by considering two botanical examples that also let us explore the proposition that scaling relations permit functional organic proportions within a relatively narrow window of absolute size. The ontogenetic and intraspecific scalings of leaves serves as an example of the former. Figure 5.7 plots lamina surface ares S versus petiole length L for thirty-four mature leaves from the sugar maple, *Acer saccharum*. Least squares regression of these data shows that $S = 0.147\ L^{1.05}$ ($r^2 = 0.78$; $\alpha_{RMA} = 1.19 \pm 0.07$), indicating that the intraspecific relation between leaf surface area and petiole length is anisometric and positive. That the morphology of each mature leaf represents the conclusion of a specific ontogenetic pattern characterized by its own scaling relation is shown by plotting S versus L measured for three leaves as they develop and increase in size until they achieve their adult form. Figure 5.7 provides these data and reveals that the "ontogenetic" values of α_{RMA} increase with the final adult size of leaves (measured in terms of either S or L). The smallest of the three leaves has a scaling exponent of 0.949 ± 0.003, the

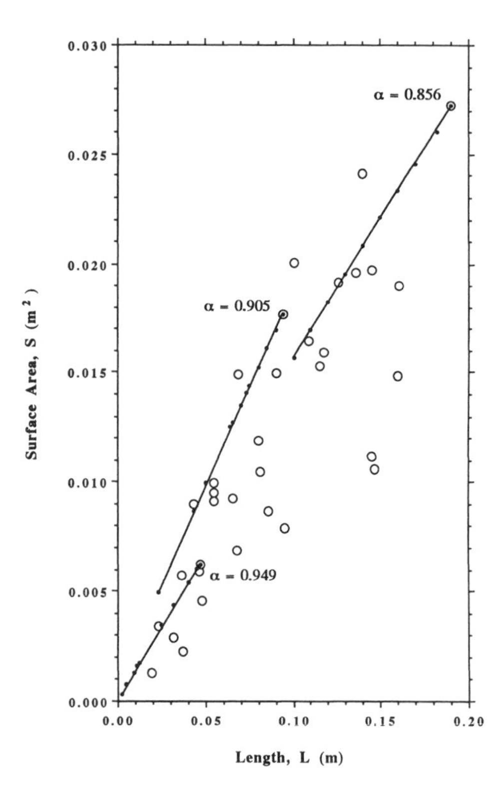

Figure 5.7. Log-log (base 10) plot of lamina surface area S against petiole length L of mature leaves (open circles) and plot of S against L for three developing leaves (closed symbols connected by LS regression curves) from the sugar maple, *Acer saccharum*. Scaling exponent obtained from RMA regression analyses is provided for each developing leaf.

intermediate leaf has an ontogenetic exponent of 0.905 ± 0.002, and the largest mature leaf among the three has a scaling exponent equal to 0.856 ± 0.002. Based on their 95% confidence intervals, these three scaling exponents differ significantly from one another; all three ontogenetic scaling exponents differ from the intraspecific scaling exponent 1.19 ± 0.07.

The tendency of intrataxonomic scaling exponents to be negatively correlated with the mean adult size of different taxa is illustrated by the relation between the scaling exponent for plant height H and stem diameter D plotted as a function of the mean plant height $\overline{H}$. From table 5.1 we see that over 50% of the total variance of plant height and stem diameter is contributed by families nested in orders and orders nested in classes. Consequently, species within genera and genera within families contribute additional, but not independent, data points for these two biological variables. Thus we can determine the scaling exponents for family mean values of plant height versus family mean values of stem diameter and plot these exponents against family mean height. Figure 5.8 provides

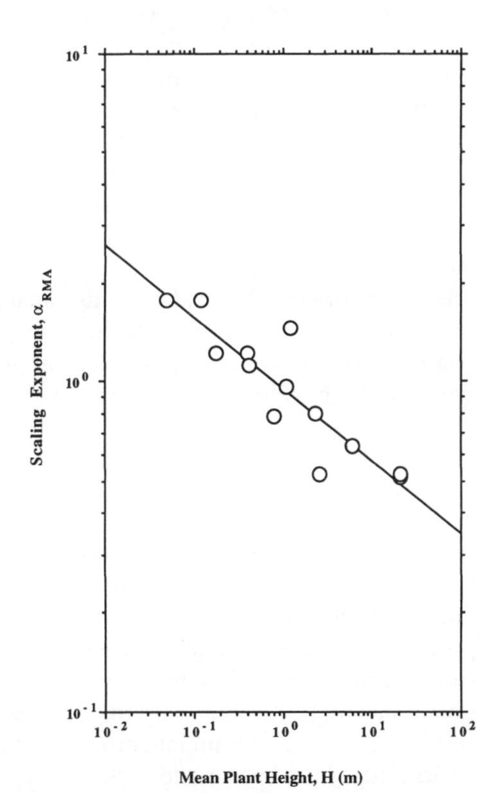

Figure 5.8. Log-log (base 10) plot of scaling exponent obtained from RMA regression analysis of family data α_{RMA} against family mean height H. Solid diagonal line is the regression curve obtained from LS regression analysis.

this comparison for nine families of angiosperms (distributed in seven orders of monocots and dicots), three families of gymnosperms, and one family of pteridophytes. Accordingly, this figure provides a broad intertaxonomic comparison for which data-point inflation and the "phyletic effect" have been more or less controlled. Least squares regression of these data shows that $\alpha \propto \overline{H}^{-0.218}$ ($r^2 = 0.80$; $\alpha_{RMA} = -0.244 \pm 0.08$), indicating that, on the average, the scaling exponent for mean height decreases roughly as the ¼ power of the family mean plant size ($\overline{H}$). The paucity of data points on which this scaling relation is based warrants concern over the precise functional (mathematical) relation between α_{RMA} and $\overline{H}$ but does not detract from the general observation that a negative correlation is obtained from these intertaxonomic scaling exponents versus size.

The most general explanation for the relations shown in figures 5.7 and 5.8 is that each scaling relation is functional only within a relatively narrow window of absolute size. Indeed, these two scaling relations were purposely selected because they were previously considered in great quantitative detail (see chap. 3). In general terms, we know that the petiole mechanically operates as a cantilevered beam whose vertical tip deflection depends on beam length and the magnitude of the distal loading. From the mechanical interrelations among petiole length, lamina weight, lamina surface area, and vertical tip deflection, the ontogenetic scaling relations among these parameters for each leaf are expected to scale in reference to final leaf size so as to obtain a negative correlation between α_{RMA} and increasing mature leaf size. In terms of the vertical axes of plants that mechanically operate as columns supporting the loadings produced by other plant organs, the load-carrying capacity of this type of biomechanical device is dictated in part by column length and axial girth. The scaling relation between these two biological variables must avoid or at least minimize elastic buckling. For comparatively short plants, length and axial diameter may scale as predicted by geometric similitude ($L \propto D^{1.00}$; see chap. 3). As we have seen, the scaling of height and diameter of the comparatively small sporophytes of mosses is very nearly isometric (see chap. 3). Extending this scaling into the larger size range of dicot herbs would yield vertical axes too slender to mechanically support the proportionally larger and more massive branched sporophyte. The intertaxonomic reduction in the scaling exponent shown in figure 5.8 demonstrates a proportional increase in stem girth with respect to shoot height. This provides an increased safety factor in that it diminishes the probability of elastic buckling. Another way to deal with the mechanical problems of vertical growth is determinate growth in height. Determinate shoots can avoid elastic buckling provided their final size is below the critical buckling height. In the absence of indeterminate growth in length, alternative developmental modifications are required to sustain the continued vertical growth of plants. Dicot and gymnosperm tree species increase the girth of their trunks by producing annual growth layers of secondary xylem in a manner that mechanically compensates for yearly increments in overall height and vertical biomass. This method of dealing with self-loading automatically reduces the scaling exponent of height with respect to stem diameter. Consequently, the tendency for the scaling exponent of plant height to decrease with increasing plant size shown in figure 5.8 is compatible with engineering theory and affords circumstantial evidence that adjustments in scaling parameters are required when terrestrial plants occupy different domains of absolute plant size. It is noteworthy that projecting the scaling relation attained by tree species into the smaller size realm of herbaceous plants is unnecessary from a mechanical perspective, since this produces plants with mechanical support members that are substantially "overbuilt" in terms of the actual loadings they must support.

The proposition that changes in scaling parameters attend changes in the size range of organisms does not require a negative correlation between scaling exponents and absolute size. It merely argues that the functional domain permitted by a particular scaling relation is size dependent such that changes in scaling parameters will attend phyletic changes in size. Clearly, how and whether scaling parameters change depends on the particular relation we consider. Indeed, a particular scaling relation may appear not to change over a broad range of size because a panoply of other variables influencing functional relations are ignored. Figure 5.9 illustrates this point in terms of the scaling of plant height with respect to stem diameter. Species mean values are plotted in this figure, rather than family mean values of height and diameter, because our concern is not with the precise numerical values of scaling parameters but rather with the superimposition of anatomical and developmental evolutionary innovations that attend phyletically quantitative differences in height and stem diameter. Perhaps the most apparent feature of the interspecific scaling relation shown in figure 5.9 is the comparatively abrupt "break" in slope that denotes the advent of elastic buckling (indicated by a downward-pointing arrow) for the tallest plants lacking large amounts of secondary xylem in their stems. The tallest of these plants are species of palm that undergo modest elastic buckling when they reach their critical buckling heights. Further increases in stem length result in proportionally greater bending. The largest specimens of these species thus do not grow taller above their critical buckling height but rather "bow" to their mechanical loadings. The tallest palm specimens appear to demonstrate the limiting height to which plants can grow vertically in the absence of a vascular cambium capable of substantially thickening the girth of vertical stems. Plants obviously have extended beyond this "limiting" condition through the evolution of a vascular cambium. The vascular cambium appears to have evolved independently in a number of plant groups—for example, the lycopods, horsetails, ferns, progymnosperms, pteridosperms, conifers, and dicots—suggesting that this developmental innovation conferred selective advantages in terms of the elevation of leaves and reproductive organs.

The second feature shown in figure 5.9 is the apparently uniform scaling relation for plant height below the size range of the tallest palm species. Although intertaxonomic scaling exponents for mosses, pteridophytes, and herbaceous dicots differ somewhat, a surprisingly uniform trend in the scaling of "nonwoody" plant height becomes apparent. Indeed, in the absence of additional information we might be inclined to consider the data points below and including the limiting condition of elastic buckling as belonging to a continuous scaling relation defined by

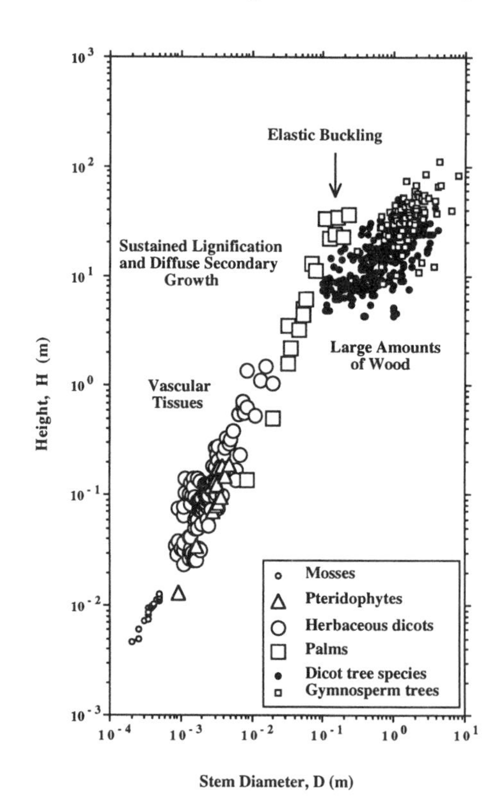

Figure 5.9. Log-log (base 10) plot of plant height H against stem diameter D for various plant species (see inset).

geometric similitude. Nonetheless, numerous anatomical and developmental innovations are required to obtain this "single" scaling relation. The smallest vertical terrestrial sporophytes, such as those of moss species, lack vascular tissues; the plants occupying the intermediary size range between mosses and the tallest palms possess primary vascular tissues mechanically configured in a variety of ways as well as expressing a comparatively limited capacity to produce secondary xylem; and the very tall palm species mechanically attain their vertical posture by the diffuse secondary growth of nonvascular tissue and the sustained lignification of primary vascular cell types in the vascular strands that anastomose throughout their stems. It should be evident that what appears to be a single scaling relation for "nonwoody" plants is in reality a consequence of the superimposition of at the very minimum three scaling relations, all reflecting significant phyletic differences in anatomy and developmental choreography.

The final point that must be made regarding figure 5.9 is that without benefit of the fossil record we have no temporal ordering principle for inferring how long or in what sequence different plant groups have occupied their present size ranges. In this regard the various plant groups shown are not ordered in terms of their chronological order of appearance in the fossil record, nor do they reflect past size ranges. Among the seed plants, the most ancient groups are the gymnosperms, which currently occupy the uppermost range of plant size. Whereas the tallest extant "nonwoody" plants are the palms, these arborescent monocots traditionally are phyletically derived from a dicot ancestral group. It also is noteworthy that the herbaceous dicot growth habit frequently is derived from an ancestral shrub growth habit. Note that the individual scaling relations shown in figure 5.9 overlap. The tallest palms actually exceed the shortest but nonetheless record-size dicot tree species. By the same token, the upper size range of herbaceous dicot species overlaps with the lower size range of palms. And finally, although not shown by these data, the tallest moss sporophytes extend well into the lower size range of extant pteridophytes. Had we included the estimated heights and stem diameters of fossil pteridophyte species in this data set, the size ranges of different plant lineages would have overlapped significantly more. One trunk of the fossil lycopod genus *Lepidodendron* is reported to measure 35 m in length up to the level where it first branched to produce a "crown" of twigs, leaves, and reproductive organs; the trunk of the fossil horsetail genus *Calamites* is reported to have reached a more modest, albeit still impressive 10 m (Stewart 1983, 103 and 159). These and other examples drawn from the fossil record indicate that the size range at present occupied by major plant lineages is not indicative of the size range occupied in the past.

Clearly, in the absence of information regarding the histories of lineages, adaptationist explanations for why different plant groups occupy the size ranges they do based on present-day anatomical and developmental features are largely fatuous.

I conclude my treatment of heterauxesis and interspecific allomorphosis with an attempt to mathematically derive an interspecific scaling relation from the scaling of ontogeny based on a simple quantitative-genetic model. A variety of previously treated scaling relations is at our disposal, but the one I shall focus on is the scaling of reproductive biomass M to stem diameter D. The relation between these two biological variables was presented in chapter 4 (see fig. 4.20). We saw that the interspecific scaling relation was $M = 0.12\ D^{2.90}$ ($r^2 = 0.94$, $N = 16$ species; $\alpha_{RMA} = 2.99 \pm 0.05$), meaning that, on the average, reproductive biomass scales roughly as the cube power of stem diameter. We also saw that this scaling relation was the statistical summation of sixteen intraspecific relations, each falling into one of only two general intraspecific categories distinguishable by the numerical value of α_{RMA}. One of these categories was described by $\alpha_{RMA} = 4.63 \pm 0.04$, while the other was described by $\alpha_{RMA} = 2.08 \pm 0.04$. Figure 5.10 illustrates the relation between these two intraspecific scaling categories and the interspecific scaling relation by plotting data from four of the total of sixteen species. My objective here is to derive a simple heuristic model for the genetic variation of M and D that can explain how these two sets of ontogenetic scaling relations obtain the interspecific scaling relation shown in figure 5.10.

I begin by noting that plants like the moss *Polytrichum* and the pteridophyte *Psilotum* bear their reproductive organs flush to their vegetative axes. Plants like these attain very high scaling exponents for their reproductive biomass (on the average, $\alpha_{RMA} = 4.63 \pm 0.04$). In contrast, plants like the extinct pteridphyte *Psilophyton* and the extant gymnosperm *Larix* bear their reproductive organs on lateral branches. These plants have comparatively much lower scaling exponents (on the average, $\alpha_{RMA} = 2.08 \pm 0.04$). Based on this observation, I will hypothesize that the intraspecific scalings of M depend upon whether reproductive organs directly develop from meristems that also give rise to vegetative axes (e.g., *Polytrichum* and *Psilotum*), in contrast to plants that produce reproductive organs on specialized branches that develop from meristems secondarily derived from the apical meristems that produce the principal vegetative branches (e.g., *Psilophyton* and *Larix*). To model both categories of development, I shall assume that only three traits are involved: trait A, which defines early growth in stem diameter, trait B, which will determine later growth in stem diameter, and trait C, which dictates the total biomass M of mature reproductive organs measured distal to mature stem diameter ("body size"), which equals $A + B$. In the case of plants that develop their reproductive organs from meristems derived "directly" from those producing vegetative branches, I assume that M evolves as a correlated response to early stem growth A. By contrast, of those plants that bear reproductive organs "indirectly" on lateral specialized branches, I will assume that M has evolved as a correlated response to later growth in stem size B. For both categories of plants, variance in trait B will be held within comparatively narrow limits because the numerical value of this biological variable dictates a variety of mechanical and physiological features other than M, such as total plant height, the ability to support photosynthetic organs, and the like.

To simulate the evolutionary consequences of these assumptions on the scaling of M with respect to D, we assign initial values to A, B, and C for the "ancestral condition." In a logarithmic scale, these values will be taken as zero ($A = 1$, $B = 1$, and $C = 1$). Each evolutionary step in our simulation is as follows:

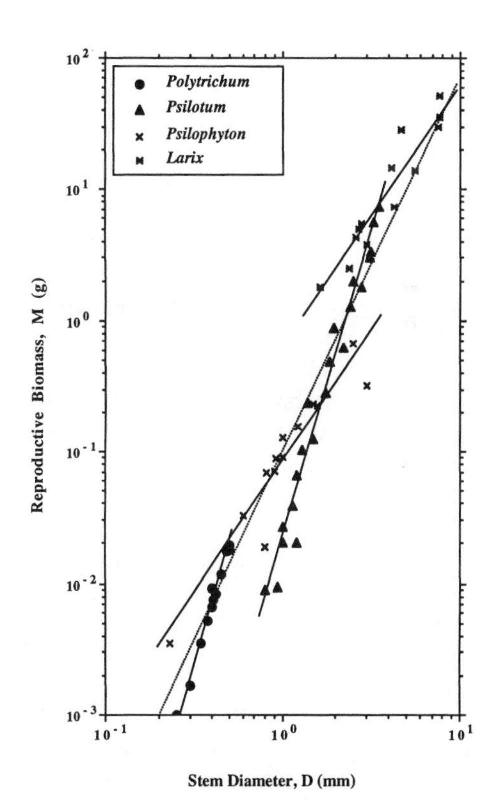

Figure 5.10. Log-log (base 10) plot of reproductive biomass M against stem diameter D measured proximal to M for four species (see inset). Solid and dotted lines are the regression curves obtained from LS regression.

1. A random number r is selected from a normal distribution such that $3 \leq r \leq -3$.
2. A is assigned a new value $A^* = A + r$.
3. B is assigned a new value $B^* = B + \mathrm{r}$:
   if $B^* > 1$, then $B^* = 1$.
   if $B^* < 1$, then $B^* = -1$.
4. In the case where M depends on early stem growth, $C = 4.63\ A^*$.
5. In the case where M depends on late stem growth, $C = 2.08\ (A + 2B)$.

Steps (1) and (2) correspond to random selection on early stem growth in diameter. Step (3) imposes a limit on subsequent stem growth in size. Steps (4) and (5) simulate the two correlated evolutionary responses of total reproductive biomass to stem growth. Step (4) simulates the response of plants that bear reproductive organs directly on their principal vegetative axes, for which the intraspecific scaling exponent equals 4.63. Step (5) simulates the response of plants that produce reproductive organs on specialized stems, for which the intraspecific scaling exponent equals 2.08.

Figure 5.11 plots the results of these simulations. As in the empirical relation, the simulated interspecific scaling exponent is obtained from two categories of intraspecific scaling relations differing in their scaling exponents. These two categories are shown for four simulated intraspecific comparisons. Two simulated "species" have scaling exponents equal to 4.63 (data points shown as closed circles); two simulated "species" have scaling exponents equal to 2.08 (data points shown as open circles). Clearly, these intraspecific scaling exponents are the direct result of steps (4) and (5) and therefore warrant little amazement. The scaling exponent for the reduced major axis regression of the simulated interspecific comparison (shown as a dotted line), however, equals 2.95. This exponent agrees remarkably well with that empirically obtained for the interspecific comparison among sixteen species for which $\alpha_{RMA} = 2.99$ (see figs. 4.20 and 5.10). Accordingly, the assumptions that random selection operates on early stem growth in diameter and that functional limitations on subsequent stem growth in size (i.e., steps 1–3) are sufficient to model the consequences of intraspecific (ontogenetic) scaling relations on the interspecific (allomorphosis) scaling relation for total reproductive biomass.

Clearly, this simulation is naive—it depicts the evolutionary relationship between ontogeny and interspecific allomorphosis in a terribly oversimplified manner. For example, reproductive biomass was assumed to be a neutral trait. This likely is not true for most plants over very long periods of evolutionary time. By the same token, natural selection likely does not operate directly on stem diameter ("body size") but rather works on a wide range of morphological, anatomical, and physiological traits. True, we can view body size as a crude index of the complex way natural selection operates on these many size-dependent biological variables. But this perspective does not mitigate our failure to distinguish the effects of selection on these traits and the covariance among these traits. The simulation presented here, however, does serve a useful pedagogic purpose by demonstrating that interspecific scaling relations can be modeled in terms of the effects of quantitative-ontogenetic variations. By doing so, we begin to see how different patterns of genetic and ontogenetic variation can achieve an interspecific scaling relation. An interesting analogue to this is seen in the zoological literature treating the relation between placental mammal body size and brain size. Interspecifically, brain weight scales roughly as the ¾ power of mammal body size, yet within individual groups of mammals, the scaling exponent tends to gravitate to 0.37 (Gould 1966; Martin 1981; Armstrong 1983). The interspecific scaling of placental mammal brain weight has been modeled in much the same way as I have treated plant reproductive biomass (Lande 1979; Atchley

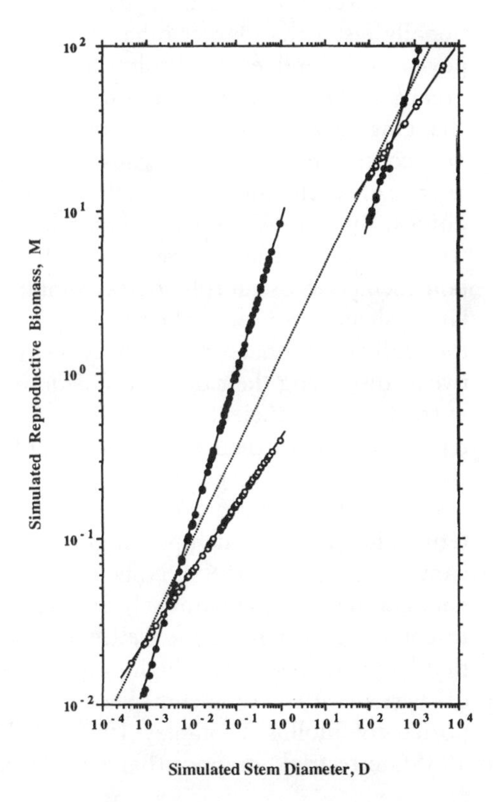

Figure 5.11. Log-log (base 10) plot of simulated reproductive biomass M against simulated stem diameter D proximal to M. Solid lines are regression curves obtained from LS regression of each simulated scaling relation; dotted diagonal line is the regression curve for the LS regression of all four scaling relations.

et al. 1984; Leamy 1988; Riska 1989)—that is, directional selection operating on one trait (body size) can be shown to result in a correlated response in another trait (brain weight). Such correlated (ontogenetic) response models offer great hope that we will eventually be able to understand the biological significance of differences between ontogenetic and interspecific scaling relations.

5.4 On Being Small or Large

Depending upon the environment where an organism lives, there are opposing advantages to small or large size, and plant size within individual lineages has historically been modified by changing environmental conditions. Here I shall explore the notion that intralineage size tends to diminish over time through the progressive elimination of larger species.

Among the advantages of being small is comparatively rapid growth. For example, previous interspecific comparisons among unicellular and multicellular plants and animals show that growth rate scales roughly as the ¾ power of body size. Thus small plants and animals grow proportionally faster than larger organisms (see chap. 1). Notably too, the ¾-power "rule" influences absolute as well as relative growth rates—larger organisms take longer to reproduce. In habitats characterized by rapidly changing local environmental conditions, therefore, rapid growth rate may confer an advantage because individuals can complete their reproductive life cycles during relatively brief episodes of favorable growth conditions. By contrast, other scaling relations suggest that benefits accrue with increasing absolute size—even when individuals differing in size obtain metabolic or morphological similitude. Tall terrestrial plants cast larger shadows on their shorter neighbors and thereby deprive them of the sunlight essential for growth. Larger plants also tend to be more effective at dispersing their diaspores, acquire and conserve water and other nutrients more efficiently, and are more insulated from rapid changes in temperature than their smaller morphological counterparts. In general terms, therefore, larger organisms tend to metabolically perceive their environment as fine-grained such that their physiochemical responses to external alterations are dampened in comparison with those of smaller organisms (see Levins 1968). Nonetheless, larger organisms tend to occupy habitats that are comparatively, stable, in part because they take longer to complete their reproductive life cycles. It is not surprising therefore, that large organisms have high probabilities of extinction during geological periods of environmental instability. Indeed, evolutionary trends reported for mollusks (Stanley 1979; Hallam 1990), arthropods (Briggs 1985), terrestrial tetrapods (Bakker 1977), and terrestrial plants (Tiffney and Niklas 1985) suggest that large size is an adaptive response to favorable environmental conditions over long periods of geological history whereas environmental stress causes a phyletic decrease in size. The similarity between the evolutionary trends observed for ecologically and phyletically diverse plants and animals presumably result from similar size-dependent relations to resource utilization, probability of extinction, population density, and growth rate.

That the physical environment influences plant size and that the physical environment and size range of individual plant lineages have changed over evolutionary history are well known. Perhaps the best-documented case of changes in the size range of a plant lineage in response to global environmental conditions is for the arborescent Carboniferous lycopods called lepidodendrids (see DiMichele, Phillips, and McBrinn 1991; DiMichele and Bateman 1992; Phillips and DiMichele 1992). The antecedents of this group were comparatively diminutive Devonian plants that lacked a vascular cambium and possessed a rhizomatous growth habit. The taxonomic radiation of lepidodendrids was confined to the ecologically stable and geographically expansive paleotropical wetlands of Euramerica (DiMichele, Mahaffy, and Phillips 1979; Bateman, Di Michele, and Willard 1992). Fossil specimens of the arborescent lycopod genus *Lepidodendron*, which achieved heights of over 30 m, grew in geographically extensive forestlike communities whose fossilized remains largely comprise today's economically important coal resources. The closure of the Carboniferous in the late Stephanian C ($\approx 325 \times 10^6$ years ago) was attended by the ecological collapse of the wetland habitats in which the lepidodendrids grew. This deterioration in habitat saw a concomitant decline in the numbers of subordinate taxa as well as individual plants within species. In part this appears to be a consequence of the dependency on freestanding water for the completion of the life cycle—the diaspores of some lepidodendrids (called aquacarps; see Phillips and DiMichele 1992, 584–85) were adapted for dispersal over the water surface, whereas the completion of the life cycle of most other less reproductively specialized lycopod species depended on hydrated conditions (see Phillips 1979). The taxonomic and numerical reduction in arborescent lycopods was attended by a reduction in the height of successively surviving taxa. The evolutionary history of the lycopods provides an unusually detailed case history for the way the size range of a particular plant lineage has initially expanded and subsequently contracted. Similar trends occur during the evolutionary history of the horsetails and eusporangiate fern lineages, whose mean height declined (after reaching a maximum, typically in the early half of the history of each lineage) attended by dramatic environmental changes. Thus, the Carboniferous arborescent Calamitaceae and the eusporangiate fern *Psaronius* were substantially larger than their phyletic predecessors or followers.

Evidence for the effect on size of habitat deterioration or episodic short-duration environmental instability also is available from data obtained for extant nonwoody and woody vascular plants. Figure 5.12A shows the frequency distribution of the log-transformed data for genus mean height. A total of 1,133 genera found in the northeastern United States and adjacent Canada are represented (data from Gleason 1968; aquatic plants and vines have been excluded because measurements of height for these organisms are ambiguous). These data indicate that there are substantially fewer tall than short genera. Indeed, the frequency distribution is strongly skewed to the right even when genus mean height is log-transformed (skewness = 0.71 ± 0.13). The pronounced tail toward

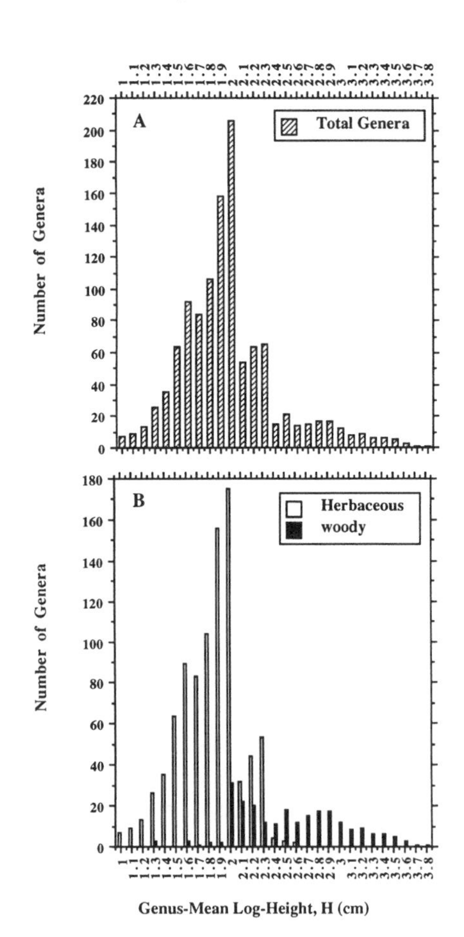

Figure 5.12. Number of genera *N* plotted as a function of log-transformed genus mean height *H* for a total of 1,133 genera found in the northeastern United States and adjacent Canada. Data from Gleason 1968 (genera dominated by aquatic species or species with liana growth habits excluded). (A) Frequency diagram for all genera. (B) Frequency diagram for herbaceous and woody genera in total data set.

taller genera is occupied principally by plants with a woody growth habit, as shown in figure 5.12B. Thus, in terms of the present genus composition of this flora, there is an evident bias favoring small nonwoody plants. A number of factors are responsible for this frequency distribution, among which the episodic glaciation in northern America and adjacent Canada likely has played an important role. Another is the phyletic history of the various lineages represented in northeastern America and adjacent Canada. Figure 5.13 provides the frequency histograms of log mean height for genera within the Compositae and Ranunculaceae, as well as for lycopod, horsetail, and fern genera (data from Gleason 1968). The pteridophytes are the most ancient group of plants, whereas the Compositae is the most evolutionary recent taxon among the three nonwoody plant groups considered in this figure. A comparison among the three histograms shows that, on the average, genera within the Compositae are taller than ranunculaceous genera, while pteridophytic genera are, on the average, shorter than those in either of the two flowering plant families. Similar analyses of other flowering plant families reveals that average height of nonwoody

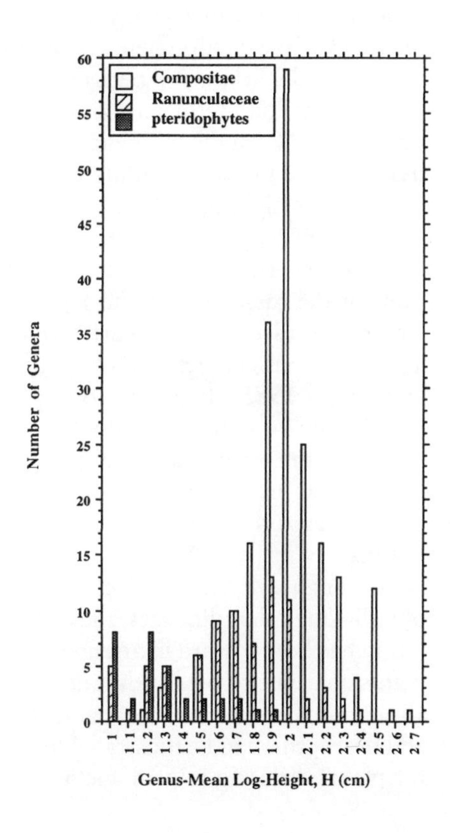

Figure 5.13. Number of genera N plotted as a function of log-transformed mean height H of genera in the Compositae (= Asteraceae), Ranunculaceae, and pteridophytes (ferns, *Equisetum*, and *Lycopodium*) found in the northeastern United States and adjacent Canada. Data from Gleason 1968.

genera tends to be inversely proportional to the relative evolutionary age of the family in which genera are nested.

The apparent inverse relation between mean size and the geological age of plant lineages can be explained in the following way. On the average, the rates of plant maturation and growth are inversely correlated with plant size, while larger plants also have a higher reproductive rate, so both the rate of growth of plant populations and population density are inversely correlated with mean plant size. By contrast, on the average, the utilization of available resources by an individual plant as well as the biomass allocation to the reproduction of the individual is positively correlated with size. Thus larger plants tend to occupy more space, use more resources, and produce larger diaspores than do smaller plants. In theory, therefore, a phyletic gain in size will occur during geological periods characterized by environmentally undisturbed conditions, whereas the probability of extinction will increase with plant size during geological periods of environmental stress. Clearly, plant taxa occupying the lower size range with a lineage can provide founder species for new taxa whose subsequent evolution can repeat the general historical pattern of gain in size provided stable environmental conditions resume. The frequency of environmental disturbance experienced by a lineage will play a pivotal role in whether the long-term phyletic trend is one of increasing or decreasing size occurs. When the frequency of disturbance is low, the general phyletic trend will be in the ascending mode of plant size. Conversely, when the frequency is high, the general trend will be in the descending mode. This explanation is compatible with a number of paleobotanical observations. For example, rhizomatous pteridophyte species tend to have longer durations in the fossil record and higher probabilities of survival than their larger arborescent counterparts, which generally fail to pass through geological periods characterized by dramatic ecological changes (see Tiffney and Niklas 1985). It should come as no surprise, therefore, that the size distribution of extant pteridophytes from North America and Canada, which are the remnants of the most ancient groups of vascular plants, has a nearly flat histogram throughout the size range of nonwoody plants but fails to trespass into the range of more geologically recent woody angiosperms.

Although size tends to decline after reaching a maximum within most of individual plant lineages, the collective tendency of vascular plant evolution has been toward increasing plant size. It is also clear that very small plants continue to persist over geological history. Consequently, the overall size range of plants has expanded through evolutionary history. In general terms, this can be illustrated by means of the record stem diameter D for plants differing in geologic age. Recall that plant height tends to increase in proportion to stem diameter, and therefore D provides a crude estimate of the maximum height to which extinct plants grew. Figure 5.14, which shows D versus the various geologic periods within the Phanerozoic—specifically, from the Upper Silurian ($\approx 424 \times 10^6$ years ago) to the Recent. As may be seen, on the average, maximum plant size tends to increase throughout the Phanerozoic. Plant size increased sharply in the early Paleozoic (during the late Silurian to the Carboniferous). This trend has been noted by others (e.g., Zimmermann 1930; Chaloner and Sheerin 1979) and will be discussed in some detail later on. For now, however, note that maximum plant size vacillates during the Permo-Triassic ($\approx 280 \times 10^6$ to 200×10^6 years ago), reaches a local maximum during the Middle Eocene ($\approx 45 \times 10^6$ years ago), and sharply rebounds once again after the Upper Miocene ($\approx 10 \times 10^6$ years ago). These broad

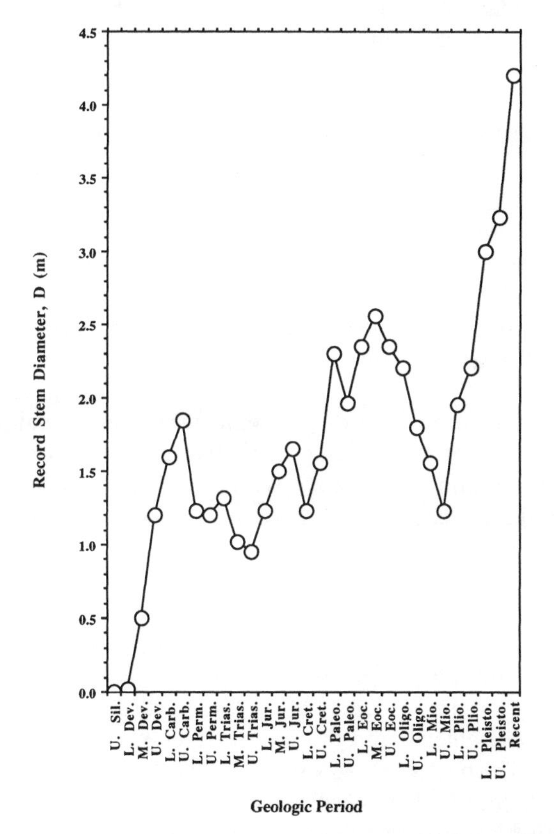

Figure 5.14. Record stem diameter D for fossil vascular plants plotted as a function of geologic period. Duration of geologic periods not drawn to scale. Data from a survey of the primary paleobotanical literature.

evolutionary changes in size undoubtedly reflect a number of factors, such as major floristic turnover and the effects of global climatic conditions on individual plant groups. Nonetheless, the general trend toward increasing plant size is undeniable. The Paleozoic flora was dominated by pteridophytes whose arborescent members reached their greatest size in the Carboniferous, only to be eradicated during the global deterioration of wetland communities attending the closure of this geologic period. Although comparatively small pteridophytes persist to the Recent, in the Mesozoic the pteridophyte-dominated flora of the early Paleozoic was replaced by one composed principally of gymnosperms. In turn, in the Cenozoic this gymnosperm-dominated flora was replaced by one dominated by angiosperms. It is noteworthy that extant gymnosperms, specifically the conifers, hold the current world record for plant height and stem diameter. The ascent of plant size during the latter part of the Cenozoic is mainly the consequence of increasing conifer size, thereby providing the principal exception to the generalization that geologic age and plant size (for an individual lineage) are inversely proportional.

The comparatively sharp increase in maximum plant size during the early Paleozoic noted earlier is of particular interest because it coincides with the initial radiation of terrestrial vascular plants. The behavior of a group of organisms recently introduced to an unoccupied landscape warrants particular concern over the interpretation of apparent directional trends. Although natural selection is most often implied or explicitly assumed as the mechanism responsible for increasing size or anatomical complexity, alternative models to natural selection exist (e.g., stochastic, diffusion, etc.). These models provide a well-reasoned concern for interpreting every trend a priori as the consequence of directional selection. In the context of the initial radiation of terrestrial plants, it is evident that comparatively small and morphologically simple organisms were poised on the edge of an accessible and unoccupied domain of size and shape. Biological systems like these are predictably expansive over time and can be described as a process of diffusion, one of whose manifestations is an increase in variance (Stanley 1973; Fisher 1991; Gould 1990). For this reason I shall dwell temporarily on the early colonization of land by plants to examine the alternative hypotheses of adaptive evolution and diffusive evolution in the context of three biological variables: plant height, tracheid diameter, and the volume fraction of xylem in stems.

Figure 5.15 quantifies changes in plant size during the early Paleozoic by plotting the logarithm of record stem diameters D (in units of m) for different major lineages as a function of geologic time. The relation between these two variables is exponential ($r^2 = 0.82$), revealing that, on the average, maximum plant height increased over time. Figures 5.16 and

5.17 show that this overall gain in plant size was attended by an increase in tracheid diameter and the volume fraction of xylem in stems. Figure 5.16 plots tracheid diameter (in μm) versus geologic age for representative Upper Silurian and Devonian plant remains. Recalling that the capacity to transport water is proportional to the fourth power of the diameter of capillary-like cells (see chap. 1), it is reasonable to argue that the increase in the upper limit of tracheid diameters conferred an increased hydraulic capacity to transport water. Clearly this capacity depends on the amount of xylem tissue in a stem as well as on the diameters of its constituent cells. Figure 5.17 reveals that the volume fraction of xylem tissue in cross sections through fossilized stems as well as the relative hydraulic efficiency of individual xylem cells increased during the Devonian (see Niklas 1986a, b). The mechanical implications of a phyletic increase in

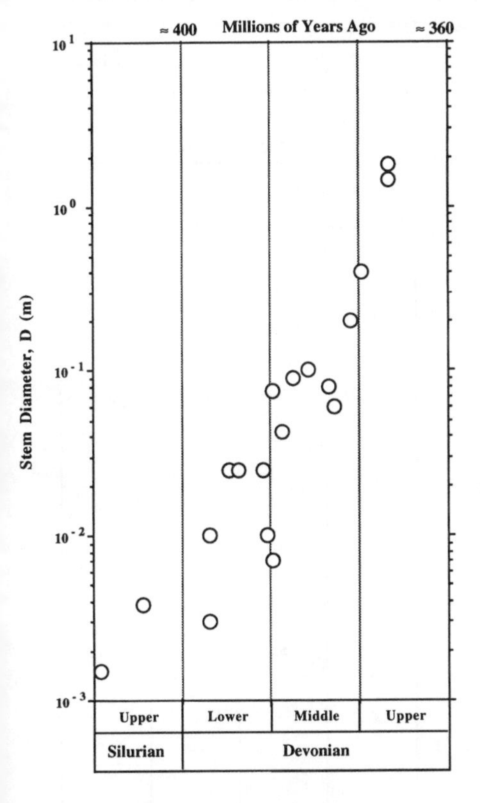

Figure 5.15. Log (base 10) stem diameter D plotted as a function of geologic time for early Paleozoic plants.

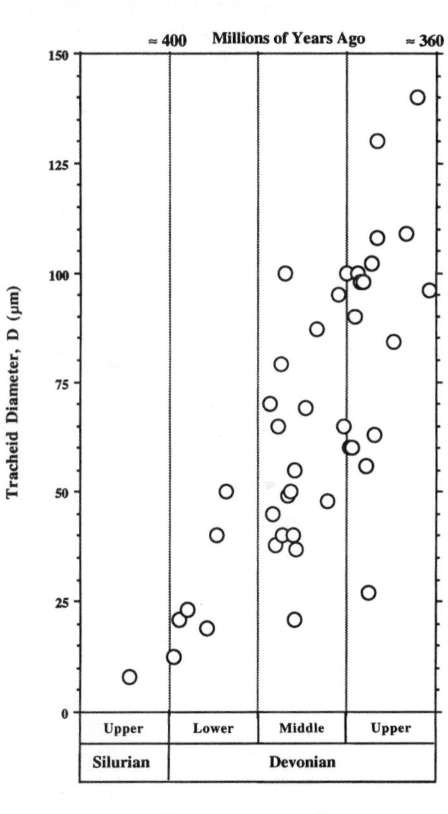

Figure 5.16. Tracheid diameter D plotted as a function of geologic time.

the cross-sectional area of xylem are rather obvious and need not be emphasized here. Although the increase in maximum plant size through the Devonian can be explained by recourse to a nonadaptive "diffusion" model in which initially small and morphologically and anatomically simple plants radiated into an essentially biologically barren landscape, the simultaneous evolutionary changes in a constellation of many functionally important anatomical, morphological, and reproductive features suggest that selection pressures operated on the ability of plants to absorb, transport, and retain water as well as to intercept sunlight and disperse spores. To be sure, the "diffusion" and "natural selection" models are not mutually exclusive. Each organism is the collection of numerous biological features, and some may be subject to natural selection while others are free to vary in a more or less stochastic way.

Before leaving the Devonian, we can explore how nested analysis of variance can be used to reveal the way the taxonomic level experiencing phyletic changes in size or anatomy shifts over evolutionary time. Figure 5.18 plots the percentage variance in plant height, tracheid diameter, and

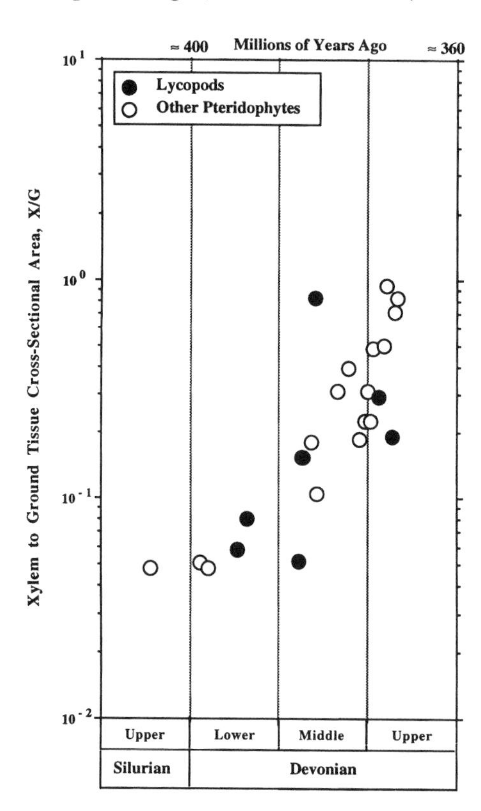

Figure 5.17. Log (base 10) of the quotient of the cross-sectional areas of xylem and ground tissue *X/G* plotted as a function of geologic time.

xylem cross-sectional area contributed by the sum of the variance obtained for species nested in genera and that contributed by genera nested in families. Before we attend to this figure, it is important to realize that the taxonomy of many Devonian plants above the family level and below the genus level is currently poorly understood and likely does not reflect a natural classification. Consequently figure 5.18 must be examined as an intellectual exercise rather than a legitimate analysis, because the taxonomic structure of the data set does not reflect the real phylogenetic relations among these Devonian plant remains, and therefore the fundamental premise of nested analysis of variance in the context of scaling analysis is violated. With this caveat in mind, the data indicate that the percentage variances of tracheid diameter and xylem cross-sectional area, on the average, decline during the late Silurian and the Devonian. If this is a legitimate feature of Devonian plant evolution, then it suggests that the taxonomic level in which evolutionary changes occurred progressively shifted from the species and genus levels to the family and higher taxonomic levels. Clearly, even if we had at our disposal a natural system of classifi-

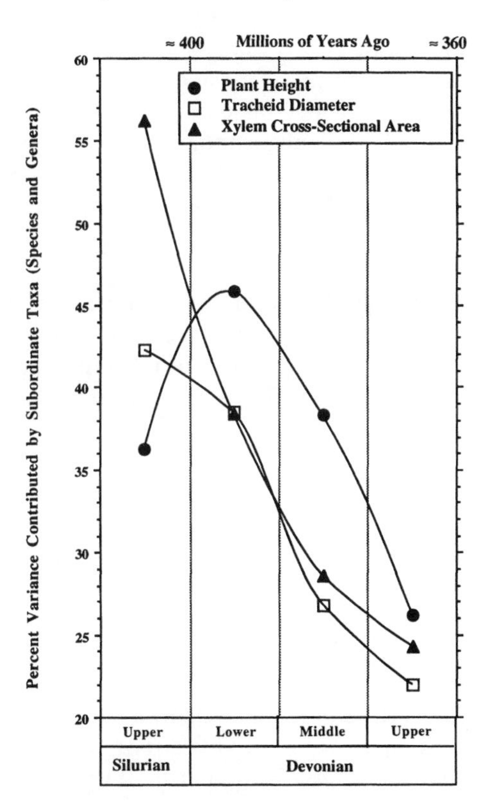

Figure 5.18. Changes in the sum of species and genus variances in morphological and anatomical variables (see inset) plotted as a function of geologic time.

cation for Devonian plant remains, this shift could be an artifact of the capacity for greater taxonomic "resolution" with decreasing geological age—younger plant remains tend to be more morphologically and anatomically complex and therefore offer more numerous characters with which to distinguish subordinate taxa. In this regard, one might argue that the temporal pattern in the percentage variance of plant size at the species and genus taxonomic levels should be less prone to this potential artifact. In this context, although the percentage variance of plant size appears to have much less "directionality" across the Silurian-Devonian boundary, this biological variable nonetheless declines throughout the Devonian. In general terms, therefore, a nested analysis of variance approach to Devonian plant remains suggests that evolutionary changes, which undeniably occur within populations, operate so that the variance in biological variables was initially high within subordinate taxonomic levels and subsequently increased at higher taxonomic levels.

Finally, we must be sensitive to the fact that all evolutionary innovations occur at the level of the species and that these innovations are used by systematists to establish higher taxonomic groups. Consequently, although we may speak of "shifts in the relative contributions of taxonomic levels to the total variance observed for particular biological variables within a clade," this phraseology does not imply that supraspecific taxa evolve from equivalent or higher taxonomic levels—clearly genera do not evolve from genera, nor families from families, in any meaningful biological sense. The paradigm of evolutionary biology is that the taxonomic level at which evolutionary innovations arise will always be the species. Accordingly, temporal shifts in the distribution of the variance of a biological variable among taxonomic levels reflect how far evolutionary innovations have established morphological and anatomical departures from the ancestral condition defining the founding species of a clade and how these departures have been used by systematists to recognize subordinate taxa nested in higher taxonomic levels.

5.5 The Scaling of the Plant Life Cycle

By virtue of its life cycle, each plant species consists of two types of organisms, each with its own distinctive ontogeny, development, maturation, and reproductive function in the life history of the species. Thus each phase in the life cycle is characterized by its own scaling relations for size, expression of multicellularity, reproductive effort, and growth rate. Accordingly, the life cycle of a species affords two occasions for ontogenetic and scaling modification, and each occasion has the potential to yield a new type of organism in terms of life history. Clearly, the scalings of the two phases in the life cycle are linked by their ecological consequences on life history. Thus, although ontogenetic modifications of the two phases in the life cycle are potentially independent, they must permit the completion of the cycle in order to perpetuate a life history variant through evolutionary time. We will explore the plant life cycle in terms of how compliant ontogenetic modifications of the two phases have influenced plant history.

Figure 5.19 diagrams three versions of the plant life cycle. The possibil-

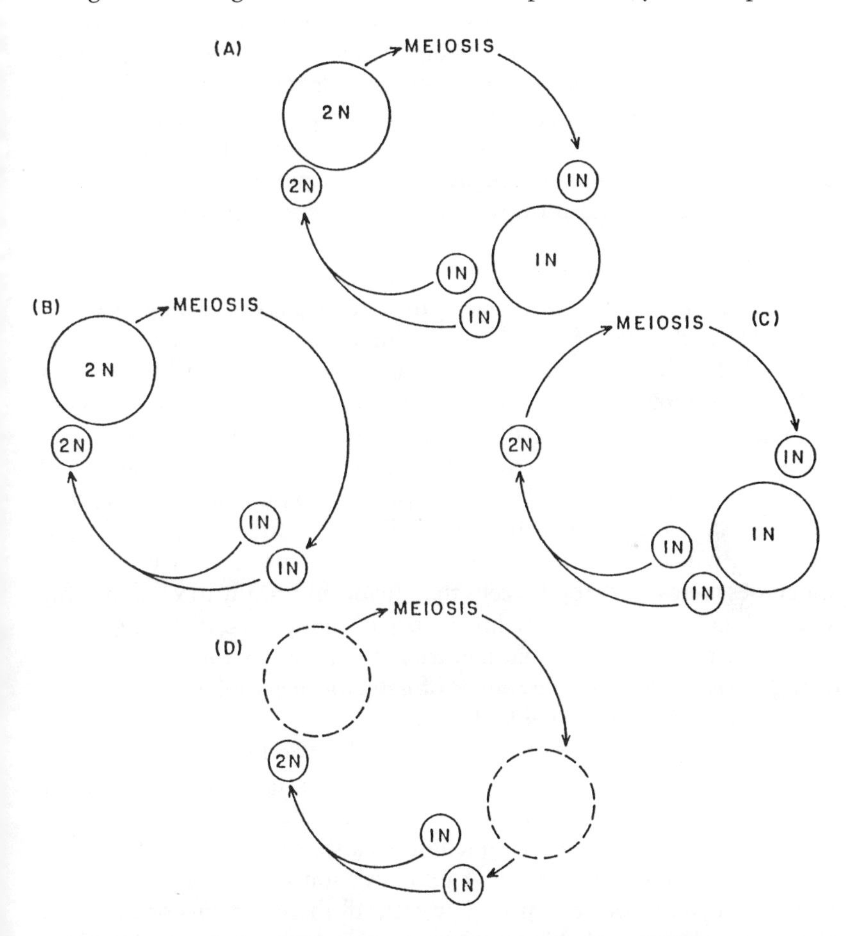

Figure 5.19. Schematic versions of the plant life cycle. Multicellular individuals are denoted by large circles, unicellular individuals by smaller circles. (A) Diplobiontic ("alternation of generations" version with two multicellular generations. (B) Haplobiontic-diploid version with a diploid (2*N*) multicellular generation. (C) Haplobiontic-haploid version with a haploid (1*N*) multicellular generation. (D) Aspects of the plant life cycle shared by versions A–C (dashed circles denote relative positions of multicellular individuals).

ities of polyploidy and asexual reproduction have been ignored, although neither is uncommon and they are even the rule for some species (see Bell 1989). Figure 5.19A diagrams what is called the *alternation of generations* or, more technically, the diplobiontic version of the plant life cycle, which involves the alternation of two multicellular generations or phases: the sporophyte that produces spores by meiotic cell division and the gametophyte that produces gametes (either egg or sperm or both) by mitotic cell division. In general terms, the onset of sexual maturity for the sporophyte and the gametophyte occurs with the initiation of sporogenesis and gametogenesis. A change in the timing of sporogenesis or gametogenesis relative to the somatic maturation of the sporophyte or gametophyte can alter the life history of a species enough to warrant recognizing a new group of plants. For example, a delay in the timing of sporogenesis or acceleration in the tempo of vegetative growth can lead to the elaboration of sporophytic size and form relative to that of the conspecific gametophyte.

The life cycle versions diagrammed in figure 5.19B and C are called haplobiontic life cycles. Only one multicellular phase occurs in each reproductive cycle. One of these cycles is similar to the life cycle of a typical mammal (fig. 5.19B); it is called the haplobiontic-diploid life cycle. A multicellular diploid individual produces gametes by meiotic cell division. The zygote resulting from fertilization undergoes repeated mitotic cell division to produce the multicellular individual in the life cycle. The second type of haplobiontic life cycle is called the haplobiontic-haploid life cycle (fig. 5.19C). A multicellular haploid individual produces gametes by mitotic cell division; then the gametes fuse to form zygotes, each of which undergoes meiosis to produce cells that divide mitotically to yield the only multicellular individual in the life history of the species. In both of these life cycles, the onset of sexual maturity is denoted by the advent of gametogenesis. A delay in the timing of gametogenesis relative to somatic growth can yield a larger haploid individual.

In general terms, each of the two haplobiontic life cycles represents one aspect of the diplobiontic life cycle. In each, only one multicellular individual makes an appearance. The multicellular individual in the haplobiontic-diploid life cycle functions as a sporophyte. The multicellular individual in the haplobiontic-haploid version assumes the function of a gametophyte. Upon reflection, however, all three versions of the plant life cycle are distinguishable regarding where and to what extent the condition of multicellularity is expressed (fig. 5.19D). In the diplobiontic variation, both phases in the life cycle are multicellular organisms. In the case of the two haplobiontic versions of the plant life cycle, multicellularity is expressed in only one of the two phases. From this point of view, the zygote is a unicellular sporophyte and the gamete is a unicellular gametophyte. The intercalation of mitotic cell division between syngamy and sporogenesis transforms the zygote into a multicellular individual. Likewise, the intercalation of mitotic cell division between gametogenesis and syngamy translates the gamete into a multicellular individual. Conversely, the loss of the capacity to express the multicellular condition can convert the sporophyte and the gametophyte into the zygote and the gamete.

That the zygote and the gamete are organisms is hardly a new point of view. But this perspective has become conceptually obscured by cell-theory dogma (see Scharp 1926; Kaplan and Hagemann 1991; see also sec. 1.3). Wilhelm Hofmeister (1851), who elucidated the alternation of generations (*Generationswechsel*) based on the morphological differences between the sporophytic and gametophytic phases, and who was one of the first to vigorously oppose the cell theory, concluded that the formation of cells is a function of general growth and not its cause. Hofmeister also maintained that the increase in the mass of the individual (whether in the sporophytic or gametophytic phase) cannot be considered to be the sum of the formation of individual cells but rather is an increase in the volume of the protoplasm of the individual that may or may not express the capacity for multicellularity. Thus the zygote, meiospore, and gamete are viewed by Hofmeister as unicellular individuals whose morphogenetic capacities prefigure those of their multicellular counterparts. This conceptual distinction between the cell and the organism should have eradicated subsequent controversies regarding plant evolution, such as the two theories advanced to explain the evolution of the sporophytic phase: the homologous (or modification) theory and the antithetic (or intercalation) theory. According to the former, neither the sporophyte nor the gametophyte can be given evolutionary precedence. Both generations are viewed as evolving simultaneously. According to the antithetic theory, the sporophytic phase is interpreted as an entirely new manner of organism "intercalated" between two successive gametophytic phases in the haplobiontic-haploid version of the general plant life cycle. The advocates of the homologous theory held to their views based on the presumed improbability that a new component, the sporophyte, could be evolutionarily inserted into the life history of an organism: *ex nihilo nihil fit* (from nothing, nothing comes). Echoing Hofmeister's logical clarity, F. O. Bower (1908) countered this point of view by pointing out that the zygote is a unicellular organism that fulfills the functional role of the sporophyte: *ex aliquo aliquid* (from something, something comes). Regrettably, the unnecessary dialectic of the homologous and the antithetic theory persists even today (see Remy, Gensel, and Hass 1993).

The logical distinction between the cell and the individual organism emphasizes the bio*logical* difference between the concept of the individual plant and the life cycle of which it is a part. This distinction is important because the limits imposed on the individual do not necessarily specify the full range of limitations that operates at the level of the life history of the species it belongs to. The *vegetative* functions and adaptations that permit the survival of an individual and the *reproductive* functions and adaptations that permit the completion of a life cycle are not the same. Although a very fine line exists between vegetative and reproductive functions, as well as the physical limitations imposed on the completion of these functions (the individual sporophyte or gametophyte must survive and grow before it reaches sexual maturity and invests a portion of its metabolic reserves into the formation of spores or gametes), a counterpoint between reproductive and vegetative functions exists, as reflected by the scalings that govern these two functions. This becomes clear when we consider one of the more general evolutionary trends among plant groups, the positive scaling relation between plant size and multicellularity. This scaling relation is important in terms of the capacity of an individual plant to produce numerous spores or gametes over its lifetime. For lack of a better phrase, this may be called "reproductive redundancy." Consider that in the case of a truly unicellular organism the existence of the individual is terminated by fertilization (gametophytic phase) or meiosis (sporophytic phase). Large unicellular plants exist (e.g., *Caulerpa*), but the unicellular vegetative condition is lost with the advent of reproduction. The capacity for multicellularity permits the production of many gametes or spores during a lifetime. In this regard the scaling of plant size and cell number is interesting. On the average, larger plants tend to have more cells because cell size tends to be a highly conserved feature for each type of cell, perhaps owing to the influence that the ratio of surface area to volume has on the rates of physiological processes for which cell types have become specialized. Consequently, larger multicellular plants have a greater capacity for reproductive redundancy. However, a negative scaling relation appears to exist between the size of the sporophyte and that of its gametophyte. A gain in the ability of the sporophyte to produce many spores is in large part counterbalanced by a loss in the ability of the conspecific gametophyte to produce many gametes. Figure 5.20 illustrates this tendency in terms of the life histories of a charophycean alga (*Chara*), a moss (*Sphagnum*), a pteridophyte (*Equisetum*), a gymnosperm (*Pinus*), and an angiosperm (*Lycopersicon*). The arrangement of these plants in figure 5.18 crudely reflects the relative times of appearance of each plant group in geological history. The charophycean algae, which are presumably the ancestral plexus from which embryophytes were divided (see fig.

5.1), are the most ancient of the five plant groups represented in this figure. In ascending order, the bryophytes, pteridophytes, and gymnosperms are more evolutionarily recent, while the angiosperms constitute the most recent major plant clade to debut.

It is tempting to describe the trends diagrammed in figure 5.20 in terms of heterochrony. In general terms, the intertaxonomic procession of decreasing gametophytic size suggests a precocious onset of sexual maturity in progressively less multicellular and therefore more vegetatively juvenile plants. This trend could be ascribed to paedomorphosis, specifically progenesis, which leads to the rapid completion of sexual function relative to vegetative function (see table 5.2). More to the point, however, this presumed evolutionary series of gametophytes coincides with a reduction in the expression of the multicellular condition associated with determinate growth in size and a general loss in reproductive redundancy. For example, under normal conditions, the angiosperm megagametophyte produces a single egg and therefore normally a single embryo in its lifetime. At the other end of the spectrum, a single *Chara* gametophyte, which can be as large as a small tree in terms of biomass, can produce thousands of

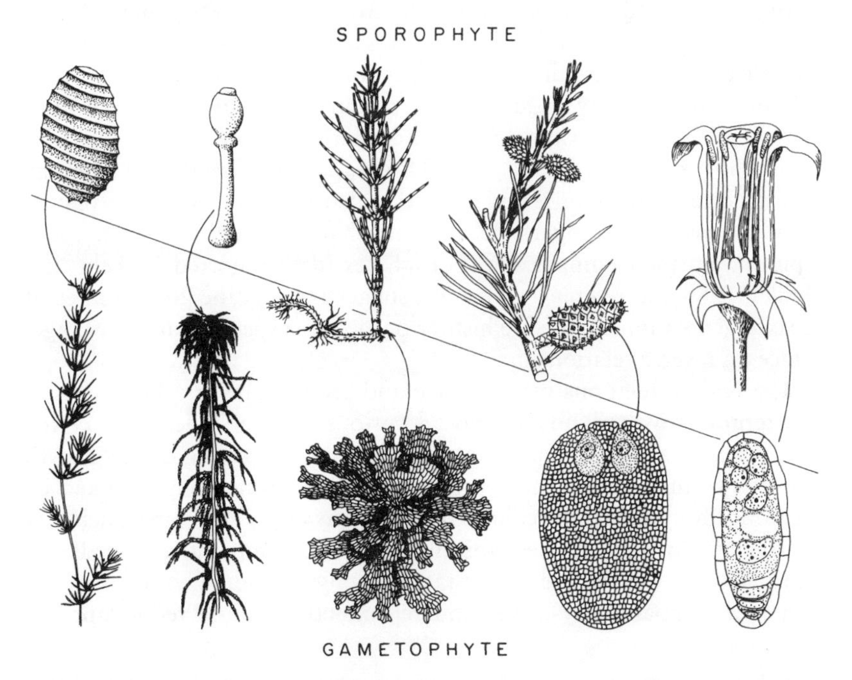

Figure 5.20 Comparative sizes of the sporophytes and gametophytes of various types of plants arranged to show the relative times when major plant groups made their evolutionary appearance. For details see text.

eggs in its lifetime by virtue of its indeterminate growth in cell number and size. Turning to the evolutionary procession of sporophytes, at least three heterochronies assigned to the class of peramorphosis are theoretically capable of yielding a descendant whose form progressively extends beyond that of its ancestor (acceleration, hypermorphosis, or predisplacement). Much more to the point, the phyletic increase in sporophytic sizes is associated with an increase in reproductive redundancy. The unicellular sporophyte of *Chara* is a zygote that normally yields a single gametophyte; the multicellular sporophyte of a typical angiosperm can produce thousands of seeds or more in a year.

The reproductive redundancy conferred by large size and multicellularity, however, appears to come at a cost in terms of growth rate. Based on broad intertaxonomic comparisons, larger unicellular and multicellular plants and animals, on the average, grow more slowly than smaller ones. This seems true for gametophytes as well as sporophytes. Figure 5.21 plots the average growth rate of the pollen tube G versus average pollen grain volume V for thirteen representative species from eight families (nested within five orders) of flowering plants (data from Fehr and Hadley 1980 and references therein). Admittedly, the scaling relation shown by these pollen grains likely reflects a phyletic effect, since eleven of the thirteen species are dicots and no gymnosperm pollen is represented. Additionally, a variety of factors other than size influence growth rates of pollen.[3] But the anisometric scaling relation shown in figure 5.21 is statistically robust ($G \propto V^{0.54}$; $r^2 = 0.96$, $\alpha_{RMA} = 0.55 + 0.005$) and may hold true in qualitative terms for the pollen of other flowering plants. Also, interspecific comparisons indicate that gametophytes grow more rapidly than their comparable sporophytes (data reported by Fehr and Hadley 1980 and references therein), suggesting that the growth rates of the two generations in the life history of the same species, on the average, evince an inverse relationship.

The scaling relations between size and growth rate and between size and reproductive redundancy have a curious interdependency, since the rate at which a species reproduces depends on the scaling of the growth rate of two individuals, the sporophyte and the gametophyte, linked by the same life cycle. These scalings tend to be inversely correlated such that the life cycle of each species typically involves a comparatively small and reproductively less redundant but fast-growing individual whose counterpart is a comparatively larger and reproductively more redundant but

3. As early as 1830 it was known that pollen grains receive nutrients from stylar tissues (Amici, cited by Cruden and Lyon 1985), while more recent studies show that stylar substrates are used in the construction of the pollen tube (see Campbell and Ascher 1975).

slower-growing individual. Since the absolute rate of the life cycle depends on the absolute as well as the relative size of both individuals, how well a life cycle can tolerate environmental disturbance depends on the ability of the larger individual in the life cycle to vegetatively survive periods of stress, because this individual is the slower growing of the two. Thus, in environments that are characteristically disturbed, comparatively small and fast-growing sporophytes that produce even smaller and faster-growing gametophytes may be favored. In stable environments, big and slow-growing sporophytes that produce comparatively large and laggard gametophytes are permissible.

5.6 The Evolution of Plant Life Histories

The sporophyte and gametophyte of a species typically assume different sizes and shapes under natural conditions of growth and development. When artificially grown under similar conditions, however, the two indi-

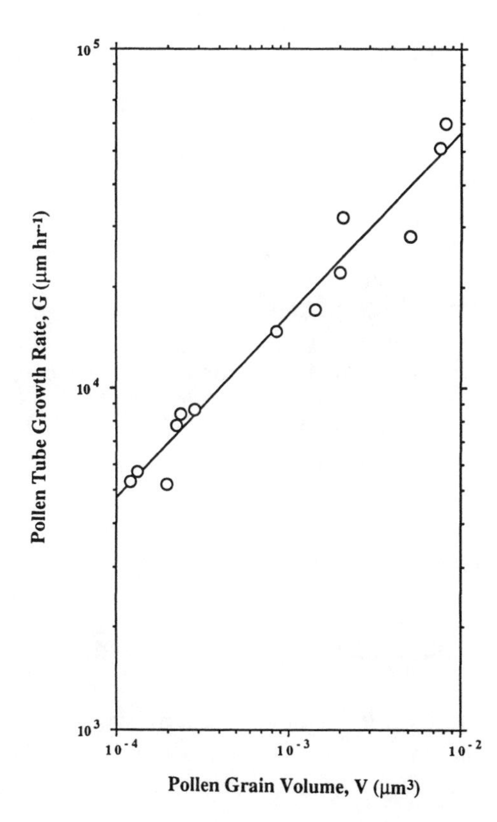

Figure 5.21. Log-log (base 10) plot of average pollen tube growth rate *G* against average pollen grain volume *V* for angiosperm species. Data from Fehr and Hadley 1980 and references therein.

viduals are capable of assuming the same size, morphology, and anatomy, thereby attesting to the receptivity of plant morphogenesis and development to environmental factors attending growth and also the intrinsic capacity of both generations in the life history to express nearly identical morphology and anatomy. For example, pollen grains can be cultured to produce massive multicellular haploid plants that look exactly like their conspecific sporophytes, even to the point of producing flowers. Figure 5.22 illustrates this for the common field mustard, *Brassica campestris.* Similar conditions (apospory and apogamy) occur naturally among bryophytes and pteridophytes, while "sporophytes" derived from haploid megagametophytic tissues of the gymnosperm *Ephedra foliata* have been cultured in the laboratory (Konar and Singh 1979). Thus the gametophytes and sporophytes composing many plant species manifest similar morphogenetic potentials, either naturally or under laboratory conditions. Nonetheless, throughout the plant kingdom a pervasive tendency exists for heteromorphic life histories (see fig. 5.20). That the sizes and shapes of the gametophytic and sporophytic phases of the same species typically differ attests to the different environmental factors that attend the growth and development of the gametophyte and sporophyte. When grown and developed within the anther sac, the microgametophytes of *Brassica,* which have an intrinsic capacity to mimic the sporophyte in size and form, assume the characteristics of normal pollen grains. It is reasonable to assume that the physical and metabolic microenvironment within the anther locule influences morphogenesis and histogenesis. Likewise, the procession of normal embryogenesis is the consequence of the physical and metabolic microenvironment supplied by the ovule. When removed from this "habitat," however, the developing zygote attains morphological and histological features uncharacteristic of normal development. Nonetheless, these features are reproducible for each set of environmental conditions.

How much physical and metabolic influence the sporophyte exerts on the development of the gametophyte and how much the gametophyte physiochemically influences the development of the sporophyte have undergone a number of phyletic modifications that have profoundly affected plant life histories. Among these modifications a few are particularly noteworthy: those attending the transition from an aquatic to a terrestrial life history, the innovation of the archegonium, and the evolution of heterospory and the seed habit. For simplicity, I shall consider these modifications as if they occurred as a series of evolutionary events.

The colonization of land by plants was a major adaptive feat requiring dramatic departures from the constellation of biological features defining the ancestral aquatic plant condition. Nonetheless, these departures had

Figure 5.22. Diploid and haploid plants of common field mustard, *Brassica campestris* (shown to the left and right, respectively).

to be compatible with the physiological requirements for growth. For all plants these requirements are radiant light energy, inorganic carbon in the form of carbon dioxide or bicarbonate ions, mineral nutrients, and water. In this sense all plants require the same resources regardless of their physical habitat. In the aquatic ecosystem, water obviously is not a limiting factor, nor typically are inorganic nutrients, since these occur in large supply in their dissolved states. For these reasons aquatic plants have a physiological advantage over their terrestrial counterparts. By contrast, from Bouguer's law we know that any optically translucent medium attenuates the intensity of light as well as altering the spectrum of radiant energy available for photosynthesis. Accordingly, the maximum light intensity and nearest spectral equivalent to sunlight in any column of water occur just below the water/air interface. Based purely on the proportional relation between the rate of photosynthesis and growth, small aquatic plants occupying the upper few centimeters of the water column or plants in the uppermost interstitial spaces in wet soil are expected to grow faster, on the average, than their more deeply submerged counterparts. From Fick's law, occupying the upper realms of a column of water or the uppermost strata of waterlogged soil is also metabolically beneficial because in general terms the flux densities of carbon dioxide and oxygen are inversely proportional to the distance between the plant and the atmosphere (see chaps. 1 and 2). Thus, living just below the water/air interface rather than deep within the water column confers considerable metabolic advantages regardless of a plant's phyletic affiliations. Provided water can be conserved, however, the physiological requirements of all plant species are better met in the air than in water because light and carbon dioxide are more available. For these reasons it is likely that the colonization of land by plants began within biologically congested communities lying just beneath the surface of freestanding bodies of fresh water, in the uppermost layers of water-saturated soils, and so on. From previous scaling relations we can infer that the growth rates of these plants would have been high because of the availability of light and atmospheric gases in these microhabitats. Also, because of the physical nature of their habitats, plants occupying shallow columns of fresh water and wet soils must be small. Consequently the surface area relative to volume of plants living in these microhabitats must have been comparatively large. Since large ratios of surface area to volume aid passive diffusion and therefore photosynthesis, once again we may speculate that the growth rates of these small plants were comparatively high.

I further hypothesize that the mutation rates of smaller and more rapidly growing plants are proportionally higher than those of larger and slower-growing plants. If true, then the paleocommunities occupied by the algae from which the land plants ultimately arose were characterized by the capacity for comparatively rapid morphological and physiological innovation. In terms of selection pressures favoring the transition from water to land, the microhabitats these plants occupied likely underwent periodic changes in water level, exposing individuals to the air. Such changes could have provided the physical mechanism selecting individuals deviating from the modal forms of their species in capacity to deal with water stress. Individuals incapable of surviving brief exposure to air would have been eliminated, leaving behind those better able to cope with limited or no direct access to water. Over evolutionary time, species with modifications permitting survival and growth during longer periods out of fresh water would have made their appearance.

The appearance of the first bona fide land plants came with the ability to successfully reproduce in the terrestrial landscape. Perhaps the most influential modification in plant life history relates to the evolutionary appearance of the archegonium, the definitive feature of the embryophytes. The archegoniate condition is characterized by oogamy, plasmodesmatal connections between the egg and adjoining gametophytic cells, and the retention of the zygote and the developing sporophyte embryo within gametophytic tissues. These three features have adaptive value for an aquatic organism living in shallow aquatic habitats or relying on liquid water for fertilization in a terrestrial landscape. The retention of nonflagellated eggs that contain a disproportionately large metabolic investment in the next generation within a larger multicellular gametophyte reduces the probability that the most expensive volume fraction of the gamete population will be lost during a brief period of water deprivation. Likewise, the developing sporophytic embryo is supplied with metabolites, kept hydrated, and physically protected by the archegonium as it develops. From the point of view of the evolution of plant life history, the advent of the archegonium placed aspects of early sporophyte development under the direct physical and metabolic influence of the gametophytic generation. For example, the thickness of the cell wall surrounding the egg is controlled by the gametophytic generation and may have been an important physical influence on the initial development of the sporophyte. In the charophycean algae, the cell wall around eggs is comparatively thick and elastic. Mechanically, therefore, these walls can inhibit cell expansion. An additional feature is that the volume of the zygote protoplast is dictated by the gametophyte and does not differ from that of the unfertilized egg. The bulk of metabolites that the zygote invests in the gametophyte are sequestered in the egg well before fertilization occurs. By comparison, the cell wall around the eggs of embryophytes is highly plastic and can accommodate the expansion of the egg and the zygote.

The physical influence of the archegonium on the developing sporophyte also cannot be neglected. For example, the mechanical and perhaps chemical influence of the calyptra, a structure derived from the archegonium and therefore from the gametophyte, is known to influence the ontogeny and development of the moss sporophyte. Premature removal of the calyptra results in abnormal development of moss sporophytes (French and Paolillo 1976).

Historically, the free-living gametophytes of embryophytes have evolved little or no ability to limit how quickly their tissues lose water to the external atmosphere. In most cases these plants survive on land by hugging their substrate, operating physiologically within the confines of the boundary layer produced by the plant and its substrate (see fig. 3.2). This microenvironment is conducive to syngamy, since flagellated sperm cells require liquid water for their survival and transport from antheridia to archegonia. In this sense the reproductive function and physical architecture of the free-living gametophytes of mosses and liverworts are compatible. By contrast, in terrestrial habitats the sporophytic phase fulfills its reproductive role outside the boundary layer created by the substrate. It must aerially disperse spores whose short-term survival after release depends neither on the water status of the parental gametophyte nor on the immediate physical microenvironment. The homoiohydric nature of the terrestrial vascular sporophyte is a logical evolutionary derivative of a reproductive role that requires a reversal of the use of the boundary layer from that of its gametophytic phase. That is, free-living gametophytes essentially reproduce in the semiaquatic microenvironment generated by their boundary layers, whereas sporophytes fulfill their reproductive role in comparatively rapidly moving air currents. The higher spores are released above the boundary layer around the gametophyte, the greater their potential for long-distance dispersal. The selective advantages of this in decreasing local population density and colonizing new habitats are obvious. The probability of intraspecific competition decreases, whereas the probability of outcrossing increases as the population density of gametophytes decreases (see chap. 4). But the taller a plant becomes, the more it extends above the protective influence of the boundary layer beneath it and the more it may lose water vapor to a moving and potentially dry atmosphere (see fig. 3.3). The morphological and anatomical adaptations that characterize the sporophytes of embryophytes (stomata, intercellular air spaces, low-resistance pathways for the flow of water) and the way this type of individual treats water deprivation are all explicable in terms of adaptive evolutionary changes in the context of the reproductive role of the sporophytic phase in the land plant life cycle.

Free-living gametophytes grow and develop in a manner largely removed from the direct physical and metabolic influence of the sporophytic generation. This isolation from the other phase in the life cycle in general affords gametophytes flexibility in sexual expression and the timing of gametogenesis. This is particularly true for homosporous plants, whose gametophytes fix carbon and acquire nutrients and water from their external environment. The condition of sexual flexibility perhaps is best reflected in the gametophytes of the horsetail, *Equisetum*. The meiospores of this plant are morphologically homosporous (Duckett 1970a). Gametophytes, however, either produce only antheridia (*male* gametophytes) or produce antheridia and then, at larger sizes, archegonia (*female* gametophytes; see Duckett 1970b). In a lengthy series of experiments, Duckett has shown that no inherent physiological differences exist between spores that produce male and female gametophytes. Both male and female gametophytes can be regenerated from isolated portions of either male or female gametophytes (Duckett 1977). All the available evidence suggests that the development of *Equisetum* gametophytes and subsequent sexual expression are mediated by environmental conditions. The development of male or female gametophytes depends on a series of threshold environmental conditions. Indeed, the ratio of male and female gametophytes from regenerated portions of either a male or a female gametophyte can be altered by changing the environmental conditions in which regenerated gametophytes are grown. The male condition results from growth under environmentally stressful conditions.

Flexibility in the sexual expression of gametophytes, however, can be indirectly influenced by the sporophyte. Gametophytic gender expression is influenced by the scaling of growth, which in turn is influenced by the size of the spore from which the gametophyte develops. In a variety of homosporous pteridophytes, principally ferns, the absolute growth rates of large spores are higher than those of smaller spores, although smaller spores grow faster for their size. Larger spores develop into plants that produce archegonia. By contrast, smaller spores develop into plants that produce antheridia (reviewed by Bell 1989). While still under direct gametophytic influence, therefore, the timing of gamete production and the ratio of male and female gametophytes within a population is indirectly under the control of the sporophytic generation because of the way protoplasm and nutrients are partitioned among developing sporocytes within the sporangia of the sporophytic individual. An extension of this phenomenology can easily be seen to yield the heterosporous life history. Precocious development and confinement of the gametophyte largely within the spore wall transfers the control of gender expression from the gametophyte to the sporophyte. The adaptive value of heterospory and free-living gametophytes mirrors the adaptive value of the archegoniate condition.

The egg-producing gametophyte is invested with proportionately larger metabolic reserves and provides physical protection supplemented by the spore wall, which is in part produced by the parental sporophytic generation.

A review of the life histories of embryophytes (see fig. 5.20) shows that the next major phyletic innovations were the thinning and essential loss of the megaspore wall around the megagametophyte and the appearance of pollination mechanisms. The former essentially "invaginates" the archegoniate condition—the developing egg-bearing gametophyte is enveloped by the sporophyte phase in the plant's life history. In these circumstances the delivery of sperm-bearing gametophytes to the sites where eggs are produced becomes essential for syngamy. Although the evolution of pollination is complex, one basic feature of interest is that the growth of megagametophytes within sporophytic tissues and the transport of microgametophytes by either abiotic or biotic means are assisted in both instances by producing comparatively small male gametophytes. And as we have been, small gametophytes grow more rapidly than larger ones. Thus, one of the benefits of this life history is a comparatively rapid completion of its sexual phase. It is somewhat ironic that the dominance of the gametophytic over the sporophytic phase was reversed during the procession of plant history modifications. A logical extension of this procession suggests that the somatic cells of both gametophytes in the seed plant habit will be lost, thereby yielding a life history occupied by a single multicellular phase, the sporophyte, that essentially reproduces like a mammal.

5.7 Summary

Three propositions were reviewed in the context of plant phyletic trends and heterochrony: (1) that a change in intraspecific scaling reflects a modification in ontogeny and development; (2) that each scaling relation is functional only within a narrow domain of size and form; and (3) that changes in scaling relations prefigure or attend the appearance of new plant groups. The first of these propositions was shown to require information on the scaling of ontogeny (heterauxesis), the scaling of static comparisons among conspecific adults drawn from the same population (intraspecific allomorphosis), and the scaling of adults drawn from different taxa (interspecific, intergeneric, etc., allomorphosis). Examples of functionally useful scaling relations were given and revealed a tendency for a negative correlation between the scaling exponent and either the size of the individual or the average size of the taxon in which the individual is nested. These negative correlations were shown to result from adjustments in scaling parameters that permit the occupancy of different domains of size and organic form. These adjustments can result from changes in either the tempo or the pattern of ontogenetic events. The term heterochrony applies to either of these types of ontogenetic modification. The terminology of heterochrony was shown to be inadequate in terms of mechanisms that can alter plant ontogeny and development. Also, the way plants grow results in ambiguity regarding how this terminology is applied.

Variations in the plant life cycle were reviewed in terms of the capacity to express multicellularity. Each plant species was shown to consist of two types of organisms (sporophyte and gametophyte), each characterized by its own distinctive ontogeny, development, maturation, and reproductive function. Thus the life cycle of a species affords two occasions for ontogenetic modification. Each has the potential to yield a new type of organism in terms of life history. The life cycle of each species typically involves a comparatively small and reproductively less redundant but fast-growing individual that is mirrored by a comparatively larger and reproductively more redundant but slower-growing one. The absolute rate of the life cycle depends on the absolute sizes as well as the relative sizes of both individuals. Environments that are characteristically disturbed favor small and fast-growing sporophytes that produce even smaller and faster-growing gametophytes. In undisturbed environments, big and slow-growing sporophytes that produce comparatively large and laggard gametophytes are permitted. Despite perhaps historically random changes in the stability of terrestrial ecosystems, the intertaxonomic procession of land plant gametophytes evinces a phyletic reduction in size that was interpreted in terms of a precocious onset of sexual maturity in progressively less multicellular and therefore more vegetatively juvenile plants. This trend coincides with a reduction in the expression of the multicellular condition associated with indeterminate growth in size and a general loss in reproductive redundancy. The evolutionary procession of sporophytes reveals a phyletic gain in size relative to the size of conspecific gametophytes. The size and form of descendant sporophytes extend beyond those of their ancestors. The tendencies were discussed in the context of changes in plant life histories and the scaling of vegetative and reproductive features.

Epilogue

This book has had two principal aims. The first has been to present the myriad types of biological phenomena that correlate with and therefore operate within the reference system of absolute as well as relative size. The second has been to expose the practical as well as philosophical difficulties that give us pause whenever we believe the behavior of living things can be reduced to numerals and mathematical symbols. In retrospect, the juxtaposition of these two aims was necessary and desirable because they superficially appear antagonistic to one another. Although Darwin paid homage to the biological correlations characterizing each living thing and, while treating the topic of morphology, explicitly noted that selection on one organic feature "pulls" on every other feature, much like someone pulling a thread in a complex tapestry, he paradoxically stressed the capacity for equivalent variation in every direction under the blind influence of selection. By hindsight, Darwin's view that variation can occur in any direction to equal degree is subject to well-reasoned criticism that curiously evaded (or at least was minimized by) one of biology's most astute observers. The correlations evident among the parts of each type of organism and the morphogenetic observance of restricted combinatorial variations of features among organisms flout the notion of equiprobable and equivalent variation for every biological variable and the corollary that all permutations of organic form are equiprobable and independent. Indeed, one of the major accomplishments of the modern synthesis, which sought to wed developmental and historical biology to genetics, was the recognition that the expression of organic form is constrained by developmental as well as genetic barriers and that the procession of superficially different morphologies in ontogeny and the neighboring morphologies in the adaptive walks of evolution are the products of *permissible* transformations within a family of forms produced by shared underlying developmental and genetic mechanisms. Upon reflection, one of the major conclusions of this book is that these mechanisms and the "push/pull" they engender, whose physical manifestations are the phenotypes we study, are undeniably responsive to physical as well as temporal scale.

Another conclusion is that the scale of dimension, like that of time, is a legitimate as well as convenient way to order ontogenetic or evolutionary transformations of form and process. From the molecular traffic within each living cell to the reappearance of numerous shapes and structures that characterize the complex river of organic evolution, size-dependent variations show that the transformations in external shape, internal structure, and chemical process of remarkably phyletically and ecologically dissimilar organisms repeatedly assume the same mathematical syntax. Curiously, the heritage of Charles Darwin often compels us to place these transformations within the vehicle of natural selection, whose operation is made manifest by *change* and whose inferred absence in the face of *no change* is often taken to imply simply heredity continuity. This inference is needless and in many instances demonstrably in error. The absence of overt change does not logically argue for the absence of natural selection. Nonetheless, a single "allomorphic rule" capable of mathematically ordering size-correlated alterations in organic form often is taken to imply the subservience of selection to, for lack of better phrase, *hereditary inertia*. Yet the equally impressive intellectual inheritance of D'Arcy Thompson's elegant and rightly famous book *On Growth and Form* shows us that biological order in an otherwise overtly chaotic inorganic world often reflects the influence of overriding physical and chemical "rules" that propel changes in organic form and process during the stately procession of ontogenetic changes in size and the seemingly unpredictable cortege of phylogeny. In this book I have implicitly argued, and here I explicitly state my view, that physiochemical "rules" contain no small portion of the algorithms for transformations in form and process relating different morphologies as members of a series permitted to change into one another. This view is a logical extension of Thompson's pivotal argument—that organisms make use of physical mechanisms whose operation extends the organism's raw genetic capacity to coordinate changes in form and structure.

Like many before me, however, I depart from Thompson's worldview of biology, in which the role of natural selection is overtly ignored. His reasons for this approach are idiosyncratic and often submerged in brilliant but to my mind deceptive metaphor. The principal reason Thompson eschews natural selection appears to revolve around the argument that, since physiochemical principles are immutable, natural selection need play no important subsequent role once organisms adapt to the workings of these principles. Clearly, organisms can rely on the immutability of physical and chemical processes, but the history of life demonstrates that living things do not repose exclusively on the principles of physics and chemistry and that natural selection must "work as hard" to maintain the absence of overt change as to create change.

In light of the complexity of organisms, scaling analysis appears a simple, often naive tool, contingent on fragile assumptions, flawed by the appearance of great precision despite the likelihood of inaccurate measurements of time and space, and confronted by a bewildering panoply of biological variables. Our understanding of size-correlated phenomena cannot be brought to satisfactory closure until we resolve the practical difficulties of amassing the requisite data for broad interspecific comparisons, determining which among the seemingly innumerable biological variables are truly relevant, discovering the relation between ontogenetic and phylogenetic transformations, finding the most reliable way to formulate and test phyletic hypothesis, and establishing the statistical foundations on which scaling relations are quantified. Rather than establishing grounds for concern, to my mind, these and other practical and theoretical difficulties afford a keen challenge, one that undoubtedly will be met by each of the biological disciplines animated to understand living things, regardless of the level of biological organization emphasized and in spite of the often disparate reasons that compel them to devote energy and time to studying organic form and process. The real task, however, is to establish and maintain communication among disciplines that profess superficially different biological Weltanschauungs. Given the diversity of size-correlated phenomena evinced by different hierarchical levels of biological organization, perhaps it is inevitable that some will view the question, What are the consequences of size on organic form and process? as presumptuous. But I think the more perceptive will see this staggering diversity as a *Luftspiegelung*—a mirage that has intellectually distorted our perception of biology as a whole because we have conveniently dissected organisms into their more easily comprehensible, albeit inanimate, parts. Unlike a mechanical device, a living thing is functional because each of its parts exists both for and as a consequence of all the others, just as each organism is the means and the end of evolutionary change. Within my *Weltanschauung* of biology lies an analogue of the Newtonian ideal of physics, the drive to account for apparently dissimilar phenomena by means of a comparatively few underlying principles. If it is true that organisms differ largely because of combinatorial variations of few features rather than as a result of an unlimited capacity to endlessly add new features to those that came before, then the diversity of life is the consequence of permutation, not simple addition. In biology as in physics, not all possible permutations are equiprobable; some are more likely to occur and remain stable than others. If so, then the familiar attempt to understand the underlying mechanisms that faithfully cause ontogeny to unroll before us must be recast to seek those patterns and forms that have been most easily "found" by organisms over the course of their evolutionary excursion through vastly different domains of size, shape, and structure. I believe, therefore, that the quest for "mechanism" can be sped on its way by means of scaling analysis.

Appendix: Methods

A.1 Introduction

This appendix is devoted to some of the methods used to determine and interpret scaling relations. Some of these methods are illustrated by employing data sets presented in previous chapters. With the advent of handheld calculators and personal computers, many of the computations illustrated by the manipulation of data need not be done manually. For pedagogical reasons, however, these manipulations demonstrate that regression analysis, correlation analysis, and analysis of variance, the principal topics dealt with here, are comparatively easy. Nonetheless, I have made no attempt to be inclusive either in analyzing data sets or, more important, in presenting all aspects of each method discussed. The techniques are dealt with only so far as they directly relate to scaling analysis, and even then not in a complete manner owing to limitations on space. Those requiring additional information should consult appropriate statistics textbooks or the citations provided in the following sections.

A.2 The Functional Bivariate Relation (Y and X)

Scaling relations are empirically determined by establishing the form and significance of the functional relation between two or more biological variables. The phrase functional relation means the *mathematical* function that lets us predict the value of one variable that corresponds to the specified value of another variable. This definition is important to bear in mind because functional relation has been used to mean a "cause and effect relation" (sensu "organic correlation" of Kermack and Haldane 1950).

A mathematical function between two variables, one of which is an independent variable, takes the general form $Y = f(X)$, where Y is the dependent variable, X is the independent variable, and f denotes the mathematical operations required to establish the equality between the two sides of the equation. The form of the functional relation between Y and X is established by means of regression analysis, which estimates the identity of f. The statistical significance of the functional relation is determined by correlation analysis. Regression analysis and correlation analysis are not the same thing, although both are used in conjunction to examine and test scaling relations and hypotheses.

The simplest functional relation between Y and X encountered in the scaling literature is given by the formula $Y = a + bX$, where a is the Y-intercept ($Y = a$, when $X = 0$) and b is the slope (the first derivative of Y with respect to X; $b = dY/dX$). In statistics, b is called the *regression coefficient*, and the function $Y = a + bX$ is called the *simple linear regression equation*. This equation establishes the "best line of fit" between Y and X variates. Since values of Y are regressed on values of X to obtain the equation $Y = a + bX$, statisticians often denote the regression coefficient as $b_{Y,X}$ to distinguish it from the case when X is regressed (legitimately) on Y (i.e., $b_{X,Y}$). This protocol is required when dealing with formal mathematical statements but is often dispensed with when the relation between the variables Y and X is made explicit with graphs and such. Difficulties sometimes arise, however, because statisticians, mathematicians, and allometrists frequently employ different notation. In mathematics, the equation for a linear function is sometimes written as $Y = b + mX$, whereas in scaling analysis the linear equation often is written as $Y = \beta + \alpha X$. This latter convention provides parity with the more frequently encountered power function equation $Y = \beta X^{\alpha}$, which obtains a linear form with log-transformation, $\log Y = \log \beta + \alpha \log X$. Unfortunately, both $Y = \alpha + \beta X$ and $Y = \alpha X^{\beta}$ are encountered in the scaling literature. In scaling analysis, regardless of the symbols used for the Y-intercept and the slope, the slope of the simple linear and power function equation is called the *scaling exponent* and the Y-intercept is called the *scaling coefficient*. Throughout this book, α denotes the scaling exponent and β represents the scaling coefficient.

Regression analysis is required to establish the form of the functional relation between Y and X, because in virtually every real case, values of Y and X do not lie perfectly along the best line of fit (the regression line). Deviation of observed values from predicted values of Y for a given value of X results from inherent natural variation in and measurement errors of the biological variable Y. The estimate of the functional relation between Y and X, therefore, does not imply that the numerical value of Y for a specified value of X precisely equals $\beta + \alpha X$ (or, in the case of the power function equation, βX^{α}). Rather, regression analysis estimates the functional relation between Y and X by establishing an expected mean for the values of Y for a specified value of X.

As noted, Y and X are called the "independent" and "dependent" variables, respectively. These terms have very precise meanings because of the assumptions made by Model I regression analysis. In Model I regression analysis, the functional (mathematical) relation between Y and X is established by minimizing the sum of the squared deviations of Y values about the regression line. For this reason, Model I regression is often called ordinary "least squares" regression analysis, henceforth denoted as LS regression. By contrast, Model II regression assumes that *Y* and *X* have error measurements. Consequently *X* is not an "independent" variable. In Model II regression this is often emphasized by abandoning the notation *Y* and *X* in favor of Y_1 and Y_2, where Y_1 is the variable plotted on the ordinate and Y_2 is the variable plotted on the abscissa of a bivariate plot. Sections A.3 and A.4 review the assumptions and methods of Model I and Model II regression analysis. For the time, it is important to know that, although Model I regression underestimates the scaling exponent, this technique has predictive value. It is the only regression technique that generates *Y*-deviation scores that are completely uncorrelated with the *X* variate. LS regression, therefore, is an appropriate technique when the objective is to estimate the mean value of *Y* for a specified *X* or to compare the standard dispersion statistics obtained from two or more samples of data. For this reason LS regression analysis is considered in some detail first. Model II regression analysis controls in part (but not fully) for variations in Y_2 and therefore typically establishes a more reliable estimate of the scaling exponent, which often is the primary object of concern in scaling analysis. For this reason, Model II regression analysis is given somewhat greater emphasis.

A.3 Model I Regression Analysis (*Y* versus *X*)

The most frequently used regression method is called Model I regression (LS regression), which makes four assumptions typically obtained by experimental data:

1. The values of the *X* variate do not randomly vary (*X* is measured without error—values are specified by the investigator).
2. The expected relation between *Y* and *X* is linear (the functional relation takes the form $Y = \beta + \alpha X$).
3. Values of *Y* for any specified value of *X* are independently and normally distributed ($Y_i = \beta + \alpha X_i + \varepsilon_i$, where *i* denotes a particular value of *X* and ε is the random deviation of the error term, which is assumed to be normally distributed with a mean value equal to zero).
4. Samples of *Y* along the regression line have a common variance σ^2 that is the variance of ε_i (the variance is independent of the magnitude of *Y* or *X*).

These four assumptions make explicit what is meant by the "independent" and "dependent" variables. LS regression assumes that the numerical values of *X* are "fixed" by the investigator. That is, each X_i is "unique" such that it has no frequency distribution, and therefore each X_i varies independently. For each value of X_i, however, values of Y_i are not fixed by the investigator but rather are assumed to vary randomly such that they have a normal distribution about X_i. Note that, although only one value of Y may be experimentally obtained for X_i, Model I regression analysis assumes that this Y_i is a sample size of one ($N = 1$) drawn at random from a population of independently and normally distributed variates.

The reason Model I regression is called "least squares" regression becomes evident from the menu required to determine the form and significance of the LS regression relation:

(1) the sum of X values, ΣX,
(2) the sum of squares of X values, ΣX^2,
(3) the sum of Y values, ΣY,
(4) the sum of squares of Y values, ΣY^2,
(5) the sum of products of X and Y values, ΣXY,
(6) the mean of X, $\overline{X} = \Sigma X/N$,
(7) the mean of Y, $\overline{Y}$,
(8) the sum of squares of Y, $\Sigma y^2 = \Sigma Y^2 - (\Sigma Y)^2/N$,
(9) the sum of squares of X, $\Sigma x^2 = \Sigma X^2 - (\Sigma X)^2/N$, and
(10) the sum of products of X and Y, $\Sigma xy = \Sigma XY - (\Sigma X)(\Sigma Y)/N$.

From this menu, we compute the scaling exponent and the scaling coefficient of the LS regression curve

(A.1) $$\alpha_{LS} = \frac{\sum xy}{\sum x^2} \text{ and}$$

(A.2) $$\beta_{LS} = \overline{Y} - \alpha_{LS}\overline{X},$$

from which we can determine the LS regression line, $Y = \beta_{LS} + \alpha_{LS}X$, whose significance is given by the correlation coefficient r_{YX}:

(A.3) $$r_{YX} = \frac{\sum xy}{\left[\left(\sum x^2\right)\left(\sum y^2\right)\right]^{1/2}}.$$

Confidence intervals for the parameters in the LS regression line are determined by the following formula:

(A.4a) $$L_1 = [P - t_{\alpha(df)}SE_P]$$

(A.4b) $$L_2 = [P - t_{\alpha(df)}SE_P],$$

where P denotes the parameter, SE is the standard error of P, and $t_{\alpha(df)}$ is the critical value of the Student's distribution for n degrees of freedom df (e.g., for 95% confidence limits, $t_{\alpha(df)} = t_{0.05(df)}$).

In scaling analysis, attention must be given to what is actually meant by "mean" Y and "mean" X. The best line of fit of an intraspecific LS regression passes through the mean values of Y and X, $\overline{Y}$ and $\overline{X}$ (see eq. A.2). As noted in the LS regression menu,

$$\overline{Y} = \frac{\sum Y}{N} \text{ and } \overline{X} = \frac{\sum X}{N}, \tag{A.5}$$

where N is the sample size—the number of individuals drawn from the population. A basic assumption of any regression analysis is that each datum represents an independent observation on the smallest sampling unit. When dealing with interspecific scaling relations, therefore, N denotes the number of species sampled, not the number of individuals. This becomes clear once it is recognized that the characterization of each species requires numerous measurements taken from conspecifics differing in size and that the means of the measurements of Y are regressed against the means of X from each species (rather than all the individual measurements) to yield the interspecific regression formula. The latter is inappropriate, since it results in data point inflation (the representation of a "single" observation by two or more data points). Species means are computed from eq. (A.5), where N now denotes the number of measurements of Y and X for each species. It is sometimes necessary to calculate weighted means for Y and X for each species and use these weighted means (denoted by the subscript w) as the variates in regression analysis. This is sometimes necessary because the frequency distributions of Y and X from each species often deviate significantly from a normal distribution. The appropriate formulas for the weighted means of Y and X obtained from each species are as follows:

$$\overline{Y}_w = \frac{\sum^N (w_i Y_i)}{\sum^N w_i} \text{ and } \overline{X}_w = \frac{\sum^N (w_i X_i)}{\sum^N w_i}, \tag{A.6a}$$

where w_i is the weighting factor whose numerical value depends on the extent to which the assumption of normal distribution is violated. These formulas militate against data point inflation in interspecific comparisons. It is evident that they yield unweighted means for each species when N is substituted for $\sum^N w_i$, where N now is the number of individuals sampled per species ($w_i = 1$). Note that interspecific regression curve passes through

$$\overline{Y} = \frac{\sum \overline{Y}_w}{N} \text{ and } \overline{X} = \frac{\sum \overline{X}_w}{N}, \tag{A.6b}$$

where N is the number of species.

Often LS regression obtains nearly equivalent coefficients of correlation, r_{YX}, for untransformed and log-log transformed data. The investigator therefore must elect $Y = \beta + \alpha X$ or $\log Y = \log \beta + \alpha \log X$ as the appropriate description of the functional relation. A useful "rule of thumb" is to elect $\log Y = \log \beta + \alpha \log X$ over $Y = \beta + \alpha X$ as the true functional relation whenever the range of Y and the range of X span two or more orders of magnitude and the correlation coefficient r_{YX} of the log-log-transformed data equals or exceeds that obtained by LS regression of the untransformed data. In the scaling literature, log-log transformation is frequently encountered because it is assumed that the biological relation between Y and X conforms to Huxley's "allometric law" ($Y = \beta X^{\alpha}$; see Huxley 1932). Log-log transformation also may be used to reduce the statistical problems resulting from a number of outliers, to fulfill the LS assumption that data have a normal frequency distribution and homoscedasticity, or because log transformation simply is the easiest way to explain proportionality.[1] Smith (1980) cautions against the uncritical use of log-transformed data. Harvey (1982) notes that the functional relation of intraspecific data generally is well estimated by regression of untransformed rather than transformed data, whereas interspecific data, which tend to be lognormally distributed, are best treated after log-log transformation (r_{YX} increases after log-log transformation of interspecific data). Clearly, although the range of size evinced by an interspecific comparison typically is larger than for an intraspecific comparison, there is no a priori reason to assume that an interspecific relation between Y and X is nonlinear, and the assumption that interspecific data have a lognormal distribution should always be tested.

Logarithmic transformation to any base is acceptable, although the base of the logarithmic transformation affects the numerical values of standard errors that are used to calculate confidence limits and a correction factor is required to deal with systematic biases introduced when data are log transformed (see eq. A.11). Common (base 10) logarithms tend to be the simplest to deal with, are most frequently encountered in the scaling literature, and are used throughout this book, with one or two exceptions clearly noted. When reporting the results from log-log-transformed data, the scaling coefficient and scaling exponent should be retransformed into the linear scale (e.g., β = antilog [$\log \beta$]). In the case

1. Note that the scaling exponent of the least squares log-log regression line

$$\alpha_{LS} = \frac{\log Y_2 - \log Y_1}{\log X_2 - \log X_1}$$

becomes $\alpha_{LS} = (Y_2/Y_1)/(X_2/X_1)$ when converted to a linear scale.

of $\overline{Y}$ and $\overline{X}$, a statement regarding the standard error of $\overline{Y}$ or $\overline{X}$ has no value, and the confidence limits L_1 and L_2 for these as well as other statistics must be computed in the logarithmic scale before they are retransformed into the linear scale. The confidence limits for mean Y or mean X may not be symmetric around the mean. The formula for the confidence intervals for any regression parameter P derived from log-log-transformed data is as follows:

$$L_1 = \text{antilog}[\log P - t_{\alpha(df)}SE_P] \tag{A.7a}$$

$$L_2 = \text{antilog}[\log P - t_{\alpha(df)}SE_P], \tag{A.7b}$$

where, once again SE is the standard error of P and $t_{\alpha(df)}$ is the critical value of the Student's distribution for n degrees of freedom df (e.g., for 95% confidence limits, $t_{\alpha(df)} = t_{0.05(df)}$).

It is important to report α, β, $\overline{Y}$, $\overline{X}$, etc., and their SE to a sufficient number of decimal places. Sokal and Rohlf (1981, 151) give the following rule of thumb: Find the decimal place of the first nonzero digit of the quotient $SE/3$ and report the statistic parameter to that decimal place and report SE to one further decimal place. For example, if $\alpha_{LS} = 2.53$ and $SE = \pm\ 0.396$, then $SE/3 = 0.396/3 = 0.132$, and therefore $\alpha_{LS} \pm SE$ becomes 2.5 ± 0.40. When applied to the scaling exponent obtained from an interspecific comparison for which the range of Y and the range of X are large, however, this rounding protocol has a significance consequence on predicting Y for a specified value of X. The maximum percentage deviation, % dev_{max}, between a predicted and an observed Y value depends on the difference in the scaling exponent $\Delta\alpha$ and the logarithm of the ratio of the maximum to the minimum "size," X_{max} and X_{max} (Prothero 1986, 262):

$$\%\ \text{dev}_{max} = 100\left(10^{(\Delta\alpha)\log\frac{X_{max}}{X_{min}}} - 1\right). \tag{A.8}$$

Consider that, for the rounding rule of thumb, $\Delta\alpha = 0.03$. Assuming the range in size spans ten decades—that is, $\log (X_{max}/X_{max}) = 10$—we see that rounding α_{LS} from 2.53 to 2.5 results in % $\text{dev}_{max} \approx 100\%$. For a size range of five decades, % $\text{dev}_{max} \approx 40\%$. Even conservative rounding of scaling exponents, therefore, can result in dramatic (percentage) deviations between predicted and observed Y values. Also, the rounding of an empirically determined scaling exponent frequently gives the illusion of a good fit with a scaling exponent predicted by a scaling hypothesis. While on the subject of scaling exponents and their standard errors, I should mention the "scaling cancellation" method of Stahl (1967), where the ratio of two power function equations is taken to determine the extent to which the ratio of the two dependent variables, Y_1 and Y_2, is size independent. Our concern here is with the standard error of the residual scaling exponent, $\alpha_R = \alpha_1 - \alpha_2$. Consider the two scaling relations $Y_1 = \beta_1 X^{\alpha_1}$ and $Y_2 = \beta_2 X^{\alpha_2}$, which obtain the ratio $Y_1/Y_2 = kX^{\alpha R}$: that is,

(A.9)
$$\frac{Y_1}{Y_2} = \frac{\beta_1}{\beta_2}\frac{X^{\alpha_1}}{X^{\alpha_2}} = \frac{\beta_1}{\beta_2} X^{\alpha_1 - \alpha_2} = kX^{\alpha R} .$$

Since the standard error of α_R equals the sum of the standard error of α_1 and the standard error of α_2, the extent to which β_1/β_2 may be considered "a constant" such that Y_1/Y_2 is independent of X (= size) depends on the standard errors of α_1 and α_2 as well as on the numerical value of αR. Provided the standard errors are high, even very small residual scaling exponents (e.g., 0.03) may yield significant size-dependent differences in Y_1/Y_2. It also is very important to note that valid mathematical manipulations of equations typically do not yield statistically valid estimates of scaling relations. Although $Y = X^{\alpha}$ mathematically obtains $Y^{\frac{1}{\alpha}} = X$, the scaling exponent of Y (when X is regressed on Y) is only the inverse of the scaling exponent obtained when Y is regressed on X when Y and X are perfectly correlated ($r_{YX} = r_{XY} = 1.0$).

Confidence in the meaningfulness of an empirically determined scaling exponent clearly depends on its standard error, whose numerical value depends on the sample size N and the deviations of Y values from the LS regression line. Prothero (1986) provides an empirically determined formula (based on computer simulations) that predicts SE based on these two features:

(A.10)
$$SE_{\text{predicted}} = \frac{0.019 \text{ MPD}}{\log\left(\frac{X_{\max}}{X_{\min}}\right) N^{1/2}},$$

where MPD is mean percentage deviation of Y values about the LS regression line. Notice that the predicted SE is inversely proportional to the logarithm of the size range. Prothero's equation, however, is appropriate only for data that evince a very high degree of correlation ($0.95 < r \leq 1.0$). Where they do, it may be used to determine whether the standard error reported for an empirically determined interspecific scaling exponent is consistent with that expected given the sample size and size range of the data set. This is particularly important when the interspecific scaling exponent is claimed to be representative for the scaling relation in general.

Recall that in regression analysis the estimated functional relation between Y and X establishes an expected mean for the values of Y for a specified value of X. A mathematical consequence of dealing with log-transformed Y values is that their *mean* is the *median* of the distribution of untransformed values of Y. Since the mean and median Y are not numerically equivalent, log transformation introduces a bias in the estimate of regression parameters. Although the error estimate introduced by this bias typically is fairly small ($\leq 10\%$), a correction factor CF for this bias, estimated from the standard error SE of log β, should be used (Sprugel 1983):

$$\text{(A.11a)} \qquad SE = 2.303\left[\frac{\sum(\log Y_i - \log Y_P)^2}{df}\right]^{1/2}$$

$$\text{(A.11b)} \qquad SE = \left[\frac{\sum(\ln Y_i - \ln Y_P)^2}{df}\right]^{1/2}$$

$$\text{(A.11c)} \qquad CF = \exp\left[\frac{SE^2}{2}\right],$$

where Y_P are predicted values of the observed values of Y_i. The numerical factor 2.33 in eq. (A.11a) is necessary when Y and X are log transformed to base 10 (common) logarithms because the value of SE depends on the logarithm base used by the investigator and because eq. (A.11c) is geared for base e logarithms.

The width of the confidence limits for the scaling coefficient and scaling exponent plays an important role in determining whether observed scaling relations show a statistically significant difference among data sets. Likewise, L_1 and L_2 are used to test whether observed scaling exponents differ from those predicted by scaling hypotheses. The width of the confidence intervals for this statistic may be reduced by reducing its standard error or by increasing sample size. Since most scaling studies are based on data gathered from natural populations, and since the investigator has no control over the standard error of data, the tactic is to increase the sample size.[2] The relevant question then becomes, How large a sample need be obtained to show that a true difference δ exists between two data sets? An appropriate formula to answer this question is

$$\text{(A.12)} \qquad n \geq 2\left(\frac{SE_T}{\delta}\right)[t_{\alpha(df)} + t_{2(1-P)(df)}]^2,$$

where n is the number of replicates, SE_T is the true standard error, and P is the specified probability that an observed difference δ is significant.

2. In experimental systems, however, Model I regression analysis assumes that the investigator has control over the variance of X and therefore can exercise control, albeit indirect, over the variance of the response variable Y.

Note that the actual values of SE_T and δ need not be known to estimate n. Rather, the ratio of the true standard error and the percentage of difference one desires to detect may be estimated to calculate n.

As noted, the statistical significance of the functional relation between Y and X is established by r_{YX} (see eq. A.3). The methods of correlation analysis are well known and require little further comment here except in the case of the correlation coefficient of what is known as a *part-whole* regression analysis. This type of regression involves regressing data obtained from measurements of a part of an organism Y_1 against the size of the whole organism Y_2. An example of a part-whole regression is the regression of leaf weight against the weight of a whole shoot. Another example is the regression of total seed weight against the weight of a fruit. Part-whole regressions sometimes are considered to yield "illegitimate" results because they typically attain very high correlation coefficients. The correlation coefficient of a part-whole regression is not spurious, however, but rather is a consequence of the way the dependent and independent variables are defined:

$$r_{12} = \frac{s_1 + r_{01}s_0}{(s_0^2 + 2r_{01}s_0s_1 + s_1^2)^{1/2}}, \tag{A.13}$$

where r_{12} is the correlation coefficient of the regression of the part against the whole, s_1 is the standard deviation of the part, s_2 is the standard deviation of the whole, s_0 is the standard deviation of the whole minus the part (i.e., $Y_0 = Y_2 - Y_1$), and r_{01} is the correlation coefficient obtained from the regression of the whole minus the part Y_0 against the part Y_1 (see Sokal and Rohlf 1981, 578). Note that when r_{01} equals zero (when no correlation exists between Y_0 and Y_1), eq. (A.13) becomes $r_{12} = [1 + (s_0^2/s_1^2)]^{-1/2}$ (i.e., the correlation between Y_1 and Y_2 depends only on the standard deviation of Y_0 and the standard deviation of Y_1). Sokal and Rohlf (1981) note that $r_{12} = [1 + 1]^{-1/2} = 1/\sqrt{2} \approx 0.71$ when these two standard deviations equal one another. Thus r_{12} can be very high when these two variables are equally variable and evince no correlation with one another. Clearly, whenever the results of part-whole regression and correlation analyses are presented, all the statistical parameters shown on the right-hand side of eq. (A.13) should be reported.

A.4 Model II Regression Analysis (Y_1 versus Y_2)

The techniques employed by LS regression analysis are rarely appropriate for scaling analysis because "size" is subject to natural variation and measurement error and therefore is not an independent variable. Bacon

(1953) presents the equations for LS regression when there is dispersion in both variables and when both variables have the same units of measurement. Unfortunately, these equations have a limited range of application to scaling analysis. Also, it often is desirable to emphasize that "size" is not fixed by or under the direct control of the investigator, as well as that "size" is not the "cause" of the independent variable. For these reasons, Model II regression analysis often must be employed in scaling analysis even when the two variables are measured in the same units.

Two approaches have been suggested to overcome the bias introduced by measurement errors such that no independent variable exists (i.e., Y_1 and Y_2). These approaches are called major axis (or principal axis) regression and reduced major axis regression. Henceforth I shall denote these approaches as MA regression and RMA regression, respectively. When both Y_1 and Y_2 are in the same units of measurement or when the units of Y_1 and Y_2 differ but are predetermined by the investigator (i.e., nonarbitrary measurement units), the slope of the major axis of the bivariate sample may be the preferred method. The approach is to compute the elliptical confidence distribution of the variates and to use the slope of the major axis of the elliptical confidence distribution as the scaling exponent (the major axis of the elliptical confidence distribution is the principal regression line). As in LS regression, the criteria for establishing the major axis through bivariate Y_1 and Y_2 data are to minimize the sum of squares of the deviations of observed Y values from the major axis and to fit the major axis through $\overline{Y}_1$ and $\overline{Y}_2$. In LS regression the deviations of observed values of Y are *parallel* to the Y-axis of the bivariate plot. In MA regression, the deviations of observed Y_1 values are *perpendicular* to the regression line established by the major axis regression line (i.e., the technique considers the joint deviations of Y_1 and Y_2).

Denoting the variable plotted on the ordinate as Y_1 and the variable plotted on the abscissa as Y_2, the menu for major axis regression is as follows:

(1) the sum of Y_2 values, ΣY_2,
(2) the sum of squares of Y_2 values, ΣY_2^2,
(3) the sum of Y_1 values, ΣY_1,
(4) the sum of squares of Y_1 values, ΣY_1^2,
(5) the sum of products of Y_2 and Y_1 values, $\Sigma Y_2 Y_1$,
(6) the mean of Y_2, $\overline{Y}_2 = \Sigma Y_2/N$,
(7) the mean of Y_1, $\overline{Y}_1$,
(8) the sum of squares of Y_1, $\Sigma y_1^2 = \Sigma Y_1^2 - (\Sigma Y_1)^2/N$,
(9) the sum of squares of Y_2, $\Sigma y_2^2 = \Sigma Y_2^2 - (\Sigma Y_2)^2/N$, and
(10) the sum of products of Y_1 and Y_2, $\Sigma y_2 y_1 = \Sigma Y_2 Y_1 - (\Sigma Y_2)(\Sigma Y_1)/N$.

The MA regression menu is identical to the menu for LS regression up to and including this step. It continues:

(11) the sample variance of Y_1, $s_1^2 = \Sigma y_1^2/(N-1)$,
(12) the sample variance of Y_2, $s_2^2 = \Sigma y_2^2/(N-1)$,
(13) the covariance of Y_1 and Y_2, $s_{12} = \Sigma Y_1 Y_2/(N-1)$.

We then need to compute the slope for the major axis (which is the scaling exponent) and the slope of the minor axis of the elliptical confidence intervals:

(A.14a) $$\alpha_{major} = \frac{s_{12}}{(\lambda_1 - s_1^2)}$$

(A.14b) $$\alpha_{minor} = \frac{-1}{\alpha_{major}},$$

where

(A.15) $$\lambda_1 = \frac{1}{2}\{s_1^2 + s_2^2 + [(s_1^2 + s_2^2)^2 - 4(s_1^2 s_2^2 - s_{12}^2)]^{1/2}\}.$$

Note that the slope of minor regression axis equals the reciprocal of the slope of the major axis. In mathematical terms, the major and minor axes of the MA regression line are eigenvectors. The regression formula for the major and the minor axes is as follows:

(A.16a) $$Y_1 = \overline{Y}_1 + \alpha_{major}(Y_2 - \overline{Y}_2)$$

(A.16b) $$Y_1 = \overline{Y}_1 + \alpha_{minor}(Y_2 - \overline{Y}_2).$$

Finally, the confidence intervals for the major axis (the scaling exponent) are given by the following formula:

(A.17a) $$L_1 = \tan\left(\arctan \alpha_{major} - \frac{1}{2}\arctan 2\sqrt{C}\right)$$

(A.17b) $$L_1 = \tan\left(\arctan \alpha_{major} + \frac{1}{2}\arctan 2\sqrt{C}\right).$$

Perhaps intuitively, we see that the shape of the ellipse obtained from the minor and major axes is an important concern. It is given by the formula $[\lambda_1/(s_1^2 + s_2^2 - \lambda_1)]^{1/2}$. When the ellipse is very narrow, the variance about the minor axis is very small and the bivariate data are very highly correlated. That is, most of the variance is accounted for by the major axis such. When the shape approaches a circle, the variance about the minor axis increases, and when the shape is circular, no correlation between Y_1 and Y_2 exists. The technique of MA regression is used in multivariate analysis (e.g., principal component analysis) where the confidence intervals of the n-dimensional space occupied by data points are described by hyperellipsoids. The principal axis of a hyperellipsoid defines the orthogonal axis that accounts for the greatest variation in the data, the second axis accounts for the next greatest variation, and so on.

One limitation of MA regression is that it is sensitive to the scale of measurement and can yield different results when the axes of the bivariate plot are rotated. RMA regression overcomes the scale dependence of the MA regression technique by standardizing the variables Y_1 and Y_2 before the scaling exponent is computed. The scaling exponent determined by RMA regression is given by the following formula:

$$\text{(A.18)} \qquad \alpha_{\text{RMA}} = \left(\frac{\sum y_1^2}{\sum y_2^2}\right)^{1/2} = \frac{SE_{Y_1}}{SE_{Y_2}}.$$

The numerical value of α_{RMA} will always be greater than that of the LS scaling exponent because

$$\text{(A.19)} \qquad \alpha_{\text{RMA}} = \frac{\alpha_{\text{LS}}}{r_{12}},$$

where r_{12} is the correlation coefficient determined from LS regression of Y_1 against Y_2. The scaling coefficient can easily be determined because the reduced major axis must pass through the mean values of Y_1 and Y_2, just as in LS regression. Thus,

$$\text{(A.20)} \qquad \beta_{\text{RMA}} = \overline{Y}_1 - \alpha_{\text{RMA}}\,\overline{Y}_2\,.$$

The standard error of α_{RMA} and the standard error of β_{LS} are the same as those of α_{LS} and β_{LS}. The SE of α_{LS} therefore can be used to compute the confidence intervals for α_{RMA}. For this reason, it is often convenient to obtain r_{12}, α, $\overline{Y}_1$, and $\overline{Y}_2$ from LS regression and then determine $\alpha_{\text{RMA}} \pm SE$ and β_{RMA}, particularly since the LS regression line must be used anyway to determine the joint function between Y_1 and Y_2.

At this point it is useful to draw comparisons among LS, MA, and RMA regressions. For illustration the relation between plant height H and stem diameter D for nonwoody plants (mosses, pteridophytes, herbaceous angiosperms, and palms) will be considered. These data were used to adduce the scaling relation for the height of 190 plant species in chapter 3 (see fig. 3.32, where LS regression curves are provided). Here the same data set will be used to compute the form and significance of the functional relation when log H (in m) is regressed against log D (in m) by the three regression techniques.

In terms of LS regression, log $H = Y$ and log $D = X$. The log-log-transformed data from nonwoody plants obtain the following statistics: $\Sigma X = -543.580$, $\Sigma X^2 = 1498.117$, $\Sigma Y = -208.448$, $\Sigma Y^2 = 334.008$, $\Sigma XY = 91.218$, $\overline{X} = -2.6260$, and $\overline{Y} = -1.0071$. The sum of squares of Y is $\Sigma y^2 = 124.1020$, the sum of squares of X is $\Sigma x^2 = 70.6813$, and the sum of the product of X and Y is $\Sigma xy = 91.21787$. The quotient of the sum of products and the sum of squares obtains the LS scaling exponent $\alpha_{LS} = \Sigma xy / \Sigma x^2 = 1.29058$. From eq. (A.2), the scaling coefficient is calculated as follows:

$$\log \beta_{LS} = -1.0071 - 1.29058\,(-2.6260) = 2.3820.$$

Thus the LS regression line is log $H = 2.3820 + 1.2906$ log D. Finally, from eq. (A.3), the significance of this regression is statistically robust, since $r_{YX} = \Sigma xy / [(\Sigma x^2)(\Sigma x^2)]^{1/2} \approx 0.9740$ ($N = 190$).

Turning to MA regression, the sample variances and covariance for log $H = Y_1$ and log $D = Y_2$ are computed as follows:

$$s_1^2 = \frac{\sum y_1^2}{N-1} = \frac{124.102}{206} = 0.602437$$

$$s_2^2 = \frac{\sum y_2^2}{N-1} = \frac{70.6813}{206} = 0.3431068$$

$$s_{12} = \frac{\sum y_1 y_2}{N-1} = \frac{91.2179}{206} = 0.4428155.$$

Inserting these values into eq. (A.15), we find

$$\lambda_1 = \frac{1}{2}\left\{\begin{matrix} 0.6024370 + 0.3431068 + \\ \left[\begin{matrix}(0.6024370 + 0.3431068)^2 - \\ 4\,(0.2067002 - 0.1960856)\end{matrix}\right]^{1/2}\end{matrix}\right\} = 0.934181$$

and conclude that $\alpha_{major} = 1.33481$ (based on eq. A.14a). Finally, from eq. (A.16a), we obtain the MA regression line for the major axis: log $H = 2.4986 + 1.3348$ log D.

The RMA regression line is comparatively easy to determine. From LS regression, we know that $\alpha_{LS} = 1.29058$ and $r_{YX} \approx 0.9740$. Thus, from eq. (A.19) we find that $\alpha_{RMA} = 1.29058/0.9740 \approx 1.3251$. Note that eq. (A.18) permits us to check this value:

$$\alpha_{RMA} = (\Sigma y_1^2 \,/\, \Sigma y_2^3)^{1/2} = (124.1020/70.68125)^{1/2} = 1.32507 \approx 1.3251.$$

Finally, by means of eq. (A.20), we obtain the scaling coefficient:

$$\log \beta_{RMA} = -1.0071 - 1.3251(-2.6260) = 2.4726,$$

from which we find the RMA regression line: $\log H = 2.4726 + 1.3251 \log D$. In summary:

$$\text{LS regression: } \log H = 2.3820 + 1.2906 \log D$$
$$\text{MA regression: } \log H = 2.4986 + 1.3348 \log D$$
$$\text{RMA regression: } \log H = 2.4726 + 1.3251 \log D.$$

Generally, but not invariably, LS regression obtains the lowest value for the scaling exponent, whereas MA yields the highest value for α. The reduced major axis technique obtains an α value intermediate between α_{LS} and α_{MA}. The same pattern generally holds true for the numerical values of the scaling coefficient (i.e., $\beta_{LS} < \beta_{RMA} < \beta_{MA}$).

Seim and Saether (1983) draw additional useful comparisons among the LS, MA, and RMA regressions. They conclude that (1) MA regression gives consistent results when the error variance of Y_1 and Y_2 are equal, (2) RMA is useful when the sizes of the error variances are equal, and (3) LS, MA, and RMA yield nearly identical statistical conclusions when the data yield very high correlation coefficients. Not surprisingly, Seim and Saether further conclude that none of these three regression methods have universal application. More recently, Pagel and Harvey (1988) conclude that MA is preferred over RMA regression for interspecific comparisons provided deviation scores of Y_1 and Y_2 are checked by LS regression. However, MA regression is sensitive to the absolute scale in which variables are measured and often yields different conclusions when coordinate axes are rotated. By contrast, RMA regression is much less sensitive to the absolute scale of measurement and axis rotation (Rayner 1985; LaBarbera 1989). In summary, the choice of which regression method to use depends on the nature of the data. Model II regression is a subject of controversy, and therefore definitive recommendations are not currently possible, despite some authors' assertions. When the object of regression analysis is to predict Y_1 for a specified value of Y_2, LS regression may be preferred over MA or RMA regression because unbiased estimates of Y_1 for a given value of Y_2 as well as minimum width confidence limits for estimated Y_1 values generally are not obtained from the functional equations derived from Model II regressions that describe the covariation between two random (independent) variables. Even when Model II regression is chosen over LS regression, however, the investigator still must determine which method of Model II regression best overcomes the bias in the scaling exponent resulting from error measurements of size. This choice should be based on the absolute and relative sizes of the Y_1 and Y_2 error variances.

A.5 Multiple Regression (Y_0 versus $Y_1, Y_2, \ldots, Y_n$)

The principal objective of any regression analysis is to reduce the unexplained error variance (s^2_{Y}, in the case of Model I regression, and $s^2_{Y_1}$, for Model II regression). In all cases except when two variables evince perfect correlation ($r = 1.0$), some unexplained error variance remains after a bivariate regression. In theory this "residual" variance can be further reduced by regressing the unexplained deviations on other variables that are presumed to govern the operation of the system under analysis. The technique used to regress three or more variables simultaneously is called multiple regression analysis. This technique computes the least squares that best fits a linear function involving three or more variables, one of which is regressed on the other variables. In this section, the regressed variable is symbolized by Y_0. The variables Y_0 is regressed on are denoted as $Y_1, Y_2, \ldots, Y_n$, where n is the total number of variables considered in the multiple regression analysis. In scaling analysis, multiple regression analysis is used either to establish a linear prediction equation or to construct a mathematical model that purports to explain the variation of Y_0 in terms of the simultaneous variations in $Y_1, Y_2, \ldots, Y_n$. By analogy with LS regression, Y_0 may be considered the "dependent" variable and $Y_1, Y_2, \ldots, Y_n$, the "independent variables." With the exception of experimental data, however, this terminology is misleading, particularly in scaling analysis, where most if not all variables are subject to natural variation and measurement error.

Unlike previously discussed regression methods, the equation obtained from multiple regression analysis expresses not a regression *line* but rather a three-dimensional surface. Consider that in a bivariate regression, the scaling exponent is a slope defining the rate of change of Y_0 with respect to a particular "independent variable," denoted here as Y_j (i.e., $\alpha_{0,j} = \Delta Y_0/\Delta Y_j$). By contrast, in multiple regression analysis, Y_0 is simultaneously regressed on $Y_1, Y_2, \ldots, Y_n$. The result is a series of *partial scaling exponents* and collectively defines a three-dimensional surface intersecting with the ordinate axis at β (the Y_0-intercept). In statistics, these partial scaling exponents are called *partial regression coefficients*. This is best understood by noting that the multiple regression equation can be expressed in one of two forms—as a conventional equation or as a standardized equation:

(A.21a) $Y_0 = \beta + \alpha_{0,1}Y_1 + \alpha_{0,2}Y_2 + \cdots + \alpha_{0,n}Y_n$ (conventional form)

(A.21b) $y_0' = \alpha_{0,1}'y_1' + \alpha_{0,2}'y_2' + \cdots + \alpha_{0,n}'Y_n'$, (standardized form)

where α is the partial regression coefficient and α' is the standardized partial regression coefficient. The subscripts used for the partial regression coefficient and the standardized partial regression coefficients indicate the variable against which Y_0 is regressed (i.e., $\alpha_{0,1}$ is the partial regression coefficient obtained by regressing Y_0 against Y_1). It is important to know that the partial regression coefficient is obtained by presuming that all other variables are constant. For example, $\alpha_{0,1}$ denotes the relation between Y_0 and Y_1 when all other "independent variables" are held constant (i.e, $Y_2 = k, \ldots, Y_n = k$). Thus, $\alpha_{0,1}$ may be symbolized as $\alpha_{0,1:2-n}$, where the subscripts to the right of the colon indicate the "independent variables" that are held constant. This more complex notation will be employed later when we consider the relation between the standardized partial regression coefficient and the correlation coefficients between pairs of variables. For now, simply recognize that each partial regression coefficient is a slope whose units reflect the scales in which the "dependent" and "independent" variables are measured. Thus the partial regression coefficients in the conventional multiple regression equation have the units of measurements of Y_0 in their numerators and the units of Y_1 and so on in their denominators.

The dependency of the conventional multiple regression equation on the units of measurement in which the "dependent" and "independent variables" are expressed is circumvented by means of the standardized multiple regression equation. In this form, all the variables are transformed into their respective standard deviations:

(A.22) $$y_0' = \frac{Y_0 - \overline{Y}_0}{s_0} \text{ and } y_j' = \frac{Y_j - \overline{Y}_j}{s_j}.$$

Because of this normalization procedure, the Y_0-intercept, β, equals zero, thereby accounting for the fact that β does not appear in eq. (A.21b) ($\beta = 0$). To determine the numerical value of β, therefore, we need to convert the standardized regression form into the conventional form of the multiple regression equation. Much more important, unlike α, which is a rate of change expressed in terms of the original units of measurement of Y_0 and Y_j, each standardized partial regression coefficient α' is a slope expressed in the units of standard deviation of the "dependent variable" and respective "independent variable." This is particularly useful because, in most scaling relations, the "dependent" and "independent" variables are not measured in the same units and therefore each α can have a very different unit.

The standardized form of a simple linear regression permits us to further appreciate the relation between the standardized partial regression coefficient and the partial regression coefficient. Recall that each partial regression coefficient is obtained by regressing the "dependent variable" Y_0 against a specified "independent variable," denoted here by the subscript *j*. Each partial regression coefficient therefore is the slope of a simple linear regression. When expressed in its standardized form, a simple linear regression becomes $y'_0 = \alpha'_{0,1}y'_1$. Statistical theory shows that the standardized regression coefficient in this simple linear equation equals the correlation coefficient obtained when Y_1 is regressed on Y_0. That is, $\alpha'_{0,1} = r_{1,0}$ such that $y'_0 = \alpha'_{0,1}y'_1 = r_{1,0}y'_1$. When dealing with multiple regression, however, the relation between the standardized partial regression coefficient and the correlation coefficients obtained from simple linear regression of paired variables is more complex. For example, in the case of two "independent variables," Y_1 and Y_2:

(A.23a) $$\alpha'_{0,1:2} = \frac{r_{1,0} - r_{2,0}r_{1,2}}{1 - r^2_{1,2}}$$

(A.23b) $$\alpha'_{0,1:1} = \frac{r_{2,0} - r_{1,0}r_{1,2}}{1 - r^2_{1,2}}.$$

That is, $\alpha'_{0,1:2}$ is the coefficient obtained by regressing the standardized "dependent variable" against the first standardized "independent variable" when the second standardized "independent variable" is held constant (i.e., denoted by the subscript 0,1:2), whereas $\alpha'_{0,2:1}$ is the coefficient obtained by regressing the standardized "dependent variable" against the second standardized "independent variable" when the first standardized "independent variable" is held constant (i.e., denoted by the subscript 0,2:1). Substituting eq. (A.23) into eq. (A.21b), we obtain an equation for the standardized multiple regression equation in terms of coefficients of correlation:

(A.24) $$y'_0 = \alpha'_{0,1}y'_1 + \alpha'_{0,2}y'_2 = \left(\frac{r_{1,0} - r_{2,0}r_{1,2}}{1 - r^2_{1,2}}\right)y'_1 + \left(\frac{r_{2,0} - r_{1,0}r_{1,2}}{1 - r^2_{1,2}}\right)y'_2.$$

The standardized multiple regression equation can be converted into the conventional multiple regression equation, since each standardized partial regression coefficient equals its comparable partial regression coefficient multiplied by the ratio of the standard deviations of the specified "independent variable" and the "dependent variable. For example, in the case of two "independent variables" (see eq. A.23), the following holds true:

(A.25) $$\alpha'_{0,1:2} = \alpha_{0,1:2} \frac{s_1}{s_0} \text{ and } \alpha'_{0,2:1} = \alpha_{0,2:1} \frac{s_2}{s_0}.$$

Thus the partial regression coefficients are given by the formulas

(A.26) $$\alpha_{0,1:2} = \frac{r_{1,0} - r_{2,0}r_{1,2}}{1 - r^2_{1,2}} \left(\frac{s_0}{s_1}\right) \text{ and } \alpha_{0,2:1} = \frac{r_{2,0} - r_{1,0}r_{1,2}}{1 - r^2_{1,2}} \left(\frac{s_0}{s_2}\right),$$

such that the conventional form of the multiple regression equation is

(A.27) $$\begin{aligned} Y_0 &= \beta + \alpha_{0,1}Y_1 + \alpha_{0,2}Y_2 \\ &= \beta + \frac{r_{1,0} - r_{2,0}r_{1,2}}{1 - r^2_{1,2}} \left(\frac{s_0}{s_1}\right) Y_1 + \frac{r_{2,0} - r_{1,0}r_{1,2}}{1 - r^2_{1,2}} \left(\frac{s_0}{s_2}\right) Y_2. \end{aligned}$$

The numerical value of β is obtained as follows:

(A.28) $$\beta = \overline{Y}_0 - \alpha_{0,1:2}\overline{Y}_1 - \alpha_{0,2:1}\overline{Y}_2 .$$

Although the previous equations appear somewhat overwhelming, it is comparatively simple to determine the values of r and s from bivariate simple linear regression of the appropriate Y variables. Once the standardized form of the multiple regression equation is determined, it can be converted into its conventional form with equal ease. Consequently, multiple regression analysis involving only two "independent variables" is not difficult and can be done manually. As the number of variables increases, however, manual computation becomes progressively much more difficult. For example, the following formulas for the standardized partial regression coefficients apply when the "dependent variable" is regressed simultaneously against three "independent variables" (see eq. A.23):

(A.29a) $$\alpha'_{0,1:2-3} = \frac{1}{A}\left[(1 - r^2_{2,3})r_{1,0} + (r_{2,3}r_{3,0} - r_{2,0})r_{1,2} + (r_{2,3}r_{2,0} - r_{3,0})r_{1,3}\right]$$

(A.29b) $$\alpha'_{0,2:1,3} = \frac{1}{A}\left[(1 - r^2_{1,3})r_{2,0} + (r_{1,3}r_{3,0} - r_{1,0})r_{1,2} + (r_{1,3}r_{1,0} - r_{3,0})r_{2,3}\right]$$

(A.29c) $$\alpha'_{0,3:1,2} = \frac{1}{A}\left[(1 - r^2_{1,2})r_{3,0} + (r_{1,2}r_{2,0} - r_{1,0})r_{1,3} + (r_{1,2}r_{1,0} - r_{2,0})r_{2,3}\right],$$

where $A = 1 + 2(r_{1,2}r_{1,3}r_{2,3}) - r^2_{1,2} - r^2_{1,3} - r^2_{2,3}$.

The conventional partial regression coefficients are given by the following formulas (see eq. A.25):

(A.30a) $$\alpha'_{0,1:2,3} = \alpha'_{0,1:2,3}\,\frac{s_o}{s_1}$$

(A.30b) $$\alpha'_{0,2:1,3} = \alpha_{0,2:1,3}\,\frac{s_o}{s_2}$$

(A.30c) $$\alpha_{0,3:1,2} = \alpha_{0,3:1,2}\,\frac{s_o}{s_3}.$$

And finally, we can obtain β from the following equation (see eq. A.28):

(A.31) $$\beta = \overline{Y}_0 - \alpha_{0,1:2,3}\overline{Y}_1 - \alpha_{0,2:1,3}\overline{Y}_2 - \alpha_{0,3:1,2}\overline{Y}_3\,.$$

When four or more "independent variables" are involved, the investigator may wish to use a computer for multiple regression analysis.

We are now in a position to illustrate the menu for multiple regression analysis with a data set for the biomass (in g) of the perianth M_P, gynoecium M_G, and androecium M_A obtained from forty-four of the ninety species of perfect flowers considered in chapter 4 (see fig. 4.8). Our objective is to establish a predictive equation for perianth biomass based on the biomass of the gynoecium and the biomass of the androecium for log-transformed data. Thus, $\log M_P = Y_0$ and $\log M_G$ and $\log M_A$ are the "independent variables" Y_1 and Y_2, respectively.

First we compute:

$$\sum Y_0 = -72.199 \quad \sum Y_0^2 = 157.08 \quad \overline{Y}_0 = \frac{\sum Y_0}{N} = -1.6409$$

$$\sum Y_1 = -104.03 \quad \sum Y_1^2 = 283.21 \quad \overline{Y}_1 = \frac{\sum Y_1}{N} = -2.3643$$

$$\sum Y_2 = -101.07 \quad \sum Y_2^2 = 272.22 \quad \overline{Y}_2 = \frac{\sum Y_2}{N} = -2.2971.$$

The correlation coefficients for these data are

$$r_{1,2} = 0.93702 \quad r_{1,0} = 0.93059 \quad r_{2,0} = 0.96021.$$

We compute the sum of squares of Y_0

$$\sum y_0^2 = \sum Y_0^2 - \frac{(\sum Y_0)^2}{N} = 157.08 - \frac{(-72.199)^2}{44} = 38.610$$

the sum of squares of Y_1

$$\sum y_1^2 = \sum Y_1^2 - \frac{(\sum Y_1)^2}{N} = 283.21 - \frac{(-104.03)^2}{44} = 37.250$$

the sum of squares of Y_2

$$\sum y_2^2 = \sum Y_2^2 - \frac{(\sum Y_2)^2}{N} = 272.22 - \frac{(-101.07)^2}{44} = 40.058$$

the sample standard deviation of Y_0

$$s_0 = \left(\frac{\sum y_0^2}{N-1}\right)^{1/2} = \left(\frac{38.610}{43}\right)^{1/2} = 0.94758$$

the sample standard deviation of Y_1

$$s_1 = \left(\frac{\sum y_1^2}{N-1}\right)^{1/2} = \left(\frac{37.250}{43}\right)^{1/2} = 0.93074$$

and the sample standard deviation of Y_0

$$s_2 = \left(\frac{\sum y_2^2}{N-1}\right)^{1/2} = \left(\frac{40.058}{43}\right)^{1/2} = 0.96518.$$

We now obtain the standardized partial regression coefficients (see eq. A.23):

$$\alpha'_{0,1.2} = \frac{r_{1,0} - r_{2,0}r_{1,2}}{1 - r_{1,2}^2} = \frac{0.93059 - (0.96021)(0.93702)}{1 - (0.93702)^2} = 0.25292$$

$$\alpha'_{0,2.1} = \frac{r_{2,0} - r_{1,0}r_{1,2}}{1 - r_{1,2}^2} = \frac{0.96021 - (0.93059)(0.93702)}{1 - (0.93702)^2} = 0.72324.$$

Now we calculate the partial regression coefficients (see eq. A.25):

$$\alpha_{0,1.2} = \alpha'_{0,1.2}\frac{s_0}{s_1} = 0.25292\left(\frac{0.94754}{0.93074}\right) = 0.25749$$

$$\alpha_{0,2.1} = \alpha'_{0,2.1}\frac{s_0}{s_2} = 0.72524\left(\frac{0.94754}{0.96518}\right) = 0.71199.$$

The Y-intercept, β, for the conventional multiple regression equation is computed in the following way (see eq. A.28):

$$\begin{aligned}\log \beta &= \overline{Y} - \alpha_{0,1.2}\overline{Y}_1 - \alpha_{0,2.1}\overline{Y}_2 \\ &= -\ 1.6409 - 0.25749\ (-2.3643) - 0.71199\ (-2.2971) \\ &= -\ 1.6409 + 0.60878 + 1.6355 \\ &= 0.60339.\end{aligned}$$

Thus the form of the standardized multiple regression equation is as follows (see eq. A.21b):

$$Y_0 = \log \beta + \alpha_{0,1.2} Y_1 + \alpha_{0,2.1} Y_2$$

or

$$Y_0 = 0.60339 + 0.25749\ Y_1 + 0.71199\ Y_2.$$

Since $Y_0 = \log M_P$, $Y_1 = \log M_G$, and $Y_2 = \log M_A$, we find

$$M_P = 4.0123\ M_G^{0.25749}\ M_A^{0.71199}.$$

The correlation coefficient for a multiple regression equation is symbolized as R, not r. Formulas for computing R, the confidence intervals and standard errors for the partial regression coefficients, and the Y-intercept of the multiple regression equation are provided in standard statistics textbooks.

Although not frequently thought of as an example of multiple regression analysis, curvilinear regression between two variables can be evaluated by multiple regression, particularly when the curvilinear regression equation takes the form of a polynomial expansion series. This topic therefore is treated here, even though it may superficially appear to be out of place. It also affords the opportunity to emphasize that linear regression is not invariably sufficient to account for the variance of one variable with respect to another and that the presumption of a linear functional relation between two variables always should be tested. Indeed, many biological processes, such as growth, are best described by curvilinear regression equations. The logistic equation summarized in chapter 1 is a form of a curvilinear regression equation. Here we shall consider only the polynomial regression equation, which is used frequently as the best fit to empirical data evincing complex curvilinearity.

The general mathematical form of a polynomial function is:

$$Y_0 = \beta + \alpha_1 Y_1 + \alpha_2 Y_1^2 + \cdots + \alpha_n Y_1^n. \tag{A.32}$$

Although only two variables are considered in eq. (A.32), the similarities between polynomial regression and the multiple regression are self-evident. For example, curvilinear regression is a stepwise procedure. The higher-order terms shown to the right of eq. (A.32) are sequentially added in the regression analysis. The effect of these higher-order terms on explaining the variance of Y_0 in terms of Y_1 is evaluated by comparing the correlation coefficients obtained from sequential regression analysis. As higher terms are added, the power coefficients of Y_1 change in a way analogous to the way the partial regression coefficients change as more "independent variables" are added to the multiple regression analysis. In this respect the notation used in eq. (A.32) is inaccurate because values of α and β change as higher terms are added to the polynomial equation:

(A.33)
$$Y_0 = \beta_0 + \alpha_0 Y_1$$
$$Y_0 = \beta_1 + \alpha_1' Y_1 + \alpha_2' Y_1^2$$
$$Y_0 = \beta_2 + \alpha_1'' Y_1 + \alpha_2'' Y_1^2 + \alpha_3''' Y_1^3,$$

where $\beta_0 \neq \beta_1 \neq \beta_2$, $\alpha_0 \neq \alpha_1' \neq \alpha_{11}''$ and $\alpha_2' \neq \alpha_2''$.

The computation of a polynomial regression equation by multiple regression analysis is simple because each higher-order variable assumes the role of an "independent variable" (i.e., $Y_1 = Y_1$, $Y_1^2 = Y_2$, $Y_1^3 = Y_3, \cdots$, and $Y_1^n = Y_n$). With this substitution, we can calculate the standardized partial regression coefficients in the multiple regression equation that assume the roles of α values in eq. (A.33). Standard statistical techniques can be used to test for the significance of the polynomial regression equation, although the power coefficients of Y_1 are typically not tested for significance. Most computer programs geared toward statistical analysis compute polynomial regression equations. These should be used cautiously, however, since many are prone to rounding error. Also, many do not compute orthogonal polynomials, which are required whenever regressions are recomputed as terms with higher power coefficients added. Successive terms can be added to the regression equation until no improvement in fit is observed with orthogonal polynomials.

A fundamental question faces the investigator who uses multiple regression analysis: What are the predictor variables? The answer is far from simple but typically involves a "cost-profit" analysis of the data at hand. For example, in using multiple regression to obtain a predictive equation, the best set of variables is the one that is most parsimonious—that is, the smallest number of variables among those that can be selected that yields an adequate prediction. This translates into the smallest set of variables that achieves the maximum reduction in the unexplained deviations of the "dependent variable." However, these two requirements are in evident conflict, because the unexplained deviations in Y_0 decrease as the number of "independent variables" increases. The strategy therefore is to assess the point of diminishing returns obtained as more and more variables are added to the multiple regression. Three tactics have been suggested to achieve the best set of predictor variables: forward selection, backward selection, and stepwise selection (see Sokal and Rohlf 1981, 662–71). In forward selection, "independent variables" are progressively added, and their consequence on R is assessed. The selection of the variables to be added in sequence is based on their partial correlation coefficients. A partial correlation coefficient measures the correlation between any pair of variables when the other variables are held constant. We have employed first- and second-order partial correlation coefficients in the previous formulas (e.g., $r_{1,2}$ and $r_{1,2:3}$, respectively). The formulas for the second- and third-order partial correlation coefficients are as follows:

(A.34a) $$r_{1,2:3} = \frac{r_{1,2} - r_{1,3}r_{2,3}}{[(1 - r_{1,3}^2)(1 - r_{2,3}^2)]^{1/2}}$$ (second-order)

(A.34b) $$r_{1,2:3,4} = \frac{r_{1,2:3} - r_{1,4:3}r_{2,4:3}}{[(1 - r_{1,4:3}^2)(1 - r_{2,4:3}^2)]^{1/2}},$$ (third-order)

where variables 1 and 2 are regressed as 3 is held constant (second-order) or variables 1 and 2 are regressed as 3 and 4 are held constant (third-order). Equation (A.34) can be used to construct fourth-, fifth-, and so on orders of partial correlation coefficients to establish the ascending sequence of strong predictor variables used in the forward selection process. Backward selection is the reverse of the forward process. It begins with the multiple regression of all the variables believed relevant to predict the "dependent variable." The variable with the lowest partial correlation coefficient then is removed first, and this process is reiterated until the best set of predictor variables is obtained.

The forward and backward processes suffer because the effect of the addition or the removal of a variable is ingrained in the subsequent resulting multiple regression analyses. That is, either process is "linear" in its consequences on the multiple regression correlation coefficient *R*. For this reason, the stepwise process is often preferred. This tactic scans the results of backward or forward multiple regression analysis and evaluates the correlation structure of all the variables. Space does not permit an illustration of this technique, but it is important to know that regardless of which procedure an investigator follows, the subset of variables adduced as the "best" set of predictor variables is nothing more than the subset that does the "most adequate job." In any circumstances, the results of any regression analysis (whether multiple or bivariate) must be tested against theoretical expectation or additional data. Snee (1977) provides methods for testing the results of various types of regression analyses.

A.6 Analysis of Variance and Nested Analysis of Variance (ANOVA)

Analysis of variance, developed by R. A. Fisher, is the general test used to determine whether two or more samples have been drawn from the same population. It is a fundamental tool in experimental design and is important in evaluating the consequences of random effects on sample means and other statistics. At a more fundamental level, analysis of variance provides insight into the nature of variance within and among populations. The principles of analysis of variance are required if one is to proceed to an understanding of nested analysis of variance. Nested analysis of variance, also called hierarchic analysis of variance, provides a way to determine the distribution of variance among subordinate samples contained or nested within higher levels of data classification. Nested analysis of variance is used to determine the magnitude of error at various stages of an experiment or to determine the distribution of variance among hierarchical levels of data classification. The latter is particularly important to scaling analysis, whose objective is to determine whether size-dependent variations are a consequence of adaptive evolutionary changes. In this circumstance, nested analysis of variance may be used to determine the distribution of variance of characters in species nested in genera, genera nested in families, and so forth. Knowledge of the distribution of variance among taxonomic levels is important to control for the phyletic effect discussed in chapter 1. Specifically, nested analysis of variance can be used to determine the taxonomic level appropriate for scaling analysis. The following provides a brief introduction to analysis of variance and nested analysis of variance. Additional references should be consulted, however.

The principles of analysis of variance may be understood by examining the matrix in table A.1 for a continuous variable *Y*. The variates within the matrix are organized into *a* number of columns, each called a group. Within each group, *n* number of variates are arranged in rows. In the

Table A.1

	a Groups						
	1	2	3	. . .	*i*	. . .	*a*
1	Y_{11}	Y_{21}	Y_{31}	. . .	Y_{i1}	. . .	Y_{a1}
2	Y_{12}	Y_{22}	Y_{32}	. . .	Y_{i2}	. . .	Y_{a2}
3	Y_{13}	Y_{23}	Y_{33}	. . .	Y_{i3}	. . .	Y_{a3}
.	.	.	.	.	.	.	.
.	.	.	.	.	.	.	.
.	.	.	.	.	.	.	.
j	Y_{1j}	Y_{2j}	Y_{3j}	. . .	Y_{ij}	. . .	Y_{aj}
.	.	.	.	.	.	.	.
.	.	.	.	.	.	.	.
.	.	.	.	.	.	.	.
n	Y_{1n}	Y_{2n}	Y_{3n}	. . .	Y_{in}	. . .	Y_{an}
$\sum^{n} Y$	$\sum^{n} Y_1$	$\sum^{n} Y_2$	$\sum^{n} Y_3$	. . .	$\sum^{n} Y_i$	. . .	$\sum^{n} Y_a$

particular case shown, an equal number of variates is obtained for each group. Thus, the total number of variates equals na. An equal n for each a, however, is not a precondition for analysis of variance, and we shall examine the cases where groups are represented by unequal numbers of variates. For now, note that subscripts identify each variate in terms of the group it belongs to as well as locate each variate within each group. The first subscript denotes the group the variate belongs to (i.e., the column); the second subscript identifies the location of each variate in the series of measurements within each group. This notation is required because the variance within groups and the variance among groups provide separate estimates of the parametric variance of the population from which the groups (samples) are drawn.

It is generally the case that the quotient of the sum of squares Σy^2, denoted here as SS, and the degrees of freedom df equals the variance s_Y^2 of a particular sample. That is, $s_Y^2 = \Sigma y^2/(N - 1) = [\Sigma Y^2 - (\Sigma Y)^2/N]/(N - 1)$, where N is the sample size. In analysis of variance, the variance of the ith variate, Y_i, is calculated by means of an equivalent formula:

$$s_i^2 = \frac{1}{df} SS_i = \frac{1}{n-1}\left[\sum_{j=1}^{j=n}(Y_{ij} - \bar{Y}_i)^2\right]. \tag{A.35}$$

As noted, the sample size frequently differs among groups and therefore the degrees of freedom will not be equivalent among all the groups shown in the data matrix provided above. For this reason, the formulas for the variance within groups s^2_{within} and the variance among groups s^2_{among} are more complex, although mathematically homologous with eq. (A.35):

$$\begin{aligned} s^2_{\text{within}} &= \frac{1}{df} SS_{\text{within}} \\ &= \frac{1}{a(n-1)}\left[\sum_{i=1}^{i=a}\sum_{j=1}^{j=n}(Y_{ij} - \bar{Y}_i)^2\right] \end{aligned} \qquad \text{(equal } n\text{)} \tag{A.36a}$$

$$\begin{aligned} s^2_{\text{within}} &= \frac{1}{df} SS_{\text{within}} \\ &= \frac{1}{\sum_{i=1}^{i=a} n_i - a}\left[\sum_{i=1}^{i=a}\sum_{j=1}^{j=n} Y_{ij}^2 - \sum_{i=1}^{i=a}\frac{\left(\sum_{j=1}^{i=n} Y_{ij}\right)^2}{n_i}\right] \end{aligned} \qquad \text{(unequal } n\text{)} \tag{A.36b}$$

$$s^2_{\text{among}} = \frac{1}{df} SS_{\text{among}} = \frac{n}{a-1}\left[\sum_{i=1}^{i=a}(\bar{Y}_i - \bar{\bar{Y}})^2\right] \qquad \text{(equal } n\text{)} \tag{A.37a}$$

(A.37b) $$s^2_{\text{among}} = \frac{1}{df} SS_{\text{among}} \qquad \text{(unequal } n\text{)}$$

$$= \frac{1}{a-1}\left[\sum_{i=1}^{i=a} \frac{\left(\sum_{j=1}^{j=n} Y_{ij}\right)^2}{n_i} - \frac{\left(\sum_{i=1}^{i=a}\sum_{j=1}^{j=n} Y_{ij}\right)^2}{\sum_{i=1}^{i=a} n_i}\right].$$

The "grand mean" of Y (shown in eq. A.37a) computed for the entire data matrix is given by the formula $\bar{\bar{Y}} = \sum_{i=1}^{i=a} \bar{Y}_i / a$. The total variance s^2_{total} of the continuous variable Y is given by the formula

(A.38) $$s^2_{\text{total}} = \frac{1}{df} SS_{\text{total}} = \frac{1}{\sum_{i=1}^{i=a} n_1 - 1}\left[\sum_{i=1}^{i=a}\sum_{j=1}^{i=n} Y^2_{ij} - \frac{\left(\sum_{i=1}^{i=a}\sum_{j=1}^{j=n} Y_{ij}\right)^2}{\sum_{i=1}^{i=a} n_i}\right].$$

Under most conditions, the variance among groups and the total variance are computed directly, and the variance within groups is obtained by subtracting the former from the latter. To summarize up to this point, the total variance is the statistic of dispersion for the total number of observations (i.e., the product of a and n) around the grand mean—it describes the variance for the entire data matrix. The within-group variance provides the average dispersion of the items within each group around the means of groups. The among-group variance describes the dispersion of group means around the grand mean. Those familiar with analysis of variance will note that the sample variance is synonymous with the mean square, symbolized as MS. That is, $s^2_{\text{within}} = MS_{\text{within}}$ and $s^2_{\text{among}} = MS_{\text{among}}$. Likewise, the total variance is the same thing as the total mean square. In other words, $s^2_{\text{total}} = MS_{\text{total}}$.

As noted, the variance among groups and the variance within groups provide different estimates of the parametric variance σ^2 of the population from which groups are randomly drawn. Clearly, when groups are drawn at random from the same population these two estimates of the parametric mean will be nearly equivalent for very large sample sizes. They will differ, however, when groups are not drawn from the same population. The familiar F-ratio in statistics (named in honor of R. A. Fisher) is the quotient of the variance (= mean squares) among groups and the variance (= mean squares) within groups. That is, $F = s^2_{\text{among}} / s^2_{\text{within}} = MS_{\text{among}} / MS_{\text{within}}$. The F-ratio approaches unity as the sample size n of observations randomly draw from a single population increases. More specifically, the ratio of the two variances approaches $(n - 1)/(n - 3) \approx 1.0$ when n is very large. The F-ratio, or more specifically the shape of its distribution, which is described by a complicated mathematical function, is determined by two values for degrees of freedom, because we are dealing with the ratio of the variance among groups to the variance within groups. Thus, the F-ratio is formally denoted as $F_{\alpha[df_{among}, df_{within}]}$. Tests of hypotheses typically take $\alpha = 0.05$ (i.e., the "right tail" of the F-distribution is emphasized).

The results of an analysis of variance may be presented in a variety of ways, although the format shown in table A.2 is most frequently employed in computer printouts and most statistics textbooks.

Before proceeding further, let me illustrate the preliminary computations required for an analysis of variance by means of a comparatively small data set. We shall consider some data obtained for the gynoecial biomass (in mg) of perfect flowers differing in breeding systems: obligately xenogamous, facultatively xenogamous, and facultatively autogamous. These data, which were used in chapter 4, have been log transformed (see table A.3). Equation (A.37b) provides the sum of squares among groups (variance among the three groups):

$$SS_{among} = \left[\frac{(-1.0589)^2}{12} + \frac{(19.888)^2}{10} + \frac{(23.152)^2}{14}\right] - \left[\frac{(-1.0589 + 19.888 + 23.152)^2}{12 + 10 + 14}\right]$$
$$= (77.933) - (48.956)$$
$$= 28.977.$$

Equation (A.38) provides the total sum of squares:

Table A.2

Source of Variance	df	SS	MS	F-ratio	Expected Parametric Variance
Among groups	$a-1$	SS_{among}	$\frac{SS_{among}}{a-1}$	$F = \frac{MS_{among}}{MS_{within}}$	σ_2
Within groups	$\sum_{i=1}^{a} n_i - a$	SS_{within}	$\frac{SS_{within}}{\sum_{i=1}^{a} n_i - a}$		
Total	$\sum_{i=1}^{a} n_i - 1$	SS_{total}	$\frac{SS_{within}}{\sum_{i=1}^{a} n_i - 1}$		

$$SS_{\text{total}} = (4.0749 + 41.158 + 40.544) - \left[\frac{(-1.0589 + 19.888 + 23.152)^2}{12 + 10 + 14}\right]$$
$$= (85.777) - (48.956)$$
$$= 36.821.$$

And the within-group sum of squares equals the total sum of squares minus the among-group sum of squares:

$$SS_{\text{within}} = 36.821 - 28.977 = 7.8440.$$

The results of this analysis of variance are shown in table A.4. Statistical tables relating the *F*-ratio to their significance probabilities *P* are provided in most standard statistics textbooks. In the analysis of variance for perfect flowers differing in breeding system, we have an *F*-ratio that is

Table A.3

Groups ($a = 3$)		
Obligately Xenogamous ($a = 1$)	Facultatively Xenogamous ($a = 2$)	Facultatively Autogamous ($a = 3$)
−0.26761	2.1584	1.5911
−0.95861	2.1139	1.9823
0.47856	2.3979	2.1205
0.42975	1.9344	1.5910
0.11727	1.9190	1.3617
0.02938	2.5888	1.7075
1.12090	2.0863	1.7634
−0.82390	2.0043	2.6637
−0.56863	1.6434	1.8633
−0.56863	1.0413	1.6232
−0.21467		1.2552
0.16731		1.2041
		1.2787
		1.1461
Preliminary statistics:		
$n_1 = 12$	$n_2 = 10$	$n_3 = 14$
$\sum_{i=1}^{a} Y_1 = -1.0589$	$\sum_{i=1}^{a} Y_2 = 19.888$	$\sum_{i=1}^{a} Y_3 = 23.152$
$\sum_{i=1}^{a} Y_1^2 = 4.0749$	$\sum_{i=1}^{a} Y_2^2 = 41.158$	$\sum_{i=1}^{a} Y_3^2 = 40.544$

$F_{0.05[2, 33]} = 60.953$. When the numerical value of F is looked up on the appropriate table for 0.05 [2, 33], we find a highly significant ($P < 0.00001$) added variance component among breeding systems. Consequently, the null hypothesis that these three groups of data were drawn at random from the same population is rejected.

The preceding analysis of variance employed data that were purposely segregated into three groups representing different breeding systems. Therefore the assumption that samples (groups) were randomly selected was intentionally violated. This was done to help make a distinction between what is called a Model I analysis of variance and a Model II analysis of variance. The latter applies to experimental design such that variance has an *added component due to treatment*. The data for gynoecial biomass were sorted to evaluate the effects of different breeding systems on the size of the gynoecium. That is, each breeding system was considered "a treatment." The effect of a treatment i is symbolized by α_i. The consequences of treatment effects on an analysis of variance are shown formally when the generalized data matrix is recast to show the role of α_i (see table A.5). The role of treatment effects is easily illustrated. Consider the case where all groups have equal numbers of measurements. Under these circumstances, eq. (A.36a) becomes

$$\text{(A.39)} \quad s^2_{\text{within}} = \frac{1}{df} SS_{\text{within}} = \frac{1}{a(n-1)} \sum_{i=1}^{i=a} \sum_{j=1}^{j=n} [(Y_{ij} + \alpha_i) - (\overline{Y}_i + \alpha_i)]^2$$
$$= \frac{1}{a(n-1)} \sum_{i=1}^{i=a} \sum_{j=1}^{j=n} (Y_{ij} - \overline{Y}_i)^2.$$

Although the treatment effect does not alter the equation for the variance within groups, the variance among groups is altered (see eq. A.37a):

Table A.4

Source of Variance	df	SS	MS	F-Ratio
Among groups	2	28.977	$\frac{28.977}{2} = 14.488$	$\frac{14.488}{0.23769} = 60.953$
Within groups	33	7.8440	$\frac{7.8440}{33} = 0.23769$	
Total	35	36.821	$\frac{36.821}{35} = 1.0520$	

$$\text{(A.40)} \quad s^2_{\text{among}} = \frac{1}{df} SS_{\text{among}} = \frac{1}{a-1} \sum_{i=1}^{i=a} \left[(\overline{Y}_i + \alpha_i) - (\overline{\overline{Y}} + \overline{\alpha} \right]^2$$

$$= \frac{1}{a-1} \sum_{i=1}^{i=a} (\overline{Y}_i - \overline{\overline{Y}})^2 + \frac{1}{a-1} \sum_{i=1}^{i=a} (\alpha_i - \overline{\alpha})^2$$

$$+ \frac{2}{a-1} \sum_{i=1}^{i=a} (\overline{Y}_i - \overline{\overline{Y}})(\alpha_i - \overline{\alpha}).$$

Note that the first of the three terms in eq. (A.40) is the same as the variance among groups (see eq. A.37a). Two new terms, however, have been added to the variance among groups. The first of these two new terms is called the *added component due to treatment effects*. The second of these two new terms is the *covariance* between the magnitude of the treatment effects and the within-group mean of Y. The covariance equals zero when the treatment effect and group means are independent of one another. Be that as it may, since the F-ratio is the ratio of the variance among groups to the variance within groups, we see from eqs. (A.39) and (A.40) that the F-ratio is sensitive to the presence of an added component of variance due to treatment effects, which increase the variance among groups. Returning to our example of gynoecial biomass variance, we see that because the data for gynoecial biomass were specifically sorted according to three treatments—that is, breeding systems—the variance among groups was significantly increased and therefore the F-ratio was exceptionally large, leading to a rejection of the null hypothesis that the three groups of flowers were drawn at random from the same population.

Table A.5

	a Groups						
	1	2	3	...	i	...	α
1	$Y_{11} + \alpha_1$	$Y_{21} + \alpha_2$	$Y_{31} + \alpha_3$	...	$Y_{i1} + \alpha_j$	...	$Y_{a1} + \alpha_a$
2	$Y_{12} + \alpha_1$	$Y_{22} + \alpha_2$	$Y_{32} + \alpha_3$	...	$Y_{i2} + \alpha_j$	...	$Y_{a2} + \alpha_a$
3	$Y_{13} + \alpha_1$	$Y_{23} + \alpha_2$	$Y_{33} + \alpha_3$	...	$Y_{i3} + \alpha_j$	...	$Y_{a3} + \alpha_a$
⋮	⋮	⋮	⋮	⋮	⋮	⋮	⋮
j	$Y_{1j} + \alpha_1$	$Y_{2j} + \alpha_2$	$Y_{3j} + \alpha_3$	...	$Y_{ij} + \alpha_j$	...	$Y_{aj} + \alpha_a$
⋮	⋮	⋮	⋮	⋮	⋮	⋮	⋮
n	$Y_{1n} + \alpha_1$	$Y_{2n} + \alpha_2$	$Y_{3n} + \alpha_3$	...	$Y_{in} + \alpha_j$	...	$Y_{an} + \alpha_a$
$\sum^{n} Y$	$\sum^{n} Y_1 + n\alpha_1$	$\sum^{n} Y_2 + n\alpha_2$	$\sum^{n} Y_3 + n\alpha_3$	...	$\sum^{n} Y_i + n\alpha_j$	...	$\sum^{n} Y_a + n\alpha_a$

In the case of Model II analysis of variance, the added effect for each group is the consequence not of fixed treatments but rather of random effects (those over which the investigator has no control). These random effects are denoted by A_i. Their influence on the variance among groups and within groups essentially is identical to the influence of the effects of treatment in terms of the F-ratio. That is, the variance within groups remains unaffected, whereas the variance among groups becomes

$$\text{(A.41)} \quad s^2_{\text{among}} = \frac{1}{df} SS_{\text{among}} = \frac{1}{a-1} \sum_{i=1}^{i=a} \left[(\overline{Y}_i + A_i) - (\overline{\overline{Y}} + \overline{A}) \right]^2$$
$$= \frac{1}{a-1} \sum_{i=1}^{i=a} (Y_i - \overline{\overline{Y}})^2 + \frac{1}{a-1} \sum_{i=1}^{i=a} (A_i - \overline{A})^2$$
$$+ \frac{2}{a-1} \sum_{i=1}^{i=a} (\overline{Y}_i - \overline{\overline{Y}})(A_i - \overline{A})$$

Once again two new terms are added to the variance among groups. The first of these two new terms is called the *added variance component among groups*. The second new term is the *covariance*. And as before the existence of this added component to the variance among groups is detected by the F-ratio.

Although eqs. (A.40) and (A.41) are mathematically homologous, an important distinction between Model I and Model II analysis of variance exists: in the case of Model I, the differences among group means are assumed to be due to treatment effects fixed by the experimenter, and therefore analysis of variance can be used to estimate the actual difference among the group means of the Y variable, whereas in Model II the influence on variance of the added variance component among groups can only be estimated because differences among groups are the consequences of random variation. Put differently, Model I analysis of variance is concerned with the general problem of experimental design and the magnitude of experimental error at different levels of replication, whereas Model II analysis of variance is concerned with the general problem of phenotypic or genetic variation. For this reason, Model II analysis of regression is more frequently used in scaling analysis than Model I analysis.

Yet another important way of looking at the distinction between Model I and Model II analysis of variance, which will be helpful when we consider nested analysis of variance in the context of scaling studies, is as follows: All three estimates of the parametric variance (among-group, within-group, and total variance) will have the same magnitude when all groups are randomly drawn from the same population. Provided there are no treatment effects or random variation, the variance within groups and the variance among groups are estimates of the parametric variance, denoted by σ^2. However, when treatment effects or random variation components exist, the variance among groups is an estimate of

(A.42a) $$\sigma^2 + \frac{n}{a-1}\sum_{i=1}^{i=a}\alpha^2$$ (Model I ANOVA)

or

(A.42b) $$\sigma^2 + n\sigma_A^2,$$ (Model II ANOVA)

where σ_A^2 is the parametric value of added variance due to random variation and n is the number of observations per group. When unequal values of n occur among groups, the following formula may be used to compute an average n, denoted as n_0:

(A.43) $$n_0 = \frac{1}{a-1}\left(\sum_{i=1}^{i=a} n_i - \frac{\sum_{i=1}^{i=a} n_i^2}{\sum_{i=1}^{i=a} n_i}\right).$$

Thus, eq. (A.42b) becomes $\sigma^2 + n_0\sigma_A^2$, and the results of a Model II analysis of variance for unequal values of n are summarized as in table A.6.

We are now in a position to turn to nested analysis of variance, which is designed to handle cases where each group of variates is subdivided into two or more subordinate groupings or data classifications. Generally, nested analysis of variance is used to partition the total variance in a continuous variable Y among subordinate and higher levels of data classification. In scaling studies these levels of data classification are taxonomic levels (species, genera, families, etc.), and the objective of a nested analysis of variance is to determine which taxonomic level obtains an appro-

Table A.6

Source of Variance	*df*	*SS*	*MS*	*F*-Ratio	Expected *MS*
Among groups	$a - 1$	SS_{among}	$\frac{SS_{among}}{a-1}$	$\frac{MS_{among}}{MS_{within}}$	$\sigma^2 + n_0\sigma_A^2$
Within groups	$\sum_{i=1}^{a} n_i - a$	SS_{within}	$\frac{SS_{within}}{\sum_{i=1}^{a} n_i - a}$		σ^2
Total	$\sum_{i=1}^{a} n_i - 1$	SS_{total}	$\frac{SS_{within}}{\sum_{i=1}^{a} n_i - 1}$		

priate magnitude of variance for subsequent regression analysis. Recall that a fundamental assumption of regression analysis is that each datum is an independent observation taken on the smallest sampling unit. Although species are "small sampling units" in the sense of the standard taxonomic hierarchy, we have no a priori reason to assume that species are independent sampling units, because species "nested" in genera share features as a consequence of common ancestry. By examining the distribution of the total variance among subordinate and higher taxonomic levels, we gain access to how far species within and among genera provide independent observations. The procedure is to calculate the variance of species nested in genera based on species mean values of Y and then use generic means to calculate the variance of genera nested in families, and so on through higher taxonomic levels. Since, in scaling analysis subordinate groups (e.g., species) and higher groups (e.g., genera) typically are randomly selected, the nested analysis of variance is a *pure* Model II analysis of variance.

Fundamentally, no new tools are required for a nested analysis of variance, and therefore I shall proceed with an illustration of how to determine the distribution of variance among species, genera, and families (a three-way nested analysis of variance). The data set that will be used was obtained by taking measurements of gynoecial biomass from the flowers of fifty-four species distributed among nineteen genera. In turn these genera are nested within six families that belong to two orders of angiosperms. The gynoecial biomass of ten flowers was measured for each species. The data from each species were averaged to yield a mean gynoecial biomass, which was then log transformed. The data matrix obtained from this procedure is shown in table A.7. In this particular analysis, families are "groups," genera are "subgroups within groups," and species are "observations within subgroups." Thus the variance among groups is the variance among families, the variance among subgroups is the variance among genera nested within families, and the variance within subgroups is the variance among species nested within genera. This cumbersome terminology is regrettable but more or less unavoidable. In order to distinguish among groups, subgroups, and observations, the following notation will be used. Each group (= family) is denoted by i, each subgroup (= genus) is symbolized by j, and each observation (= log transformation of species mean gynoecial biomass) is distinguished as Y_{ijk}. That is, Y_{ijk} is the kth observation in the jth subgroup (= genus) of the ith group (= family). We also require symbols for the number of groups, subgroups, and observations in the nested analysis of variance: a, b, and n, respectively. Thus, the grand total of Y and the sum of squared observations are computed for three summations—the summation of observations $\sum_{k=1}^{k=n_{ij}}$, the summation of subgroups $\sum_{j=1}^{j=b_i}$, and the summation of groups $\sum_{i=1}^{i=a}$:

$$\text{Grand total } Y\text{: } \sum_{i=1}^{i=a} \sum_{j=1}^{j=b_i} \sum_{k=1}^{k=n_{ij}} Y_{ijk} = 70.756$$

$$\text{Sum of squares observations: } \sum_{i=1}^{i=a} \sum_{j=1}^{j=b_i} \sum_{k=1}^{k=n_{ij}} Y_{ijk}^2 = 145.917.$$

For convenience, the preceding data are reduced to the format shown in table A.8.

We begin the actual nested analysis of variance by first computing the sum of squared subgroup totals,

$$\sum_{i=1}^{i=a} \sum_{j=1}^{j=b_i} \frac{\left(\sum_{k=1}^{k=n_{ij}} Y_{ijk} \right)}{n_{ij}} = \frac{(-0.23746)^2}{3} + \frac{(1.97566)^2}{3} + \cdots + \frac{(1.93450)^2}{1} = 127.346,$$

the sum of squared group totals,

Table A.7

			Species			
Order	Family	Genus	1	2	3	4
1	1	1	−0.853872	0.133538	0.482873	
		2	−0.376750	0.729164	1.623249	
		3	−0.036212	1.707570	1.968482	
	2	4	−0.267606	0.117271	2.477121	
		5	−0.508638	0.187520	0.607455	
	3	6	1.724278	2.127104	3.598790	
		7	1.792390	2.113943	2.588831	
		8	0.06069	1.591064	2.663700	
	4	9	−0.958610	−0.494850	0.004321	1.000000
		10	1.113943	1.342422	1.414973	1.716003
		11	1.982271	2.004321		
		12	1.204119	1.255272	1.278753	1.602059
2	5	13	1.361727	1.462398	1.602059	
		14	1.591064	1.623249	1.623249	
		15	2.158362	2.220108	2.378397	
		16	2.397940	2.403120		
	6	17	1.707570	1.913813		
		18	1.763427	1.897627		
		19	1.93450			

$$\sum_{i=1}^{i=a} \frac{\left(\sum_{j=1}^{j=b_i} \sum_{k=1}^{k=n_{ij}} Y_{ijk}\right)^2}{\sum_{i=1}^{j=b_i} n_{ij}} = \frac{(5.37804)^2}{9} + \frac{(2.61312)^2}{6} + \cdots + \frac{(9.21694)^2}{5} = 112.751,$$

and the total variance (the total sum of squares),

$$s^2_{\text{total}} = SS_{\text{total}} = \sum_{i=1}^{i=a} \sum_{j=1}^{j=b_i} \sum_{k=1}^{k=n_{ij}} Y^2_{ijk} - \frac{\left(\sum_{i=a}^{i=a} \sum_{j=1}^{j=b_i} \sum_{k=1}^{k=n_{ij}} Y_{ijk}\right)^2}{\sum_{i=1}^{i=a} \sum_{j=1}^{j=b_i} n_{ij}}$$

$$= 145.917 - \frac{(70.756)^2}{54} = 53.2056.$$

Table A.8

Order	Family Number (a)	Genus Number (b)	Species Mean Y $\left(\sum_{k=1}^{k=n_{ij}} Y_{ijk}\right)$	Number of Species per Genus (n_{ij})	Family Mean Y $\left(\sum_{j=1}^{j=b_i} \sum_{k=1}^{k=n_{ij}} Y_{ijk}\right)$	Number of Species per Family (n_i)
1	1	1	−0.23746	3		
		2	1.97566	3		
		3	3.63984	3	5.37804	9
	2	4	2.32678	3		
		5	0.28634	3		
	3	6	7.45017	3	2.61312	6
		7	6.49516	3		
		8	4.31546	3	18.2608	9
	4	9	−0.44914	4		
		10	5.58733	4		
		11	3.98659	2		
		12	5.34021	4	14.4698	14
2	5	13	4.42619	3		
		14	4.83756	3		
		15	6.75687	3		
		16	4.80106	2	20.8216	11
	6	17	3.62138	2		
		18	3.66106	2		
		19	1.93450	1	9.21694	5

Next we compute the sum of squares within subgroups (the variance among species nested within genera), the sum of squares among subgroups (the variance among genera nested within families), and the sum of squares among groups (the variance among families):

$$s^2_{\text{within subgroups}} = SS_{\text{within subgroups}} = \sum_{i=1}^{i=a}\sum_{j=1}^{j=b_i}\sum_{k=1}^{k=n_{ij}} Y^2_{ijk} - \sum_{i=1}^{i=a}\sum_{j=1}^{j=b_i}\frac{\left(\sum_{k=1}^{k=n_{ij}} Y_{ijk}\right)^2}{n_{ij}}$$

$$= 145.917 - 127.346 = 18.5708$$

$$s^2_{\text{among groups}} = SS_{\text{among groups}} = \sum_{i=1}^{i=a}\sum_{j=1}^{j=b_i}\frac{\left(\sum_{k=1}^{k=n_{ij}} Y_{ijk}\right)^2}{n_{ij}} - \sum_{i=1}^{i=a}\frac{\left(\sum_{j=1}^{j=b_i}\sum_{k=1}^{k=n_{ij}} Y_{ijk}\right)^2}{\sum_{j=1}^{j=b_i} n_{ij}}$$

$$= 127.346 - 112.751 = 14.595$$

$$s^2_{\text{among groups}} = SS_{\text{among groups}} = \sum_{i=1}^{i=a}\frac{\left(\sum_{j=1}^{j=b_i}\sum_{k=1}^{k=n_{ij}} Y_{ijk}\right)^2}{\sum_{j=1}^{j=b_i} n_{ij}} - \frac{\left(\sum_{i=1}^{i=a}\sum_{j=1}^{j=b_i}\sum_{k=1}^{k=n_{ij}} Y_{ijk}\right)^2}{\sum_{i=1}^{i=a}\sum_{j=1}^{j=b_i} n_{ij}}$$

$$= 112.751 - \frac{(70.756)^2}{54} = 20.0397.$$

Using the subscript S for "species nested within genera," the subscript $G \subset F$ for "genera nested in families," and the subscript F for "among families," the results of the three-way nested analysis of variance up to this point are shown in table A.9. The nested analysis of variance is brought to closure once we have calculated the percentage distribution of variance among the three taxonomic levels (species, genera, and families). This requires computing the expected mean squares among groups, among subgroups, and within subgroups. As seen in the analysis of variance table, the formulas for these expected mean squares require the numerical values of the three variables, n_0, $n_{0'}$ and $(nb)_0$. The formulas for these three variables are as follows:

$$n_0' = \frac{1}{df_{\text{among groups}}} \sum_{i=1}^{i=a} \left(\frac{\sum_{j=1}^{j=b_i} n_{ij}^2}{\sum_{j=1}^{j=b_i} n_{ij}} \right) - \frac{\sum_{i=1}^{i=a} \sum_{j=1}^{j=b_i} n^2}{\sum_{i=1}^{i=a} \sum_{j=1}^{j=b_i} n_{ij}}$$

$$n_0 = \frac{1}{df_{\text{among subgroups}}} \sum_{i=1}^{i=a} \sum_{j=1}^{j=b} n_{ij} - \sum_{i=1}^{i=a} \left(\frac{\sum_{j=1}^{j=b_i} n_{ij}^2}{\sum_{j=1}^{j=b_i} n_{ij}} \right)$$

$$(nb)_0 = \frac{1}{df_{\text{among subgroups}}} \sum_{i=1}^{i=a} \sum_{j=1}^{j=b} n_{ij} - \frac{\sum_{i=1}^{i=a} \left(\sum_{j=1}^{j=b_i} n_{ij} \right)^2}{\sum_{i=1}^{i=a} \sum_{j=1}^{j=b} n_{ij}}.$$

Since

$$\sum_{i=1}^{i=a} \sum_{j=1}^{j=b_i} n_{ij} = 3 + 3 + 3 + \cdots + 1 = 54$$

$$\sum_{i=1}^{i=a} \sum_{j=1}^{j=b_i} n_{ij}^2 = 3^2 + 3^2 + 3^2 + \cdots + 1^2 = 164$$

$$\sum_{i=1}^{i=a} \left(\sum_{j=1}^{j=b_i} n_{ij}^2 \right)^2 = 9^2 + 6^2 + 9^2 + \cdots + 5^2 = 540$$

$$\sum_{i=1}^{i=a} \left(\frac{\sum_{j=1}^{j=b_i} n_{ij}^2}{\sum_{j=1}^{j=b_i} n_{ij}} \right) = \frac{3^2 + 3^2 + 3^2}{9} + \frac{3^2 + 3^2}{6} + \cdots + \frac{2^2 + 2^2 + 1^2}{5} = 17.332,$$

Table A.9

Source of Variance	*df*	*SS*	*MS*	Expected *MS*
Among groups (among families)	5	20.0397	4.00794	$\sigma_s^2 + n_o' \sigma_{GCF}^2 + (nb)_o \sigma_F^2$
Among subgroups (among genera nested in families)	13	14.595	1.12269	$\sigma_s^2 + n_o \sigma_{GCF}^2$
Within subgroups (among species nested in genera)	35	18.5708	0.53059	σ_s^2
Total	53	53.2056	1.00388	

we see that

$$n_0' = \frac{1}{5}\left(17.332 - \frac{164}{54}\right) = 2.85899$$

$$n_0 = \frac{1}{13}(54 - 17.332) = 2.82062$$

$$(nb)_0 = \frac{1}{5}\left(54 - \frac{540}{164}\right) = 10.1415.$$

The expected parametric mean squares are

$$\sigma_S^2 + n_0\,\sigma_{G\subset F}^2 + (nb)_0\,\sigma_F^2 = 4.00794$$
$$\sigma_S^2 + n_0\,\sigma_{G\subset F}^2 = 1.12269$$
$$\sigma_S^2 = 0.53059,$$

and therefore the estimated mean squares are

$$s_S^2 + 2.85899 s_{B\subset A}^2 + 10.1415 s_A^2 = 4.00794$$
$$s_S^2 + 2.82062\sigma_{B\subset A}^2 = 1.12269$$
$$s_S^2 = 0.53059.$$

We now have three equations with only two unknowns. Thus we can determine the numerical values for the estimated variance components:

$$s_{G\subset F}^2 = \frac{MS_{\text{among subgroups}} - MS_{\text{within subgroups}}}{n_0}$$
$$= \frac{1.12269 - 0.53059}{2.82062} = 0.209918$$
$$s_F^2 = \frac{MS_{\text{among groups}} - MS_{\text{within subgroups}} - n_0' s_{B\subset A}^2}{(nb)_0}$$
$$= \frac{4.00794 - 0.53059 - 2.85899(0.209918)}{10.1415} = 2.87719.$$

Since 0.209918 + 2.87719 + 0.53059 = 3.61769, the variance components expressed as percentages are as follows:

s_S^2	$= 100\% \times \frac{(0.53059)}{3.61769} \approx 15\%$	Percentage variance within subgroups (among species nested within genera)
$s_{G\subset F}^2$	$= 100\% \times \frac{(0.209918)}{3.61769} \approx 6\%$	Percentage variance among subgroups (genera nested within families)
s_F^2	$= 100\% \times \frac{(2.87719)}{3.61769} \approx 79\%.$	Percentage variance among groups (among families)

Space does not permit a treatment of the *F*-ratios for this nested analysis of variance. Tentatively, however, we conclude that the family may be the most appropriate taxonomic level for regression analysis (i.e., family mean values of gynoecial biomass afford independent data points for regression analysis). Further analysis, however, is required to consider the variance component contributed by families nested in orders. In this analysis, we would compute genus means for *Y* and consider orders as groups, families, as subgroups, and genera within families as observations. When this is done for mammals, nested analysis of variance reveals that species within genera and genera within families provide additional but nonindependent data points (see Harvey and Mace 1982; Harvey and Clutton-Brock 1985; Pagel and Harvey 1988), whereas my analyses of continuous variables for plants show that the genus and the species levels often obtain independent data points.

Determining the appropriate taxonomic level by nested analysis of variance can present difficulties, however. At a fundamental level, no taxonomic level yields truly independent data points. Additionally, the choice of taxonomic level may not be clear-cut when the percentage distribution of variance is comparatively smooth, the appropriate taxonomic level may differ for two or more variables considered important to a scaling analysis, or the variables may be represented by too few data points (i.e., taxa) for robust regression analysis. Other techniques may present better alternatives to nested analysis of variance. For example, Stearns (1983) and Wooton (1987) present a phyletic subtraction method whereby the variation in a continuous variable resulting from phyletic affiliations with higher taxonomic levels is removed from the data obtained from species. The average value of a trait for the higher taxonomic level can be subtracted from the average value measured for the subordinate level, the resulting "residuals" of the subordinate taxonomic level presumably thereby becoming free of the phyletic effect.

A.7 Timing of Ontogenetic Events

In chapter 5 I mentioned the presumed difficulty of equating the time of appearance of ontogenetic events in different but presumably phyletically related ontogenetic systems. In this section I illustrate a technique that permits direct comparisons among the temporal networks of different but related systems. This technique follows directly from systems analysis, which figures prominently in the engineering sciences.

Figure A.1 diagrams a simple ontogenetic sequence of interrelated phenotypic events A–E, configured into a temporal network of the appearance of these events. The task is to find the most probable time of appearance of an event relative to the other events. This permits direct examination of a putative descendant ontogeny to determine whether the observed time of its appearance evinces alteration. This comparison requires us to pay close attention to errors in time measurements. It also requires the selection of one event as the temporal referent. The time of appearance of *A* will be used as the arbitrary temporal referent for the other ontogenetic events shown in figure A.1. In this regard, the arrows drawn in this figure represent the difference in time between the events joined by each arrow; the symbol l_j denotes the measured time interval between events *i* and *k;* and the symbol v_j is the most probable error in the time measurement. Relative times of appearance of events are shown as follows: when l_j is positive, the event *k* at the head of the arrow is later than the point *i* at the tail of the arrow; and when l_j is negative, then event *k* at the head of the arrow is earlier than the point *l* at the tail of the arrow. Finally, the symbol h_j denotes the most probable time of appearance of event *j*.

In terms of system analysis, the known and unknown system components are as follows:

$$\begin{aligned} \text{Known } & l_j \text{ for } j = 1 \text{ to } 7 \\ \text{Unknown } & h_i \text{ for } i = 2 \text{ to } 5 \\ & v_j \text{ for } j = 1 \text{ to } 7 \end{aligned}$$

The task now is to determine the magnitudes of the unknown system components. From the notation given in figure A.1, the system components are logically restricted such that $l_j - v_j = h_k - h_i$. Because v_j is the error of time measurement, it must be subtracted from l_j to correct for the

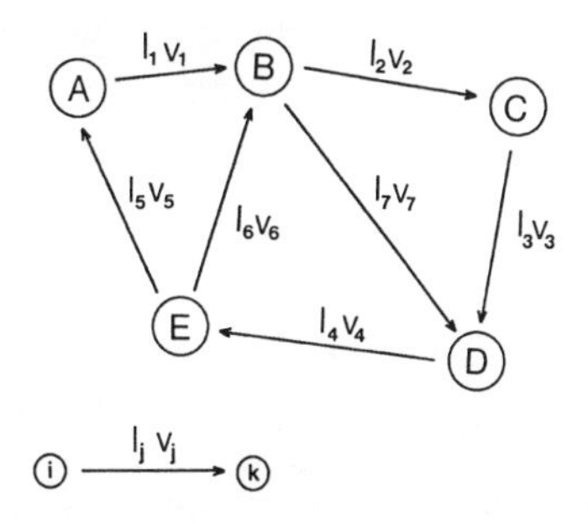

Figure A.1. Schematic of a hypothetical ontogenetic sequence of networked events *A–E*. Time of appearance of *A* used as arbitrary temporal reference for events *B–E*. Arrow indicates temporal relation between two events (e.g., $B \rightarrow C$ = event *B* occurs before event C). Time interval between neighboring events indicated by l_j (time between events *i* and *k*). Most probable error in time measurement indicated by v_j.

temporal difference between any two ontogenetic events. The following equation matrix is obtained when this is done for each of these events:

(A.44)

$$
\begin{aligned}
l_1 - v_1 &= h_2 \\
l_2 - v_2 &= -h_2 + h_3 \\
l_3 - v_3 &= -h_3 + h_4 \\
l_4 - v_4 &= -h_4 + h_5 \\
l_5 - v_5 &= -h_5 \\
l_6 - v_6 &= -h_2 + h_5 \\
l_7 - v_7 &= -h_2 + h_4
\end{aligned}
$$

Rearranging all the unknown parameters to the left-hand side yields the following equation matrix:

(A.45)

$$
\begin{array}{llllll}
v_1 & + h_2 & & & & = l_1 \\
v_2 & - h_2 & + h_3 & & & = l_2 \\
v_3 & & - h_3 & + h_4 & & = l_3 \\
v_4 & & & - h_4 & + h_5 & = l_4 \\
v_5 & & & & -h_5 & = l_5 \\
v_6 & + h_2 & & & - h_5 & = l_6 \\
v_7 & - h_2 & & + h_4 & & = l_7
\end{array}
$$

In matrix notation, eq. (A.45) is expressed as follows:

(A.46)

$$
\begin{bmatrix} v_1 \\ v_2 \\ v_3 \\ v_4 \\ v_5 \\ v_6 \\ v_7 \end{bmatrix}
+
\begin{bmatrix}
1 & 0 & 0 & 0 \\
-1 & 1 & 0 & 0 \\
0 & -1 & 1 & 0 \\
0 & 0 & -1 & 1 \\
0 & 0 & 0 & -1 \\
1 & 0 & 0 & -1 \\
-1 & 0 & 1 & 0
\end{bmatrix}
\begin{bmatrix} h_2 \\ h_2 \\ h_4 \\ h_5 \end{bmatrix}
=
\begin{bmatrix} l_1 \\ l_2 \\ l_3 \\ l_4 \\ l_5 \\ l_6 \\ l_7 \end{bmatrix}
$$

In other words,

(A.47)

$$\mathbf{V}_{(7 \times 1)} + \mathbf{A}_{(7 \times 4)} \mathbf{H}_{(4 \times 1)} = \mathbf{L}_{(7 \times 1)}.$$

Equation (A.47) cannot be used until the measurement errors v_i are fully characterized. We will assume that these errors are uniform for all measurements of time intervals l_i. That is, all values of v_i have equal accuracy. This is not an unreasonable assumption for most experimental systems. Now we may consider any unknown event, say event *B*, and assume that the probable times of appearance for events *C*, *D*, and *E* are h_3, h_4, and h_5, respectively. We make the further assumption that the time intervals between these events and event B have been measured as l_2, l_7, l_6, and l_1

(see fig. A.1). Accordingly, four different values for the relative time of appearance of event B are possible:

(A.48)
$$\begin{aligned}(h_B)_1 &= h_4 - l_2\\ (h_B)_2 &= h_4 - l_7\\ (h_B)_3 &= h_5 + l_6\\ (h_B)_4 &= l_1\end{aligned}$$

The logical way to deal with these four alternatives is to take the mean of eq. (A.48). That is, $\bar{h}_B = (1/4)[(h_B)_1 + (h_B)_2 + (h_B)_3 + (h_B)_4]$. Rearranging the terms that appear in the mean value of h_B yields the following:

(A.49)
$$4h_2 - h_3 - h_4 - h_5 = l_1 - l_2 + l_6 - l_7.$$

Applying this logic to ontogenetic events C–E yields:

(A.50)
$$\begin{aligned}&\text{Event } C \; -h_2 + 2h_3 - h_4 = l_2 - l_3\\ &\text{Event } D \; -h_2 - h_3 + 3h_4 - h_5 = l_3 - l_4 + l_7\\ &\text{Event } E \; -h_2 - h_4 - h_5 = l_4 - l_5 - l_6\end{aligned}$$

In matrix notation, eq. (A.48) becomes

(A.51)
$$\begin{bmatrix} 4 & -1 & -1 & -1\\ -1 & 2 & -1 & 0\\ -1 & -1 & 3 & -1\\ -1 & 0 & -1 & 3\end{bmatrix}\begin{bmatrix} h_2\\ h_3\\ h_4\\ h_5\end{bmatrix} = \begin{bmatrix} l_1 & -l_2 & l_6 & -l_7\\ l_2 & -l_3 & & \\ l_3 & -l_4 & +l_7 & \\ l_4 & -l_5 & -l_6 & \end{bmatrix}.$$

In other words,

(A.52)
$$\mathbf{B}_{(4\times 4)}\,\mathbf{H}_{(4\times 1)} = \mathbf{D} = {}_{(4\times 1)}.$$

It is comparatively easy to show that eqs. (A.47) and (A.52) are interrelated such that $\mathbf{B}_{(4\times 4)} = \mathbf{A}^{\mathrm{T}}{}_{(4\times 7)}\,\mathbf{A}_{(7\times 4)}$ and $\mathbf{D}_{(4\times 4)} = \mathbf{A}^{\mathrm{T}}{}_{(4\times 7)}\,\mathbf{L}_{(7\times 1)}$, where $\mathbf{A}^{\mathrm{T}}$ is the transposed matrix of $\mathbf{A}$. Thus, eq. (A.52) can be rewritten as follows:

(A.53)
$$(\mathbf{A}^{\mathrm{T}}\mathbf{A})\mathbf{H} = \mathbf{A}^{\mathrm{T}}\mathbf{L}.$$

And by grouping eqs. (A.45) and (A.51), we find that the following are true for the ontogenetic system diagrammed in figure A.1:

(A.54a) $\mathbf{V} + \mathbf{AH} = \mathbf{L}$ (system compatibility)

(A.54b) $(\mathbf{A}^{\mathrm{T}}\mathbf{A})\mathbf{H} = \mathbf{L}$ (error behavior)

(A.54c) $\mathbf{H} = (\mathbf{A}^{\mathrm{T}}\mathbf{A})^{-1}\,\mathbf{A}^{\mathrm{T}}\mathbf{L}$ (probable times of appearance)

The mathematical techniques derived from a systems analysis approach allow us to draw precise quantitative comparisons between any two ontogenies that share the same phenotypic events but for which the appearance of events is shifted in time. These techniques may appear unfamiliar to the biologist, but they figure in practical applications to engineered networks of surpassing complexity and sophistication. Although the complexity of these networked systems pales in comparison with those attacked by the biologist, the engineer and the biologist often share the same goals.

References

Adair, W. S., S. A. Steinmetz, D. M. Mattson, U. W. Goodenough, and J. E. Heuser, 1987. Nucleated assembly of *Chlamydomonas* and *Volvox* cell walls. *J. Cell Biol.* 105:2373–82.

Agustí, S. 1991. Allometric scaling of light absorption and scattering by phytoplankton cells. *Can. J. Fish. Aquat. Sci.* 48:763–67.

Albrech, P., S. J. Gould, G. F. Oster, and D. B. Wake. 1979. Size and shape in ontogeny and phylogeny. *Paleobiology* 5:296–317.

Alexander, R. McN. 1971. *Size and shape.* London: Edward Arnold.

———. 1979. *The invertebrates.* Cambridge: Cambridge University Press.

Allison, T. D. 1991. Variation in sex expression in Canada yew (*Taxus canadensis*). *Amer. J. Bot.* 78:569–78.

Amelung, E. 1893. Über mittlere Zellergrössen. *Flora* 77:176–207.

Armstrong, E. 1983. Relative brain size and metabolism in mammals. *Science* 220:1302–4.

Atchely, W. R., B. Riska, L. A. P. Kohn, A. A. Plummer, and J. J. Rutledge. 1984. A quantitative genetic analysis of brain and body size associations, their origin and ontogeny. Data from mice. *Evolution* 38:1165–79.

Atkinson, G. F. 1898. Experiments on the morphology of *Arisaema triphyllum. Bot. Gaz.* 25:1–114.

Augspurger, C. K. 1986. Morphology and dispersal potential of wind-dispersed diaspores of Neotropical trees. *Amer. J. Bot.* 73:353–63.

Augspurger, C. K., and K. P. Hogan. 1983. Wind dispersal of fruits with variable seed number in a tropical tree (*Lonchocarpus pentaphyllus:* Leguminosae). *Amer. J. Bot.* 70:1031–37.

Bachmann, K. 1983. Evolutionary genetics and the genetic control of morphogenesis in flowering plants. *Evol. Biol.* 16:157–206.

Bacon, R. H. 1953. The "best" straight line among the points. *Amer. J. Phys.* 21:428–46.

Baker, H. G. 1948. Corolla size in gynodioecious and gynomonoecious species of flowering plants. *Proc. Leeds Philos. Soc.* 5:136–39.

Bakker, R. T. 1977. Tetrapod mass extinctions—a model of the regulation of speciation rates and immigration by cycles of topographic diversity. In *Patterns of evolution as illustrated by the fossil record,* ed. A. Hallam, 439–68. Amsterdam: Elsevier.

Bange, G. G. J. 1953. On the quantitative explanation of stomatal transpiration. *Acta Bot. Neerl.* 2:225–97.

Banse, K. 1976. Rates of growth, respiration and photosynthesis of unicellular algae as related to cell size—a review. *J. Phycol.* 12:135–40.

Bateman, R. M., W. A. DiMichele, and D. A. Willard. 1992. Experimental cladistic analysis of anatomically preserved arborescent lycopods from the Carboniferous of Euramerica: An essay on paleobotanical phylogenetics. *Ann. Missouri Bot. Gard.* 79:500–559.

Bazzaz, F. A., R. W. Carlson, and J. L. Harper. 1979. Contribution to reproductive effort by photosynthesis of flowers and fruits. *Nature* 279:554–55.

Bazzaz, F. A., and J. L. Harper. 1977. Demographic analysis of the growth of *Linum usitatissimum. New Phytol.* 78:193–208.

Bell, P. R. 1989. The alternation of generations. *Adv. Bot. Res.* 16:55–93.

Beri, S. M., and S. C. Anand. 1971. Factors affecting pollen shedding capacity in wheat. *Euphytica* 20:327–32.

Bertalanffy, L. von. 1952. *Problems of life.* London: Watts.

———. 1960. Principles and theory of growth. In *Fundamental aspects of normal and malignant growth,* ed. W. W. Nowinski, 137–259. Amsterdam: Elsevier.

Berthold, G. 1882. Über die Verteilung der Algen im Golf von Neapel nebst einem Verzeichnis der bisher daselbst beobachten Arten. *Mitt. Zool. Sta. Neopol.* 3:393–536.

Bertram, J. E. A. 1989. Size-dependent differential scaling in branches: The mechanical design of trees revisited. *Trees* 4:241–53.

Berven, K. A. 1982. The genetic basis of altitudinal variation in the wood frog *Rana sylvatica.* I. An experimental analysis of life history traits. *Evolution* 36:962–83.

Beverton, R. J. H., and S. J. Holt. 1957. *On the dynamics of exploited fish population.* London: Her Majesty's Stationery Office.

Biao D., M. V. Parthasarathy, K. J. Niklas, and R. Turgeon, 1988. A morphometric analysis of the phloem-unloading pathway in developing tobacco leaves. *Planta* 176:307–18.

Bierhorst, D. W. 1977. The systematic position of *Psilotum* and *Tmesipteris. Brittonia* 29:3–13.

Bisalputra, T., and J. R. Stein. 1966. The development of cytoplasmic bridges in *Volvox aureus. Can. J. Bot.* 44:1697–1702.

Blackman, G. E. 1961. Responses to environmental factors by plants in the vegetative phase. In *Growth in living systems,* ed. M. X. Zarrow, 525–56. New York: Basic Books.

Blake, J. R., and M. A. Sleigh. 1974. Mechanics of ciliary locomotion. *Biol. Rev.* 49:85–125.

Blueweiss, L., H. Fox, V. Kudzma, D. Nakashima, R. Peters, and S. Sams. 1978. Relationships between body size and some life history parameters. *Oecologia* 37:257–72.

Blum, J. J. 1977. On the geometry of four-dimensions and the relationship between metabolism and body mass. *J. Theor. Biol.* 64:599–601.

Bookstein, F. L. 1978. *The measurement of biological shape and shape change.* Lecture Notes in Biomathematics 24. Berlin: Springer-Verlag.

Bower, F. O. 1908. *The origin of a land flora.* London: Macmillan.

Bowman, J. L., D. R. Smyth, and E. M. Meyerowitz. 1989. Genes directing flower development in *Arabidopsis. Plant Cell* 1:37–52.

Briggs, D. E. G. 1985. Gigantism in Palaeozoic arthropods. *Spec. Pap. Palaeontol.* 33:157–69.

Brokaw, C. J. 1967. Adenosine triphosphate usage by flagella. *Science* 156:76–78.

Burd, M. and T. F. H. Allen. 1988. Sexual allocation strategy in wind-pollinated plants. *Evolution* 42:403–7.

Buss, L. W. 1987. *The evolution of individuality.* Princeton: Princeton University Press.

Camp, W. H. 1932. Sex in *Arisaema triphyllum. Ohio J. Sci.* 32:147–51.

Campbell, C. S. 1982. Cleistogamy in *Andropogon* (Gramineae). *Amer. J. Bot.* 69:1625–35.

Campbell, D. R. 1989. Measurements of selection in a hermaphroditic plant: Variation in male and female pollination success. *Evolution* 43:318–34.

Campbell, R. J., and P. D. Ascher. 1975. Incorporation of radioactive label into nucleic acids of compatible and incompatible pollen tubes of *Lilium longiflorum* Thumb. *Theor. Appl. Genet.* 46:143–48.

Casper, B. B. 1987. Spatial patterns of seed dispersal and postdispersal seed predation of *Cryptantha flava* (Boraginaceae). *Amer. J. Bot.* 74:1646–55.

Chaloner, W. G., and A. Sheerin. 1979. Devonian macrofloras. In *The Devonian system,* 145–61. Special Papers in Palaeontology 23. London: Palaeontological Society.

Charlesworth, D., and B. Charlesworth. 1981. Allocation of resources to male and female functions in hermaphrodites. *Biol. J. Linn. Soc.* 15:57–74.

———. 1987. The effect of investment in attractive structures on allocation to male and female functions in plants. *Evolution* 41:948–68.

Charnov, E. L. 1982. *The theory of sex allocation.* Princeton: Princeton University Press.

Charters, A. C., M. Neushul, and D. C. Barilotti. 1969. The functional morphology of *Eisenia arborea. Proc. Int. Seaweed Symp.* 6:89–105.

Chasan, R. 1991. Modeling flowers. *Plant Cell* 3:845–47.

Chazdon, R. L. 1991. Plant size and form in the understory palm genus *Geonoma:* Are species variations on a theme? *Amer. J. Bot.* 78:680–94.

Chittenden, F. J., ed. 1931. Statistical returns of conifers. In *Conifers in cultivation,* 316–28. Proceedings of the Conifer Conference, London, 1931. London: Royal Horticultural Society.

Clay, K., and R. Shaw. 1981. An experimental demonstration of density-dependent reproduction in a natural population of *Diamorpha smallii,* a rare annual. *Oecologia* 51:1–6.

Clift, R., J. R. Grace, and M. E. Weber. 1978. *Bubbles, drops, and particles.* New York: Academic Press.

Condon, M. A., and L. E. Gilbert. 1988. Sex expression of *Gurania* and *Psiguria* (Cucurbitaceae): Neotropical vines that change sex. *Amer. J. Bot.* 75:875–84.

Cousens, R., and M. J. Hutchings, 1983. The relationship between density and mean frond weight in monospecific seaweed stands. *Nature* 301:240–41.

Cruden, R. W., and D. L. Lyon. 1985. Correlations among stigma depth, style length, and pollen grain size: Do they reflect function or phylogeny? *Bot. Gaz.* 146:143–49.

Cruden, R. W., and S. Miller-Ward. 1981. Pollen-ovule ratio, pollen size, and the ratio of stigmatic area to the pollen-bearing area of the pollinator: An hypothesis. *Evolution* 35:964–74.

Darling, M. S. 1989. Epidermis and hypodermis of the saguaro cactus (*Cereus giganteus*): Anatomy and spectral properties. *Amer. J. Bot.* 76:1698–1705.

Darwin, C. 1877. *The different forms of flowers on plants of the same species.* New York: Appleton.

Dean, T. J., and J. N. Long. 1986. Validity of constant-stress and elastic-instability principles of stem formation in *Pinus contorta* and *Trifolium pratense. Ann. Bot.* 58:833–40.

De Beer, G. 1958. *Embryos and ancestors.* Oxford: Clarendon Press.

Delesalle, V. A. 1989. Year-to-year changes in phenotypic gender in a monoecious cucurbit, *Apodanthera undulata. Amer. J. Bot.* 76:30–39.

Delf, E. M. 1932. Experiments with the stipes of *Fucus* and *Laminaria. J. Exp. Biol.* 9:300–313.

Denny, M. W. 1988. *Biology and the mechanics of the wave-swept environment.* Princeton: Princeton University Press.

Devlin, B. 1989. Components of seed and pollen yield of *Lobelia cardinalis:* Variation and correlation. *Amer. J. Bot.* 76:204–14.

DiMichele, W. A., and R. M. Bateman. 1992. Diaphorodendraceae, fam. nov. (Lycopsida: Carboniferous): Systematics and evolutionary relationships of *Diaphorodendron* and *Synchysidendron,* gen. nov. *Amer. J. Bot.* 79:605–17.

DiMichele, W. A., J. F. Mahaffy, and T. L. Phillips. 1979. Lycopods of Pennsylvania age coals: *Polysporia. Can. J. Bot.* 57:1740–53.

DiMichele, W. A., T. L. Phillips, and G. E. McBrinn. 1991. Quantitative analysis and paleoecology of the Secor Coal and Roof-Shale floras (Middle Pennsylvania, Oklahoma). *Palaios* (Society for Sedimentary Geology, Research Reports) 6:390–409.

Donhoffer, S. 1986. Body size and metabolic rate: Exponent and coefficient of the allometric equation. The role of units. *J. Theor. Biol.* 119:125–37.

Doyle, J. A., and M. J. Donoghue. 1987. The importance of fossils in elucidating seed plant phylogeny and macroevolution. *Rev. Palaeobot. Palynol.* 50:63–95.

Duckett, J. G. 1970a. Sexual behavior of the genus *Equisetum,* subgenus *Equisetum. Bot. J. Linn. Soc.* 63:327–52.

———. 1970b. Spore size in the genus *Equisetum. New Phytol.* 69:333–46.

———. 1977. Towards an understanding of sex determination in *Equisetum:* An analysis of regeneration in gametophytes of the subgenus *Equisetum. Bot. J. Linn. Soc.* 74:215–42.

Durnin, J. V. G. A., and R. Passmore. 1967. *Energy, work, and leisure.* London: Heinemann.

Duysens, L. N. M. 1956. The flattening of the absorption spectrum of suspensions as compared to that of solutions. *Biochim. Biophys. Acta* 19:1–12.

Economos, A. C. 1979. Gravity, metabolic rate, and body size of mammals. *Physiologist* 22:S-71–S-72.

Edwards, D., K. L. Davies, and L. Axe. 1992. A vascular conducting strand in the early land plant *Cooksonia. Nature* 357:638–85.

Emerson, R., and C. M. Lewis. 1943. The dependence of the quantum yield of *Chlorella* photosynthesis on wavelength of light. *Amer. J. Bot.* 30:165–78.

Engelmann, T. W. 1883. Farbe und Assimilation. *Bot. Zeit.* 41:1–13.

Eppley, R. W., and P. R. Sloan. 1965. Carbon balance experiments with marine phytoplankton. *J. Fish. Res. Bd. Canada* 22:1083–97.

———. 1966. Growth rates and marine phytoplankton: Correlation with light absorption by cell chlorophyll *a*. *Physiol. Plant.* 19:47–59.

Fehr, W. R., and H. H. Hadley. 1980. *Hybridization of crop plants.* Madison, Wis.: American Society of Agronomists and Crop Science Society of America.

Felsenstein, J. 1983. Parsimony in systematics: Biological and statistical issues. *Annu. Rev. Ecol. Syst.,* 14:313–33.

Fenchel, T. 1974. Intrinsic rate of natural increase: The relationship with body size. *Oecologia* (Berlin) 14:317–26.

Fisher, D. C. 1991. Phylogenetic analysis and its application in evolutionary paleobiology. In *Analytical paleobiology,* ed. N. L. Gilinsky and P. W. Signot, 103–22. Short Courses in Paleontology 4. Knoxville: University of Tennessee.

Fisher, R. A. 1930. *The genetical theory of natural selection.* Reprint, New York: Dover, Reprint.

Foster, A. S., and F. A. Barkley. 1933. Organization and development of foliar organs in *Paeonia officinalis. Amer. J. Bot.* 20:365–85.

Freeman, D. C., K. T. Harper, and E. L. Charnov. Sex change in plants: Old and new observations and new hypotheses. *Oecologia* 47:222–32.

Freeman, D. C., and E. D. McArthur, and K. T. Harper. 1984. The adaptive significance of sexual lability in plants using *Atriplex canescens* as a prinicpal example. *Ann. Missouri Bot. Gard.* 71:265–77.

French, C. S. 1960. The chlorophylls in vivo and in vitro. In *Encyclopedia of plant physiology,* ed. W. Ruhland, 5.1:252–97. Berlin: Springer-Verlag.

French, J. C., and D. J. Paolillo Jr. 1976. Effect of the calyptra on intercalary meristematic activity in the sporophyte of *Funaria* (Musci). *Amer. J. Bot.* 63:492–98.

Fritsch, F. E. 1965. *The structure and reproduction of the algae.* Vol. 1. Cambridge: Cambridge University Press.

Gaines, M. S., K. J. Vogt, J. L. Hamrick, and J. Caldwell. 1974. Reproductive strategies and growth patterns in sunflowers (*Helianthus*). *Amer. Nat.* 108:889–94.

Gale, M. D., and S. Youssefian. 1985. Dwarfing genes of wheat. In *Progress in plant breeding,* ed. G. S. Russell, 1:1–35. London: Butterworth.

Galilei, Galileo. 1638. *Discorsi e dimostrazioni mathematiche, intorno a due nuove scienze* (Dialogues concerning two new sciences). Trans. H. Crew and A. De Salvio. New York, 1914.

Gensel, P. G. 1992. Phylogenetic relations of the zosterophylls and lycopods: Evidence from morphology, paleoecology, and cladistic methods of inference. *Ann. Missouri Bot. Gard.* 79:450–73.

Gibbons, I. R. 1966. Studies on the adenosine triphosphate activity of 14S and 30S dynein from cilia of *Tetrahymena. J. Biol. Chem.* 24:5590–96.

Gimmler, H. C. Weiss, M. Baier, and W. Hartung. 1990. The conductance of the plasmalemma for CO_2. *J. Exp. Bot.* 41:785–95.

Gleason, H. A. 1968. *The new Britton and Brown Illustrated flora of the northeastern United States and adjacent Canada.* New York: Hafner.

Goldman, D. A., and M. F. Willson. 1986. Sex allocation in functionally hermaphroditic plants: A review and critique. *Bot. Rev.* 52:157–94.

Golub, S. J., and R. H. Wetmore. 1948. Studies of development in the vegetative shoot of *Equisetum arvense:* I. The shoot apex. II. The mature shoot. *Amer. J. Bot.* 35:755–81.

Gorham, E. 1979. Shoot height, weight and standing crop in relation to density of monospecific plant stands. *Nature* 279:148–50.

Gottlieb, L. D. 1984. Genetic and morphological evolution in plants. *Amer. Nat.* 123:681–709.

Gould, S. J. 1966. Allometry and size in ontogeny and phylogeny. *Biol. Rev.* 41:587–640.

———. 1977. *Ontogeny and phylogeny.* Cambridge: Harvard University Press.

———. 1990. Speciation and sorting as the source of evolutionary trends, or "Things are seldom what they seem." In *Evolutionary trends,* ed. J. J. McNamara, 3–27. Tucson: University of Arizona Press.

Grace, J. and J. Wilson. 1976. The boundary layer over a *Populus* leaf. *J. Exp. Bot.* 27:231–41.

Gray, B. F. 1981. On the "surface law" and basal metabolic rate. *J. Theor. Biol.* 93:757–67.

Green, P. B. 1976. Growth and cell pattern formation on an axis: Critique of concepts, terminology, and modes of study. *Bot. Gaz.* 137:187–202.

Greenhill, G. 1881. Determination of the greatest height consistent with stability that a vertical pole or mast can be made, and the greatest height to which a tree of given proportions can grow. *Proc. Cambridge Phil. Soc.* 4:65–73.

Grout, A. J. 1972. *Moss flora of North America north of Mexico.* Facsimile edition of 1936–39 ed. New York: Hafner.

Guerrant, E. O., Jr. 1982. Neotenic evolution of *Delphinium nudicaule* (Ranunculaceae): A hummingbird-pollinated larkspur. *Evolution* 36:699–712.

Günther, B. 1975. Dimensional analysis and theory of biological similarity. *Physiol. Rev.* 55:659–99.

Haber, A. H. 1962. Nonessentiality of concurrent cell divisions for degree of polarization of leaf growth: I. Studies with radiation-induced mitotic inhibition. *Amer. J. Bot.* 49:583–89.

Haber, A. H., W. L. Carrier, and D. E. Foard. 1961. Metabolic studies of gamma-irradiated wheat growing without cell division. *Amer. J. Bot.* 48:431–38.

Haber, A. H., and H. J. Luippold. 1960. Effects of gibberellin on gamma-irradiated wheat. *Amer. J. Bot.* 47:140–44.

Haig, D., and M. Westoby. 1988. Inclusive fitness, seed, resources, and maternal care. In *Plant reproductive ecology: Patterns and strategies,* ed. J. Lovett Doust and L. Lovett Doust, 60–79. New York: Oxford University Press.

Hallam, A. 1990. Biotic and abiotic factors in the evolution of early Mesozoic marine mollusks. In *Causes of evolution: A paleontological perspective,* ed. R. M. Ross and W. D. Allmon, 249–69. Chicago: University of Chicago Press.

Hallé, F. R., A. A. Oldeman, and P. B. Tomlinson. 1978. *Tropical trees and forests: An architectural analysis.* Berlin: Springer-Verlag.

Happel, J. 1973. *Low Reynolds number hydrodynamics, with special applications to particulate media.* Leiden: Noordhott International.

Harvey, P. H. 1982. On rethinking allometry. Cambridge: *J. Theor. Biol.* 95:37–41.

Harvey, P. H., and T. H. Clutton-Brock. 1985. Life history variation in primates. *Evolution,* 39:559–81.

Harvey, P. H., and G. M. Mace. 1982. Comparisons between taxa and adaptive trends: Problems of methodology. In *Current problems in sociobiology,* ed. King's College Sociobiology Group, 343–61. Cambridge: Cambridge University Press.

Harvey, P. H., and M. D. Pagel, 1991. *The comparative method in evolutionary biology.* Oxford: Oxford University Press.

Herrera, J. 1991. Allocation of reproductive resources within and among inflorescences of *Lavandula stoechas* (Lamiaceae). *Amer. J. Bot.* 78:789–94.

Heslop-Harrison, J. 1957. The experimental modification of sex expression in flowering plants. *Biol. Rev.* 32:38–90.

Hickman, J. C. 1975. Environmental unpredictability and plastic energy allocation strategies in the annual *Polygonum cascadense* (Polygonaceae). *J. Ecol.* 63:689–701.

———. 1977. Energy allocation and niche differentiation in four co-existing annual species of *Polygonum* in western North America. *J. Ecol.* 65:317–26.

Hikmat, A. A. A., B. R. Strain, and H. A. Mooney. 1972. The physiological ecology of diverse populations of the desert shrubs, *Simmondisa chinensis. J. Ecol.* 60:41–57.

Hoerner, S. F. 1965. *Fluid-dynamic drag.* Bricktown, N.J.: Hoerner Fluid Dynamics.

Hofmeister, W. 1851. *Vergleichende Untersuchungen der Keimung, Entfaltung und Fruchtbildung höherer Kryptogamen.* Leipzig: Hofmeister.

———. 1867. *Die Lehre von der Pflanzenzelle.* Handbuch der Pflanzenphysiologie, vol. 1, pt. 1. Leipzig: Wilhelm Engelmann.

Holbrook, N. M., M. D. Denny, and M. A. R. Koehl. 1991. Intertidal "trees": Consequences of aggregation on the mechanical and photosynthetic properties of sea-palms *Postelsia palmaeformis* Ruprecht. *J. Exp. Mar. Biol. Ecol.* 146:39–67.

Hossaert, M., and M. Valéro. 1988. Effect of ovule position in the pod on patterns of seed formation in two species of *Lathyrus* (Leguminosae: Papilionoideae). *Amer. J. Bot.* 75:1714–31.

Hunt, E. R., and P. S. Nobel. 1987. Allometric root/shoot relationships and predicted water uptake for desert succulents. *Ann. Bot.* 59:571–77.

Huxley, J. S. 1924. Constant differential growth-ratios and their significance. *Nature* (London) 114:895–96.

———. 1932. *Problems of relative growth.* New York: MacVeagh.

———. 1950. Relative growth and form transformation. *Proc. Roy. Soc. Lond.,* ser. B, 137:465–69.

———. 1958. Evolutionary processes and taxonomy with special references to grades. *Uppsala Univ. Årsskr.,* 21–38.

Huxley, J. S., J. Needham, and I. M. Lerner. 1941. Terminology of relative growth. *Nature* 137:780–81.

Ibarra-Manríquez, G., and K. Oyama. 1992. Ecological correlates of reproductive traits of Mexican rain forest trees. *Amer. J. Bot.* 79:383–94.

Ipson, D. C. 1960. *Units, dimensions, and dimensionless numbers.* New York: McGraw-Hill.

Jacobs, M. R. 1954. The effect of wind sway on the form and development of *Pinus radiata* Don. *Austral. J. Bot.* 2:35–51.

James, W. O. 1928. Experimental researches on vegetable assimilation and respiration: XIX. The effect of variations of carbon dioxide supply upon the rate of assimilation of submerged water plants. *Proc. R. Soc. Lond.,* ser. B, 103:1–42.

Kaiser, S. 1935. The factors controlling shape and size in *Capsicum* fruits: A genetic and developmental analysis. *Bull. Torrey Bot. Club.* 62:433–54.

Kang, H., and R. B. Primack. 1991. Temporal variation of flower and fruit in relation to seed yield in celandine poppy (*Chelidonium majus;* Papaveraceae). *Amer. J. Bot.* 78:711–22.

Kaplan, D. R. 1977. Morphological status of the shoot systems of Psilotaceae. *Brittonia* 29:30–53.

———. 1987a. The significance of multicellularity in higher plants: I. The relationship of cellularity to organismal form. *Amer. J. Bot.* (Botanical Society of America Ohio State Meeting) 74:617.

———. 1987b. The significance of multicellularity in higher plants: II. Functional significance. *Amer. J. Bot.* (Botanical Society of America Ohio State Meeting) 74:618.

Kaplan, D. R., and W. Hagemann. 1991. The relationship of cell and organism in vascular plants. *BioScience* 41:693–703.

Kaplan, R. H., and S. N. Salthe. 1979. The allometry or reproduction: An empirical view in salamanders. *Amer. Nat.* 113:671–89.

Kawano, S., and J. Masuda. 1980. The productive and reproductive biology of flowering plants: IV. Resource allocation and reproductive capacity in wild populations of *Heloniopsis orientalis* (Thund) C. tanaka (Liliaceae). *Ecologia* 45:307–17.

Kermack, K. A., and J. B. S. Haldane. 1950. Organic correlation and allometry. *Biometrika* 37:30–41.

Keyes, G. J., D. J. Paolillo, and M. E. Sorrells. 1989. The effects of dwarfing genes *Rht1* and *Rht2* on cellular dimensions and rate of leaf elongation in wheat. *Ann. Bot.* 64:683–90.

King, D. A., and O. L. Loucks. 1978. The theory of tree bole and branch form. *Radiat. Env. Biophys.* 15:141–65.

King, J. R., and T. T. Packard. 1975. Respiration and the activity of the respiratory electron transport system in marine zooplankton. *Limnol. Oceanogr.* 20:849–54.

Kirk, D. L. 1988. The ontogeny and phylogeny of cellular differentiation in *Volvox. Trends Gen.* 4:32–36.

Kirk, J. T. O. 1975. A theoretical analysis of the contribution of algal cells to the attenuation of light within natural water: II. Spherical cells. *New Phytol.* 75:21–36.

———. 1976. A theoretical analysis of the contribution of algal cells to the attenuation of light within natural water: III. Cylindrical and spheroidal cells. *New Phytol.* 76:341–58.

———. 1983. *Light and photosynthesis in aquatic ecosystems.* Cambridge: Cambridge University Press.

Kirk, J. T. O., and R. A. E. Tilney-Bassett. 1978. *The plastids.* 2d. ed. Amsterdam: Elsevier.

Klekowski, E. J., Jr. 1979. The genetics and reproductive biology of ferns. In *The experimental biology of ferns,* ed. A. F. Dyer. London: Academic Press.

Klekowski, E. J., Jr., and H. G. Baker. 1966. Evolutionary significance of polyploidy in the Pteridophyta. *Science* 153:305–7.

Klinkhamer, P. G., and T. J. De Jong. 1987. Plant size and seed production in the monocarpic perennial *Cynoglossum officinale* L. *New Phytol.* 106:773–83.

Klinkhamer, P. G. L., De Jong, T. J., and E. Meelis. 1990. How to test for proportionality in the reproductive effort of plants. *Amer. Nat.* 135:291–300.

Knight, T. A. 1811. On the causes which influence the direction of the growth of roots. *Phil. Trans. Roy. Soc. London* 1811:209–19.

Knight-Jones, E. W. 1954. Relations between metachronism and the direction of ciliary beat in Metazoa. *Q. Fl. Microsc. Sci.* 95:503–21.

Koehl, M. A. R., and S. A. Wainwright. 1977. Mechanical adaptations of a giant kelp. *Limnol. Oceanog.* 22:1067–71.

Kohn, J. R. 1989. Sex ratio, seed production, biomass allocation, and the cost of male function in *Cucurbita foetidissima* (Cucurbitaceae). *Evolution* 43:1424–34.

Kolattukudy, P. E. 1981. Structure, biosynthesis, and biodegradation of cutin and suberin. *Ann. Rev. Plant Physiol.* 32:539–67.

Konar, R. N., and M. N. Singh. 1979. Production of plantlets from the female gametophytes of *Ephedra foliata* Boiss. *Z. Pflanzenphysiol.* 95:87–90.

Kozlowski, T. T., ed. 1974. *Shedding of plant parts.* New York: Academic Press.

Kramer, G. 1959. Die funktionelle Beurteilung von Vorgängen relativen Wachstrums. *Zool. Anz.* 162:243–66.

Kramer, H. 1946. Heat transfer from spheres to flowing media. *Physica* 12:61–80.

Kramer, P. J. 1983. *Water relations of plants.* New York: Academic Press.

Kramer, P. J., and T. T. Kozlowski. 1979. *Physiology of woody plants.* New York: Academic Press.

Kreith, F. 1965. *Principles of heat transfer.* Scranton, Pa.: International Textbook Company.

LaBarbera, M. 1986. The evolution and ecology of body size. In *Patterns and processes in the history of life, 1986,* ed. D. M. Raup and D. Jablonski, 69–98. Dahlem Konferenzen. Berlin: Springer-Verlag.

———. 1989. Analyzing body size as a factor in ecology and evolution. *Ann. Rev. Ecol. Syst.* 20:97–117.

Lande, R. 1979. Quantitative genetic analysis of multivariate evolution, applied to brain:body size allometry. *Evolution* 33:402–16.

Lanner, R. M. 1982. Adaptations of whitebark pine for seed dispersal by Clark's nutcracker. *Can J. For. Res.* 12:391–402.

Leamy, L. 1988. Genetic and maternal influences on brain and body size in randombred house mice. *Evolution* 42:42–53.

Lee, T. D. 1988. Patterns of fruit and seed production. In *Plant reproductive ecology: Patterns and strategies,* ed. J. Lovett Doust and L. Lovett Doust, 179–202. Oxford: Oxford University Press.

Lenton, J. R., P. Heddon, and M. D. Gale. 1987. Gibberellin insensitivity and development in wheat—consequences for development. In *Hormone action in plant development,* ed. G. V. Hoad, J. R. Lenton, M. B. Jackson, and R. K. Atkins, 145–60. London: Butterworth.

Les, D. H., and D. J. Sheridan. 1990. Biochemical heterophylly and flavonoid evolution in North American *Potamogeton* (Potamogetonaceae). *Amer. J. Bot.* 77:453–65.

Levins, R., 1968. *Evolution in changing environments.* Princeton: Princeton University Press.

Lipkin, Y. 1979. Quantiative aspects of seagrass communities, particularly of those dominated by *Halophila stipulacea,* in Sinai (northern Red Sea). *Aquatic Bot.* 7:119–28.

Lloyd, D. G. 1974. Mating systems and genetic load in pioneer and non-pioneer Hawaiian Pteridophyta. *Bot J. Linn. Soc.* 69:23–35.

———. 1980. Sexual strategies in plants: I. An hypothesis of serial adjustment of maternal investment during one reproductive session. *New Phytol.* 86:69–79.

Lloyd, D. G., and K. S. Bawa. 1984. Modification of the gender of seed plants in varying conditions. In *Evolutionary biology,* ed. M. K. Hecht, B. Wallace, and G. T. Prance, 17:255–338. New York: Plenum.

Lord, E. M. 1979. The development of cleistogamous and chasmogamous flowers in *Lamium amplexicaule* (Labiatae): An example of heteroblastic inflorescence development. *Bot. Gaz.* 140:39–50.

———. 1980. Physiological controls on the production of cleistogamous and chasmogamous flowers in *Lamium amplexicaule* L. *Ann. Bot.,* 44:757–66.

———. 1981. Cleistogamy: A tool for the study of floral morphogenesis, function and evolution. *Bot. Rev.,* 47:421–49.

Lotka, A. J. 1956. *Elements of mathematical biology.* New York: Dover.

Lovett Doust, J., and P. B. Cavers. 1982. Biomass allocation in hermaphroditic flowers. *Can. J. Bot.* 60:2530–34.

Lumer, H. 1936. The relation between b and k in systems of relative growth functions of the form $y = bx^k$. *Amer. Nat.* 70:188–91.

Lysova, N. V., and N. I. Khizhnyak. 1975. Sex differences in trees in the dry steppe. *Sov. J. Ecol.* 6:522–47.

Maekawa, T. 1924. On the phenomena of sex transition in *Arisaema japonica. J. Coll. Agric. Hokkaido Imp. Univ.* 13:217–305.
Mariana Colombo, P., M. Orsenigo, A. Solazzi, and C. Tolomio. 1976. Sea depth effects on the algal photosynthetic apparatus. IV. Observations on the photosynthetic apparatus of *Halimeda tuna* (Siphonales) at sea depths between 7 and 16 m. *Mem., Biol. Marine Oceanogr.* 6:197–208.
Marshall, D. L. 1986. Effect of seed size on seeding success in three species of *Sesbania* (Fabaceae). *Amer. J. Bot.* 73:457–64.
Martin, R. D. 1981. Relative brain size and basal metabolic rate in terrestrial vertebrates. *Nature* 293:57–60.
Martínez, E., and B. Santelices. 1992. Size hierarchy and the −3/2" power law" relation in the coalescent seaweed. *J. Phycol.* 28:259–64.
Matlack, G. R. 1987. Diaspore size, shape, and fall behavior in wind-dispersed plant species. *Amer. J. Bot.* 74:1150–60.
Maynard Smith, J. 1978. *The evolution of sex.* Cambridge: Cambridge University Press.
Maynard Smith, J., and R. Holliday. 1979. Preface. In *The evolution of adaptation by natural selection,* ed. J. Maynard Smith and R. Holliday. London: Royal Society.
Mazer, S. J. 1987. Parental effects on seed development and seed yield in *Raphanus raphanistrum:* Implications for natural and sexual selection. *Evolution* 41:355–71.
Mazer, S. J., A. A. Snow, and M. L. Stanton. 1986. Fertilization dynamics and parental effects upon fruit development in *Raphanus raphanistrum:* Consequences for seed size variation. *Amer. J. Bot.* 73:500–511.
McAdams, W. H. 1954. *Heat transmission.* New York: McGraw-Hill.
McKone, M. J. 1989. Intraspecific variation in pollen yield in bromegrass (Poaceae: *Bromus). Amer. J. Bot.* 76:231–37.
McMahon, T. A. 1973. Size and shape in biology. *Science* 179:1201–4.
———. 1975. Allometry and biomechanics: Limb bones in adult ungulates. *Amer. Nat.* 109:547–63.
———. 1980. Food habits, energetics, and the population biology of mammals. *Amer. Nat.* 116:106–24.
McMahon, T. A., and R. E. Kronauer. 1976. Tree structure: Deducing the principle of mechanical design. *J. Theor. Biol.* 59:443–66.
McNamara, K. J. 1986. A guide to the nomenclature of heterochrony. *J. Paleontol.* 60:4–13.
Miller, E. C. 1938. *Plant physiology.* 2d. ed. New York: McGraw-Hill.
Miller, J. S. 1981. Pre-partum reproductive characteristics of eutherian mammals. *Evolution* 35:1149–63.
Mione, T., and G. J. Anderson. 1992. Pollen-ovule ratios and breeding system evolution in *Solanum basarthrum* (Solanaceae). *Amer. J. Bot.* 79:279–87.
Mishler, B. D., and S. P. Churchill. 1984. A cladistic approach to the phylogeny of the bryophytes. *Brittonia* 36:406–24.
———. 1985. Transition to a land flora: Phylogenetic relationships of the green algae and bryophytes. *Cladistics* 1:305–28.

Molau, U. 1991. Gender variation in *Bartsia alpina* (Scrophulariaceae), a subarctic perennial hermaphrodite. *Amer. J. Bot.* 78:326–39.

Molau, U., B. Eriksen, and J. Teilmann Knudsen. 1989. Predispersal seed predation in *Bartsia alpina. Oecologia* 81:181–85.

Morel, A., and A. Bricaud. 1981. Theoretical results concerning light absorption in discrete medium, and application to specific absorption of phytoplankton. *Deep-Sea Res.* 28:1375–93.

Morison, J. R., M. P. O'Brien, J. W. Johnson, and S. A. Schaaf. 1950. The forces exerted by surface waves on piles. *Petrol. Trans. AIME* 189:149–57.

Mullin, M. M., P. R. Sloan, and R. W. Eppley. 1966. Relationship between carbon content, cell volume, and area in phytoplankton. *Limnol. Oceanogr.* 11:307–11.

Munk, W. H., and G. A. Riley. 1952. Absorption of nutrients by aquatic plants. *J. Marine Res.* 11:215–40.

Munro, H. N. 1969. Evolution of protein metabolism in mammals. In *Mammalian protein metabolism,* ed. H. N. Munro, 3:133–82. New York: Academic Press.

Niklas, K. J. 1986a. Computer-simulated plant evolution. *Sci. Amer.* 254: 78–86.

———. 1986b. Evolution of plant shape: Design constraints. *Trends Ecol. Evol.* 1:67–72.

———. 1986c. Computer simulations of branching-patterns and their implications on the evolution of plants. *Lect. Math. Life Sci.* 18:1–50.

———. 1989. The cellular mechanics of plants. *Amer. J. Bot.* 77:344–49.

———. 1990. Biomechanics of *Psilotum nudum* and some early Paleozoic vascular sporophytes. *Amer. J. Bot.* 77:590–606.

———. 1992. *Plant biomechanics: An engineering approach to plant form and function.* Chicago: University of Chicago Press.

———. 1993a. Influence of tissue density-specific mechanical properties on the scaling of plant height. *Ann. Bot.* 72:173–79.

———. 1993b. The scaling of plant height: A comparison among major plant clades and anatomical grades. *Ann Bot.* 72:165–72.

———. 1993c. Allocation of organ biomass in perfect and imperfect flowers. *Ann. Bot.* 71:475–83.

———. 1993d. The allometry of plant reproductive biomass and stem diameter. *Amer. J. Bot.* 80:461–67.

———. 1994. Size-dependent variations in plant growth rates and the "3/4-power rule." *Amer. J. Bot.* 81:134–45.

Niklas, K. J., and D. R. Kaplan. 1991. Biomechanics and the adaptive significance of multicellularity in plants. In *The unity of evolutionary biology,* ed. E. C. Dudley, 489–502. Proceedings of the Fourth International Congress of Systematics and Evolutionary Biology. Portland, Ore.: Dioscorides Press.

Niklas, K. J., and V. Kerchner. 1984. Mechanical and photosynthetic constraints on the evolution of plant shape. *Paleobiology* 10:79–101.

Niklas, K. J., and D. J. Paolillo Jr. 1990. Biomechanical and morphometric differences in *Triticum aestivum* seedlings differing in *Rht* gene-dosage. *Ann. Bot.* 65:365–77.

Nobel, P. S. 1974. Boundary layers of air adjacent to cylinders: Estimation of effective thickness and measurement on plant materials. *Plant Physiol.* 54:177–81.

———. 1975. Effective thickness and resistance of the air boundary layer adjacent to spherical plant parts. *J. Exp. Bot.* 26:120–30.

———. 1983. *Biophysical plant physiology and ecology.* New York: Freeman.

Norberg, R. Å. 1988. Theory of growth geometry of plants and self-thinning of plant populations: Geometric similarity, elastic similarity, and different growth modes of plant parts. *Amer. Nat.* 131:220–56.

Oka, H.-L., and H. Morishima. 1967. Variations in breeding systems of a wild rice, *Oryza perennis. Evolution* 21:249–58.

Okubo, A., and S. A. Levin. 1989. A theoretical framework for data analysis of wind dispersal of seeds and pollen. *Ecology* 70:329–38.

Oltmanns, F. 1892. Über die Kultur und Lebensbedingungen der Meeresalgen. *Jb. Wiss. Bot.* 23:349–440.

Pagel, M. D., and P. H. Harvey. 1988. Recent developments in the analysis of comparative data. *Quart. Rev. Biol.* 63:413–40.

Parker, J. 1949. Effects of variation in the root-leaf ratio on transpiration rate. *Plant Physiol.* 24:739–43.

Parodi, E. R., and E. J. Cáceres. 1991. Variation in number of apical ramifications and vegetative cell length in freshwater populations of *Cladophora* (Ulvophyceae, Chlorophyta). *J. Phycol.* 27:628–33.

Parsons, T. R., K. Stephens, and J. D. H. Strickland. 1961. On the chemical composition of eleven species of marine phytoplankton. *J. Fish. Res. Bd. Canada* 18:1001–16.

Pearl, R. 1925. *The biology of population growth.* New York: Knopf.

Pearman, G. I., H. L. Weaver, and C. B. Tanner. 1972. Boundary layer heat transfer coefficients under field conditions. *Agr. Meteor.* 10:83–92.

Peters, R. H. 1983. *The ecological implications of body size.* Cambridge: Cambridge University Press.

Peters, R. H., S. Cloutier, D. Dubé, A. Evens, P. Hastings, H. Kaiser, D. Kohn, and B. Sarwer-Foner. 1988. The allometry of the weight of fruit on trees and shrubs in Barbados. *Oecologia* (Berlin) 74:612–16.

Phillips, T. L. 1979. Reproduction of heterosporous arborescent lycopods in the Mississippian-Pennsylvanian of Euramerica. *Rev. Palaeobot. Palynol.* 27:239–89.

Phillips, T. L., and W. A. DiMichele. 1992. Comparative ecology and life-history biology of arborescent lycopods in Late Carboniferous swamps of Euramerica. *Ann. Missouri Bot. Gard.* 79:560–88.

Plitmann, U., and D. A. Levin. 1983. Pollen-pistil relationships in the Polemoniaceae. *Evolution* 37:957–67.

Poethig, R. S. 1987. Clonal analysis of cell lineage patterns in plant development. *Amer. J. Bot.* 74:581–94.

Pomeroy, K. B., and D. Dixon. 1966. These are the champs. *Amer. For.* 72:14–35.

Popp, J. W., and J. A. Reinartz. 1988. Sexual dimorphism, biomass allocation and clonal growth of *Xanthoxylum americanum. Amer. J. Bot.* 75:1732–41.

Primack, R. B. 1987. Relationships among flowers, fruits, and seeds. *Ann. Rev. Ecol. Syst.* 18:409–30.

Prothero, J. 1986. Methodological aspects of scaling in biology. *J. Theor. Biol.* 118:259–86.

Putz, F. E., P. D. Coley, K. Lu, A. Montalvo, and A. Aiello. 1983. Uprooting and snapping of trees: Structural determinants and ecological consequences. *Can. J. For. Res.* 13:1011–20.

Raff, R. A., and G. A. Wray. 1989. Heterochrony: Developmental mechanisms and evolutionary results. *J. Evol. Biol.* 2:409–34.

Rashevsky, N. 1960. *Mathematical biophysics.* 3d ed. New York: Dover.

Raup, D. M., and J. J. Sepkoski Jr. 1984. Periodicity of extinction in the geologic past. *Proc. Nat. Acad. Sci. USA* 81:801–5.

Raven, J. A. 1984. Physiological correlates of the morphology of early vascular plants. *Bot. J. Linn. Soc.* 88:105–26.

———. 1985. Comparative physiology of plant and arthropod land adaptation. *Phil. Trans. R. Soc. Lond.*, ser. B, 309:273–388.

Raven, J. A., J. Beardall, and H. Griffiths. 1982. Inorganic C-sources for *Lemanea, Cladophora,* and *Ranunculus* in a fast-flowing stream: Measurements of gas exchange and of carbon isotope ratio and their ecological implications. *Oecologia* 53:68–78.

Rayner, J. M. V. 1985. Linear relations in biomechanics: The statistics of scaling functions. *J. Zool.* 206:415–39.

Reed, D. C., M. Neushul, and A. W. Ebeling. 1991. Role of settlement density on gametophyte growth and reproduction in the kelps *Pterygophora californica* and *Macrocystis pyrifera* (Phaeophyceae). *J. Phycol.* 27:361–66.

Reed, H. S. 1927. Growth and differentiation in plants. *Quart. Rev. Biol.* 2:79–101.

Reiss, M. J. 1989. *The allometry of growth and reproduction.* New York: Cambridge University Press.

Remy, W., P. G. Gensel, and H. Hass. 1993. The gametophyte generation of some early Devonian land plants. *Int. J. Plant Sci.* (formerly Bot. Gaz.) 154:35–58.

Rich, P. M. 1986. Mechanical architecture of arborescent rain forest plams. *Principes* 30:117–31.

———. 1987. Mechanical structure of the stem of arborescent palms. *Bot. Gaz.* 148:42–50.

Rich, P. M., K. Helenurm, D. Kearns, S. R. Morse, M. W. Palmer, and L. Short. 1986. Height and stem diameter relationships for dicotyledonous trees and arborescent palms of Costa Rican tropical wet forest. *Bull. Torrey Bot. Club* 113:241–46.

Richards, F. J. 1959. A flexible growth function for empirical use. *J. Exp. Bot.* 10:290–300.

Ridley, M. 1983. *The explanation of organic diversity: The comparative method and adaptations for mating.* Oxford: Oxford University Press.

Riska, B. 1989. Composite traits, selection response, and evolution. *Evolution* 43:1172–91.

Riska, B., and W. R. Atchely. 1985. Genetics of growth predict patterns of brain-size evolution. *Science* 229:668–71.

Ritland, C., and K. Ritland. 1989. Variation of sex allocation among eight taxa of the *Mimulus guttatus* species complex (Scrophulariaceae). *Amer. J. Bot.* 76:1731–39.

Robertson, T. B. 1923. *The chemical basis of growth and senescence.* Philadelphia: Lippincott.

Royal Horticultural Society. 1932. *Conifers in cultivation.* London: Royal Society of London.

Ryder, J. A., 1893. The correlations of the volume and surfaces of organisms. *Contr. Zool. Lab. Univ. Pa.* 1:3–36.

Sachs, J. 1893. Über einige Beziehungen der specifischen Grösse der Pflanzen zu ihrer Organisation. *Flora* 77:46–81.

Samson, A. D., and K. S. Werk. 1986. Size-dependent effects in the analysis of reproductive effort in plants. *Amer. Nat.* 127:667–80.

Sarpkaya, T., and M. Isaacson. 1981. *Mechanics of wave forces on offshore structures.* New York: Van Nostrand-Reinhold.

Scharp, L. W. 1926. *An introduction to cytology.* New York: McGraw-Hill.

Schleiden, M. J. 1838. Beiträge zur Phytogenesis. *Arch. Anat. Physiol.* 1838:137–77.

Schlesinger, D. A., L. A. Molot, and B. J. Shuter. 1981. Specific growth rates of freshwater algae in relation to cell size and light intensity. *Can. J. Fish. Aquat. Sci.* 38:1052–58.

Schmidt-Nielsen, K. 1984. *Scaling: Why is animal size so important?* Cambridge: Cambridge University Press.

Schnitzer, M. 1978. Humic substances: Chemistry and reactions. In *Soil organic matter,* ed. M. Schnitzer and S. U. Khan, 1–64. Amsterdam: Elsevier.

Schoute, J. C. 1912. Über das Dickenwachstum der Palmen. *Ann. Jardin Bot. Buitenzorg,* 2d ser., 11:1–209.

Schumacher, G. J., and L. A. Whitford. 1965. Respiration and P^{32} uptake in various species of freshwater algae as affected by a current. *J. Phycol.* 1:78–80.

Schwann, T. 1839. *Mikroskopische Untersuchungen über die Übereinstimmung in der Struktur und dem Wachstume der Tiere und Pflanzen.* Ostwalds' Klassiker der Exakten Wissenschaften 76. Leipzig: Wilhelm Engelmann.

Schwendener, S. 1874. *Das mechanische Prinzip im anatomischen Bau der Monocotylen.* Leipzig: Wilhelm Engelmann.

Seim, E., and B.-E. Saether. 1983. On rethinking allometry: Which regression model to use? *J. Theor. Biol.* 104:161–68.

Shipley, B., and J. Dion. 1992. The allometry of seed production in herbaceous angiosperms. *Amer. Nat.* 139:468–83.

Shuter, B. J. 1978. Size dependence of phosphorus and nitrogen subsistence quotas in unicellular microorganisms. *Limnol. Oceanogr.* 23:1248–55.

Sinnott, E. W. 1921. The relation between body size and organ size in plants. *Amer. Nat.* 55:385–403.

———. 1936. A developmental analysis of inherited shape differences in cucurbit fruits. *Amer. Nat.* 70:245–54.

———. 1939. A developmental analysis of the relation between cell size and fruit size in cucurbits. *Amer. J. Bot.* 26:179–89.

———. 1960. *Plant morphogenesis.* New York: McGraw-Hill.

Smith, R. C., and K. S. Baker. 1981. Optical properties of the clearest natural waters (200–800 nm). *Appl. Opt.* 20:177–84.

Smith, R. J. 1980. Rethinking allometry. *J. Theor. Biol.* 87:97–111.

Snedecor, G. W., and W. G. Cochran. 1980. *Statistical methods.* 7th ed. Ames: Iowa State University Press.

Snee, R. D. 1977. Validation of regression models: Methods and examples. *Technometrics* 19:415–28.

Social register of big trees. 1966. *Amer. For.* 72:15–35.

———. 1971. *Amer. For.* 77:25–31.

Sokal, R. R., and F. J. Rohlf. 1981. *Biometry: The principles and practice of statistics in biological research.* San Francisco: Freeman.

Soltis, D. E., and P. S. Soltis. 1992. The distribution of selfing rates in homosporous ferns. *Amer. J. Bot.* 79:97–100.

Soltis, D. E., P. S. Soltis, and R. D. Noyes. 1988. An electrophoretic investigation of intragametophytic selfing in *Equisetum arvense. Amer. J. Bot.* 75: 231–37.

Soltis, P. S., and D. E. Soltis. 1988. Estimated rates of intragametophytic selfing in lycopods. *Amer. J. Bot.* 75:248–56.

Spence, D. H. N., R. M. Campbell, and J. Chrystal. 1973. Specific leaf areas and zonation of freshwater macrophytes. *J. Ecol.* 61:317–37.

Spencer, H. 1868. *The principles of biology.* New York.

Sprugel, D. G. 1983. Correcting for bias in log-transformed allometric equations. *Ecology* 64:209–10.

Stahl, W. R. 1967. Scaling of respiratory variables in mammals. *J. Appl. Physiol.* 22:453–60.

Stålfelt, M. G. 1956. Morphologie und Anatomie des Blattes als Transpirationsorganen. *Encycl. Plant Physiol.* 3:324–41.

Stanley, S. M. 1973. An explanation for Cope's rule. *Evolution* 27:1–26.

———. 1979. *Macroevolution: Pattern and process.* San Francisco: Freeman.

Stanton, M. L., and R. E. Preston. 1988. Ecological consequences and phenotypic correlates of petal size variation in wild radish, *Raphanus sativus* (Brassicaceae). *Amer. J. Bot.* 75:528–39.

Stark, N. 1970. Water balance of some warm desert plants in a wet year. *J. Hydrol.* 10:113–26.

Starr, R. C. 1984. Colony formation in algae. In *Encyclopedia of plant physiology,* ed. H. -F. Linskens and J. Heslop-Harrison, 17:261–90. Berlin: Springer-Verlag.

Stearns, S. C. 1983. The influence of size and phylogeny on life-history patterns. *Oikos* 41:173–87.

Steenbergh, W. F., and C. H. Lowe. 1977. *Ecology of the saguaro.* Vol. 2, *Reproduction, germination, establishment, growth, and survival of the young plant.* Scientific Monograph Series 8. Washington, D.C.: National Park Service.

Stewart, W. N. 1983. *Paleobotany and the evolution of plants.* Cambridge: Cambridge University Press.

Strathmann, R. R. 1967. Estimating the organic carbon content of phytoplankton from cell volume or plasma volume. *Limnol. Oceanogr.* 12:411–18.

Sudo, S. 1980. Some anatomical properties and density of the stem of coconut (*Cocos nucifera*) with consideration for pulp quality. *IAWA Bull.,* n.s., 1:161–71.

Swain, T. 1975. Evolution of flavonoid compounds. In *The flavonoids,* ed. J. B. Harborne, T. J. Mabry, and H. Mabry, 1096–1129. New York: Academic Press.

Tam, C. K. N., and A. C. Patel. 1979. Optical absorption coefficients of water. *Nature* 280:302–4.

Taylor, T. N., and E. L. Taylor. 1993. *The biology and evolution of fossil plants.* Englewood Cliffs, N.J.: Prentice-Hall.

Tenopyr, L. A. 1918. On the constancy of cell shape in leaves of varying shape. *Bull Torrey Bot. Club* 45:51–76.

Tiffney, B. H., and K. J. Niklas. 1985. Clonal growth in land plants: A paleobotanical perspective. In *Population biology and evolution of clonal organisms,* ed. J. B. Jackson, L. W. Buss, and R. E. Cook, 35–66. New Haven: Yale University Press.

Timoshenko, S. P., and J. M. Gere. 1961. *Theory of elastic stability.* New York: McGraw-Hill.

Thompson, B. K., J. Weiner, and S. I. Warwick. 1991. Size-dependent reproductive output in agricultural weeds. *Can. J. Bot.* 69:442–46.

Thompson, D. A. W. 1917. *On growth and form.* Cambridge: Cambridge University Press.

Thompson, K., and D. Rabinowitz. 1989. Do big plants have big seeds? *Amer. Nat.* 133:722–28.

Turrell, F. M. 1961. Growth of the photosynthetic area of *Citrus. Bot. Gaz.* 122:284–98.

Vogel, S. 1981. *Life in moving fluids.* Boston: Willard Grant.

———. 1983. The lateral thermal conductivity of leaves. *Can. J. Bot.* 62:741–44.

Vogel, S., and C. Loudon. 1985. Fluid mechanics of the thallus of an intertidal red alga, *Halosaccion glandiforme. Biol. Bull.* 168:161–74.

Vries, A. Ph. de. 1974. Some aspects of cross-pollination in wheat (*Triticum aestivum* L.): 3. Anther length and number of pollen grains per anther. *Euphytica* 23:11–19.

Wagner, G., W. Haupt, and A. Laux. 1972. Reversible inhibition of chloroplast movement by cytochalazin B in the green alga *Mougeotia. Science* 176:808–9.

Watson, M. A., and B. B. Casper. 1984. Morphogenetic constraints on patterns of carbon distribution in plants. *Ann. Rev. Ecol. Syst.* 15:233–58.

Wayne, R., and M. P. Staves. 1991. The density of the cell sap and endoplasm of *Nitellopsis* and *Chara. Plant Cell Physiol.* 32:1137–44.

Weiner, J. 1988. The influence of competition on plant reproduction. In *Plant reproductive ecology: Patterns and strategies,* ed. J. Lovett Doust and L. Lovett Doust, 228–45. Oxford: Oxford University Press.

Weller, D. E. 1987. A reevaluation of the −3/2 power rule of plant self-thinning. *Ecol. Monogr.* 57:23–43.

Westlake, D. F. 1967. Some effects of low-velocity currents on the metabolism of aquatic macrophytes. *J. Exp. Bot.* 18:187–205.

Westoby, M., E. Jurado, and M. Leishman. 1992. Comparative evolutionary ecology of seed size. *Trends Evol. Ecol.* 7:368–72.

Weyl, H. 1952. *Symmetry.* Princeton: Princeton University Press.

Wheeler, W. N., and M. Neushul. 1981. The aquatic environment. In *Physiological plant ecology,* ed. O. L. Lange, P. S. Nobel, C. B. Osmond, and H. Ziegler, 1:229–47. Encyclopedia of plant physiology, n.s., vol. 12A. New York: Springer-Verlag.

White, J. 1980. Demographic factors in populations of plants. In *Demography and evolution in plant populations,* ed. O. T. Solbrig, 21–48. Oxford: Blackwell.

White, J., and J. L. Harper. 1970. Correlated changes in plant size and number in plant populations. *J. Ecol.* 58:467–85.

White, P. S. 1983. Corner's rules in eastern deciduous trees: Allometry and its implications for the adaptive architecture of trees. *Bull. Torrey Bot. Club* 110:203–12.

Whittaker, R. H., and G. M. Woodwell. 1968. Dimension and reproduction relations of trees and shrubs in the Brookhaven Forest, New York. *J. Ecol.* 56:1–25.

Wiemann, M. C., and G. B. Williamson. 1989. Wood specific gravity gradients in tropical dry and montane rain forest trees. *Amer. J. Bot.* 76:924–28.

Williams, R. B. 1964. Division rates of salt marsh diatoms in relation to salinity and cell size. *Ecology* 45:877–80.

Winn, A. A. 1991. Proximate and ultimate sources of within-individual variation in seed mass in *Prunella vulgaris* (Lamiaceae). *Amer. J. Bot.* 78:838–44.

Wooton, J. T. 1987. The effects of body mass, phylogeny, habitat, and trophic level on mammalian age at first reproduction. *Evolution* 41:732–49.

Wylie, R. B. 1951. Principles of foliar organization shown by sun-shade leaves from ten species of deciduous dicotyledonous trees. *Amer. J. Bot.* 38:355–61.

Yatsuhashi, H., and M. Wada. 1990. High-influence rate responses in the light-oriented chloroplast movement in *Adiantum* protonema. *Plant Sci.* 68:87–94.

Yoda, K., T. Kira, H. Ogawa, and H. Hozumi. 1963. Self-thinning in overcrowded pure stands under cultivated and natural conditions. *J. Biol. Osaka City Univ.* 14:107–29.

Yokozawa, M. and T. Hara. 1992. A canopy photosynthesis model for the dynamics of size structure and self-thinning in plant populations. *Ann. Bot.* 70:305–16.

Zeide, B. 1987. Analysis of the 3/2 power law of self-thinning. *For Sci.* 33:517–37.

Zimmer, E. A., R. K. Hamby, M. L. Arnold, D. A. LeBlanc, and E. C. Theriot. 1989. Ribosomal RNS phylogenies and flowering plant evolution. In *The hierarchy of life,* ed. B. Fernholm, K. Bremer, and H. Jörnvall, 205–34. Amsterdam: Elsevier (Biomedical Division).

Zimmermann, M. H. 1983. *Xylem structure and the ascent of sap.* Berlin: Springer-Verlag.

Zimmermann, W. 1930. Der Baum in seinen phylogenetischen Werden. *Ber. Bot. Ges.* 48:34–49.

Author Index

Subject Index